3000 800068 19292
St. Louis Community College

D0212252

Biomes and Ecosystems

Biomes and Ecosystems

EDITOR
Robert Warren Howarth
Cornell University

ASSOCIATE EDITOR
Jacqueline E. Mohan
University of Georgia, Athens

Volume 1
Overviews

SALEM PRESS
A Division of EBSCO Publishing
Ipswich, Massachusetts

GREY HOUSE PUBLISHING

Cover photo: Rock Islands, Koror, Palau, Pacific. © Casey Mahaney

Biomes and Ecosystems, 2013, published by Grey House Publishing, Inc., Amenia, NY, under exclusive license from EBSCO Publishing, Inc.

The paper used in these volumes conforms to the American National Standard for Permanence of Paper for Printed Library Materials, X39.48-1992 (R1997).

LIBRARY OF CONGRESS CATALOGING-IN-PUBLICATION DATA

Biomes and ecosystems / Robert Warren Howarth, general editor ; Jacqueline E. Mohan, associate editor.
volumes cm.
Includes bibliographical references and index.
ISBN 978-1-4298-3813-9 (set) -- ISBN 978-1-4298-3814-6 (volume 1) -- ISBN 978-1-4298-3815-3 (volume 2) -- ISBN 978-1-4298-3816-0 (volume 3) -- ISBN 978-1-4298-3817-7 (volume 4) 1. Biotic communities. 2. Ecology. 3. Ecosystem health. I. Howarth, Robert Warren. II. Mohan, Jacqueline Eugenia
QH541.15.B56B64 2013
577.8'2--dc23

2013002800

ebook ISBN: 978-1-4298-3818-4

First Printing

PRINTED IN THE UNITED STATES OF AMERICA

Produced by Golson Media

Contents

Publisher's Note

In conceiving, editing, and reading *Biomes and Ecosystems,* the natural question arises: What is the difference between a biome and an ecosystem? In short, an ecosystem is much smaller than a biome. Conversely, a biome can be thought of as many similar ecosystems throughout the world grouped together. An ecosystem can be as large as the Sahara Desert, or as small as a puddle or vernal pool. The distinction is worth further investigation. According to the *Merriam-Webster* dictionary, the definitions of the two terms are as follows:

Biome: Largest geographic biotic unit, a major community of plants and animals with similar requirements of environmental conditions. It includes various communities and developmental stages of communities and is named for the dominant type of vegetation, such as grassland or coniferous forest. Several similar biomes constitute a biome type; for example, the temperate deciduous forest biome type includes the deciduous forest biomes of Asia, Europe, and North America.

Ecosystem: Complex of living organisms, their physical environment, and all their interrelationships in a particular unit of space. An ecosystem's abiotic (nonbiological) constituents include minerals, climate, soil, water, sunlight, and all other nonliving elements; its biotic constituents consist of all its living members. Two major forces link these constituents: the flow of energy and the cycling of nutrients.

Still there is debate among scientists about specific regions of the Earth, whether they are technically a biome or ecosystem. The North Atlantic Ocean, for example, can be termed either way. In organizing this book the editors have targeted a consensus and have produced a work that presents five major categories of biomes as Overviews (Part 1) and some 500 ecosystems as Articles (Part 2).

Especially targeted toward high-school students, this outstanding reference work is edited to tie into the high-school curriculum, making the content readily accessible as well to patrons of public, academic, and university libraries. Pedagogical elements include a Topic Finder, Chronology, Resource Guide, Glossary, and thorough index. However, in the end, as the General Editor states in his Introduction, our writers use the terms ecosystems and biomes somewhat interchangeably in Part 2.

Scope of Coverage

This academic, multiauthor reference work serves as a general and nontechnical resource for students and teachers to understand the importance of biomes and ecosystems; to appreciate the study of ecology and how it affects life around the world; to learn of the flora and fauna in biomes and ecosystems; and to initiate educational discussion brought forth by the specific social and topical articles presented in the work.

After the frontmatter, the book opens with eight illustrations by Jared T. Williams, depicting different ecosystems around the world and showing the flora and fauna that live in these ecoregions.

In Part 1: Overviews, the book presents 39 conceptual essays organized into five biome categories: Marine and Oceanic Biomes (9); Inland Aquatic Biomes (7); Desert Biomes (6); Forest Biomes (9); and Grassland, Tundra, and Human Biomes (8).

In Part 2: Articles, the work offers 506 entries, organized in an A-to-Z presentation, on specific ecosystems around the world, ranging from Acacia-Commiphora Bushlands and Thickets to Zapata Swamp. These ecosystem articles are categorized in the Topic Finder as Coral Reefs (35); Estuaries (30); Intertidal Zones (8); Lakes, Ponds, and Pools (Lentic) (35); Oceans (9); Seas and Bays (30); Streams and Rivers (Lotic) (76); Wetlands (40); Coastal Deserts (7); Mid-Latitude Deserts (7); Paleodeserts (1); Polar Deserts (3); Rain Shadow Deserts (3); Trade Wind Deserts (12); Coniferous Forests (15); Mediterranean Forests (13); Montane Forests (14); Plantation Forests (10): Subtropical Forests (13); Temperate Forests (30); Tropical Rainforests (29); Desert and Xeric Grasslands (15); Flooded Grasslands (8); Human Ecosystems (9); Montane Grasslands (13); Temperate Grasslands (12); Tropical and Subtropical Grasslands (16); and Tundra Grasslands (13).

Article Length and Format

Articles and essays in the book range in length from 500 to 3,500 words. Each ecosystem article in Part 2 is first presented with the biome category to which it belongs (for example, Marine and Oceanic Biomes), its geographic location, and an article summary. Each article is then followed by "Further Reading" sources that include bibliographic citations. Many articles are richly illustrated with photos and captions. Finally, each article is signed by the contributor to the book.

Frontmatter and Backmatter

Volume 1 of *Biomes and Ecosystems* begins with "About the Editor" and then presents the introduction to the book. The "Contents," repeated in all four volumes, features all the articles in alphabetical order with page numbers as they are listed in the book. The "Topic Finder" shows all the articles organized by category to enable readers to find related articles by topic. The "List of Contributors" presents all the writers for the book along with their academic or institutional affiliations.

The backmatter of the book at the end of Volume 4 has the "Chronology," a timeline of major milestones in the study of biomes and ecosystems. A "Glossary" provides definitions for terms encountered in the articles. Next is the "Resource Guide" for further research that includes books that are major works in ecology as well as current editions of new works, journals in related fields, and Internet sites that pertain to the subject. Lastly, a comprehensive subject index references all concepts, terms, events, persons, places, and other topics of discussion.

About the Editors

Robert Warren Howarth, Ph.D. (General Editor) is the David R. Atkinson Professor of Ecology and Environmental Biology at Cornell University and teaches undergraduate and graduate classes in ecology and oceanography, with an emphasis on biogeochemistry and ecosystem biology. He also teaches the core graduate course in biogeochemistry for the program in Biogeochemistry and Environmental Complexity. He received his bachelor's degree in biology from Amherst College and a doctorate in biological oceanography from the Massachusetts Institute of Technology and the Woods Hole Oceanographic Institution.

A contributor in media outlets like the *Los Angeles Times*, National Public Radio, and the *New York Times*, Howarth has also contributed dozens of academic articles to government and nongovernmental organizations, publications, and encyclopedias, including the *Encyclopedia of Biodiversity*, *Encyclopedia of Inland Waters*, *Frontiers in Ecology & Environment*, *Nature*, *Scientific American*, and the *Journal of Environmental Quality*.

Howarth also chairs the International SCOPE Biofuels Project, directs the Agriculture, Energy & the Environment Program (AEEP, formerly AEP) at Cornell University, and represents the State of New York on the science and technical advisory committee of the Chesapeake Bay Program. He is the founding editor of the journal *Biogeochemistry* (editor-in-chief, 1983–2004).

Jacqueline E. Mohan, Ph.D. (Associate Editor) is an assistant professor of terrestrial ecosystem ecology and biogeochemistry at the Odum School of Ecology at the University of Georgia in Athens. She received her S.B. degree in biological chemistry from the University of Chicago, an M.E.M. in Resource Ecology from Duke University, and a Ph.D. in biology and ecology from Duke. She was a postdoctoral fellow at Harvard University and the Marine Biological Laboratory in Massachusetts. She has received teaching awards from the University of Georgia, and the Buell Award from the Ecological Society of America. She serves as an academic editor with the scientific journal *PLoS ONE*, and has published recent papers on forest responses to warming climates, and tree and poison ivy responses to rising levels of atmospheric carbon dioxide.

Introduction

The Earth holds an astounding diversity of life forms, often studied and celebrated as individual species. Those species do not live in isolation but rather are parts of ecosystems and biomes, units of biological organization that are larger than individual organisms. Just as tissues and organs are made up of cells, and organisms are composed of tissues and organs, ecosystems and biomes are made up by organisms together with the physical environment in which they live. Biomes and ecosystems and their astounding diversity are the focus of this book.

The term *ecosystem* was first coined by the British botanist A. G. Tansley back in 1935. A commonly used definition comes from the writings of American ecologist Raymond Lindeman in 1942: "... the system composed of physical-chemical-biological processes active within a space-time unit." That is, an ecosystem is composed of all the organisms—plants, animals, and microbes—and their physical environment in a defined area, such as an entire lake or forest. Culturally, the concept of an ecosystem is much older, and terms in common language such as *tundra, pampas, chaparral, swamp*, and *marsh* recognize the existence of biological units at a scale higher than that of the individual organism or species. Ecosystems represent perceptions of systems in the natural world that may be inherent in human experience.

Biomes are large areas on the Earth's surface that are characterized by a particular type of ecosystem, often described in terms of the dominant vegetation. Biomes are defined by the climate of a region, so that for example deserts exist in areas of very low moisture, grasslands occur in regions with somewhat more moisture, and forests dominate in wetter regions. The functional characteristics of organisms that dominate in a particular biome type are similar across widely disparate parts of the planet, even though these organisms often have very different genetic make up. For instance, the leaves of trees in mediterranean biomes have hard waxy coatings to help conserve water in these relatively dry areas, yet the tree species of the mediterranean biome of California, of Chile, and of the actual shores of the Mediterranean Sea have very different ancestral histories and have independently evolved this waxy-leaf characteristic in response to the climate.

In this book, we use the terms *biomes* and *ecosystems* rather interchangeably, as the concepts have much in common: both consist of all the organisms together in one location with their physical-chemical environment. Ecologists tend to think of ecosystems as being somewhat smaller in scale than biomes, but there is not a sharp dividing line between the two.

Within these volumes we deal with the terms *anthropogenic* and *human*, favoring *human* as in "human ecosystems," since that label may be better understood by a larger audience. *Anthropogenic* is an adjective defined by Merriam-Webster as "relating to, or resulting from the influence of human beings on nature" and is often used within the ecology curriculum. In most cases, we have chosen "human-induced" to describe the effects of humans on biomes and ecosystems.

Biomes and ecosystems are of inherent intellectual interest to ecologists, but nonscientists too are fascinated by many types of these systems—such as coral reefs, the savannas of Africa, and the dark redwood forests of California—because of their beauty and the wonder they inspire. Ecosystems are also a critical level of biological organization for environmental management. The effects of acid rain on lakes and the fish and other organisms in lakes first became clear through studying lakes as whole ecosystems, and ecologists are observing some of the earliest signs of disruption from global climate change at the scale of whole ecosystems. In this book, we have strived both to describe the beauty, wonder, and diversity of biomes and ecosystems and the sensitivity of these systems to damage from global change and other human-caused insults.

The book has two major parts: a series of overview essays on the major types of biomes and ecosystems around the world and then a series of place-based articles on particular biomes and ecosystems of interest. The overview essays fall into a set of broad divisions: marine and oceanic biomes, inland aquatic biomes, deserts, forests, and grasslands, tundra, and human-dominated systems. Within each of these, we have a series of further divisions, for example dividing the inland aquatic biomes into individual articles on freshwater lakes, saline lakes, great lakes, ponds, streams, rivers, and wetlands. Similarly, forests are divided into boreal forests, Mediterranean forests, montane forests, plantation forests, subtropical forests, temperate coniferous forests, temperate deciduous forests, tropical rainforests, and tropical seasonal forests. Our goal is to present the reader with the basic defining characteristics of the world's major types of biomes and ecosystems, and an understanding of the physical environment that defines these different major types of systems.

The place-based articles then provide greater detail on particular ecosystems. For example, we have articles on individual coral reefs, including the Great Barrier Reef of Australia, the Mesoamerican Barrier Reef off the shore of Honduras and Belize in the Caribbean Sea, the Florida Keys Coral Reef, and the coral reefs of the Red Sea.

The articles are designed to stand alone, so the interested reader can pick up the book and gain detailed information on any one of these reefs, or read the overview article on coral reefs, or read articles on many different coral reefs to understand the amazing diversity across these systems. For rivers, we have more than 75 articles representing most of the better known rivers across the planet. The more than 30 desert articles include coastal deserts, such as Baja California; mid-latitude deserts, such as the Mojave in the American southwest; the Nebraska sand hills as an example of a paleodesert; polar deserts, such as the dry valley deserts in Antarctica; rain shadow deserts, such as part of Patagonia in South America; and trade wind deserts, such as the Great Victoria Desert of Australia. Suggestions for further reading are provided for each article, so the book can be an excellent starting point for more in-depth study of a particular ecosystem.

Whether this is your first introduction to biomes and ecosystems or you are an experienced student or expert in this area, we trust you will find the book valuable, informative, and entertaining. We hope you agree that the authors have done an excellent job of presenting the extraordinary diversity of life on Earth through this unique lens of the diversity of biomes and ecosystems.

Robert Warren Howarth

Topic Finder

Part 1: Overviews

Part 2: Articles

Buru Rainforest
Cardamom Mountains Rainforests
Central American Atlantic Moist Forests
Christmas and Cocos Islands Tropical Forests
Congolian Lowland Forests
Congolian Swamp Forests
Costa Rican Seasonal Moist Forests
Ghats Montane Rainforests, Northwestern and Southwestern
Hawaiian Tropical Moist Forests
Irrawaddy Moist Deciduous Forests
Java-Bali Rainforests
Kakamega Rainforest
Kalimantan (Indonesia) Rainforests
Luzon Rainforest
Marajó Varzea Forests
Monte Alegre Varzea Forests
New Guinea Freshwater Swamp Forests
Niger Delta Swamp Forests
Papuan Rainforests, Southeastern
Peninsular Malaysian Rainforests
Purus Varzea Forests
Queensland Tropical Rainforests
Serra do Mar Coastal Forests
Solomon Islands Rainforests
Vanuatu Rainforests

Grassland, Tundra, and Human Ecosystems

Desert and Xeric Grasslands

Anatolian Steppe, Central
Araya and Paria Xeric Scrub
Colorado Plateau Shrublands
Galápagos Islands Xeric Scrub
Junggar Basin Semi-Desert
Kalahari Xeric Savanna
Leeward Islands Xeric Scrub
Namibian Savanna Woodlands
Oman, Gulf of, Desert and Semi-Desert
Persian Gulf Desert and Semi-Desert
San Lucan Xeric Scrub
Snake-Columbia Shrub Steppe
Tamaulipan Matorral
Windward Islands Xeric Scrub
Wyoming Basin Shrub Steppe

***Flooded Grasslands** (see also Wetlands)*

Mexican Wetlands, Central
Nenjiang River Grassland
Niger Delta Flooded Savanna, Inner
Nile Delta Flooded Savanna
Orinoco Wetlands
Rann of Kutch Seasonal Salt Marsh
Tigris-Euphrates Alluvial Salt Marsh
Zambezian Flooded Grasslands

Human (Anthropogenic) Ecosystems

Ecotourism
Fisheries
Logging Industry
Novel Ecosystems
Rural Areas
Slums
Suburban Areas
Urban Areas
Water Management

Montane Grasslands

Andean Páramo Grasslands
Andean Puna Grasslands, Central
Campos Rupestres Montane Savanna
Córdoba Montane Savanna
Ethiopian Montane Moorlands
Highveld Grasslands
Himalayan Alpine Shrub and Meadows, Eastern
Himalayan Alpine Shrub and Meadows, Western
Kinabalu Montane Alpine Meadows
Pamir Alpine Desert and Tundra
Qilian Mountains Subalpine Meadows
Tibetan Plateau Alpine Steppe, Central
Zacatonal

Temperate Grasslands

Argentine Monte
Australian Alps Montane Grasslands
California Central Valley Grasslands
Edwards Plateau Savanna
Guyanan Savanna
Kazakh Steppe
Middle East Steppe
Mongolian-Manchurian Grassland
Montana Valley and Foothill Grasslands
Northwestern Mixed Grasslands

Tropical and Subtropical Grasslands

Tundra Grasslands

List of Contributors

Angela Abela
University of Malta
Scott R. Abella
University of Nevada, Las Vegas
Carlos Abrahams
Baker Consultants
Christina Alba
Colorado State University
Matthew Alexander
Southern Methodist University
Gideon Alfred
College of African Wildlife Management
John All
Western Kentucky University
Nisreen H. Alwan
American University of Beirut
James T. Anderson
West Virginia University
John A. Anton
City University of New York
Alexandra M. Avila
Universidad San Francisco de Quito
Peter Baas
Odum School of Ecology
Matt Bahm
Sequoia and Kings Canyon National Parks
Kevin Bakker
University of Michigan
April Karen Baptiste
Colgate University
Shaunna Barnhart
Penn State University
John H. Barnhill
Independent Scholar
Andrew M. Barton
University of Maine at Farmington
Melanie Bateman
National Science Foundation
Marylin Bejarano
Universidad Nacional Autonoma de Mexico
Hannah Bement
Yale University
Okeyo Benards
Pwani University College
Lori Beshears
Stephen F. Austin State University
Medani P. Bhandari
Syracuse University
Rituparna Bhattacharyya
Alliance for Community Capacity
Justin Bohling
Penn State University

Christina Bonanni
American University
Temitope Israel Borokini
National Centre for Genetic Resources and Biotechnology (Nigeria)
Sarah Boslaugh
Keenesaw University
Damon C. Bradbury
University of California
Blake Broaddus
Stephen F. Austin State University
Victoria M. Breting-Garcia
Independent Scholar
Eric Burr
Methow Conservancy
Daren C. Card
Independent Scholar
Kylie Carman-Brown
Australian National University
Lynn Carpenter
University of Michigan
Amancay A. Cepeda
ProCAT Colombia/Internacional
Sabuj Kumar Chaudhuri
Basanti Devi College
Zhiqiang Cheng
Ohio State University
Jill M. Church
D'Youville College
Michelle Cisz
Michigan Technological University
Liane Cochran-Stafira
Saint Xavier University
Bernard W. T. Coetzee
Stellenbosch University
Itay Cohen
Hebrew University
Patrick J. Comer
NatureServe
Juanita M. Constible
Alliance for Climate Protection
Justin Corfield
Independent Scholar
Sandy Costanza
D'Youville College
Tatiana Coutto
European Union Institute for Security Studies
Joel Covington
Stephen F. Austin State University
David R. Coyle
University of Georgia
Gareth Davey
Hong Kong Shue Yan University
Israel Del Toro
University of Massachusetts, Amherst
Dominick A. DellaSala
Geo Institute
Gisela Dionísio
Universidade de Aveiro
Michael Dixon
U.S. Fish and Wildlife Service
Cliff S. Dlamini
University of Swaziland
David A Douglass
Stephen F. Austin State University
Harold Draper
Independent Scholar
Rocio R. Duchesne
Montclair State University
Frauke Ecke
Swedish University of Agricultural Sciences
Connie S. Eigenmann
Fort Hays State University
Peter Elias
University of Lagos
Robert D. Ellis
Florida State University
Joanna C. Ellison
University of Tasmania
Lilly Margaret Eluvathingal
Florida International University
Kathleen A. Farley
San Diego State University
Pedro Maria de Abreu Ferreira
Universidade Federal do Rio Grande do Sul
Fabíola Ferreira Oliveira
University of Brasilia
M. Drew Ferrier
Hood College
Judy Flemons
Stephen F. Austin State University
Máximo Florín
University of Castilla-La Mancha

Catherine G. Fontana
Yale University
William Forbes
Stephen F. Austin State University
Deborah Foss
University of La Verne
Paul T. Frankson
University of Georgia
Angelina M. Freeman
Environmental Defense Fund
Daniel A. Freiss
National University of Singapore
Jesse Fruchter
Southern Illinois University, Carbondale
Stephen M. Funk
Nature Heritage
Gillian Galford
Woods Hole Research Center
Kamal J. K. Gandhi
University of Georgia
Pedro P. Garcillán
Centro de Investigaciones Biologicas del Noroeste
José Nuno Gomes-Pereira
University of the Azores
F. Javier Gómez
Autonomous University of Barcelona
Mauricio González
Sierra to Sea Institute
Maria Jose González-Bernat
Marine Biology Association
José F. González-Maya
ProCAT Colombia/Internacional
Kyle Gracey
Global Footprint Network
Marcus W. Griswold
University of Maryland Center for Environmental Science
Lucia M. Gutierrez
Marine Biology Association of Guatemala
Jessica Haapkylä
James Cook University
Degi Harja
World Agroforestry Centre
Kaleshia Hamilton
Stephen F. Austin State University
Ingrid Hartmann
United Nations Development Programme
Matt W. Hayward
Wildlife Conservancy
Jason A. Helfer
Knox College
Lori A. Hennings
Oregon Metro
Julia-Maria Hermann
Technische Universität München
Ruth M. Higgins
University of the Azores
Annelie Hoepker
Sierra to Sea Institute
Paul Holloway
University of Texas
Jeffrey C. Howe
U.S. Fish and Wildlife Service
Falk Huettmann
University of Alaska, Fairbanks
Claire L. Hudson
The Society for Ocean Sciences
Andrew Hund
Case Western Reserve University
Nasseer Idrisi
Center for Marine and Environmental Studies
Bila-Isia Inogwabini
Swedish University of Agricultural Sciences
Zeehan Jaafar
University of Singapore
Angus C. Jackson
North Highland College
Nicholas James
Open University
Jeffrey C. Jolley
Columbia River Fishery Program
Clara B. Jones
Community Conservation
Sanghoon Kang
University of Houston, Clear Lake
Kimberly M. Kellett
Odum School of Ecology
Julie Kenkel
Northern Arizona University
Shafkat Kahn
University of Georgia

Jesse Killingsworth
Stephen F. Austin State University
Alex W. Kisingo
College of African Wildlife Management
Lee Wei Kit
National University of Singapore
Jason Krumholz
University of Rhode Island
Bill Kte'pi
Independent Scholar
Manoj Kumar
Forest Research Institute, Dehradun
Eric Landen
Independent Scholar
James S. Latimer
U.S. Environmental Protection Agency
David M. Lawrence
Virginia Commonwealth University
Adam Leavesley
Parks and Conservation Service, Australian Capital Territory
Richard E. Lee, Jr.
Miami University, Oxford, Ohio
Fern Raffela Lehman
University of Georgia
Christopher A. Lepczyk
University of Hawaii
Connie Levesque
Independent Scholar
Manja Leyk
Spring Hill College
Liang Liang
University of Kentucky
Yanna Liang
Southern Illinois University Carbondale
Jennifer K. Lipton
Central Washington University
Mary Logalbo
West Multnomah Soil and Water Conservation District
Reynard Loki
Creative Development Systems
Maricar Macalincag
Michigan State University
Megan Machmuller
Odum School of Ecology
James Mammarella
Independent Scholar
Yasmin Mannan
Colgate University
Brigitte Marazzi
University of Arizona
Erika Marín-Spiotta
Independent Scholar
Micaela E. Martinez-Bakker
University of Michigan
Dino J. Martins
State University of New York, Stony Brook
Muraree Lal Meena
Banaras Hindu University
Patricio Mena-Vásconez
Wageningen University and Research Centre
Nicole Menard
Three Rivers Park District
Julia Miller
Macquarie University
Nina Mindawati
Research and Development Centre for Forest Productivity
Susan Moegenburg
University of Vermont
Jacqueline E. Mohan
University of Georgia
Lucía Morales-Barquero
University of Bangor
Raquel Moreno-Peñaranda
United Nations University
Muhammad Aurang Zeb Mughal
University of Durham
Magdalena Ariadne Kim Muir
Arctic Institute of North America
John Mull
Weber State University
Jaime Murillo-Sanchez
Sierra to Sea Institute
Brett P. Murphy
South Dakota State University
Evans Mwangi
University of Nairobi
Ghazala Nasim
University of the Punjab

Gonzalo Navarro
Catholic University of Cochabamba, Bolivia
Kristine Nemec
University of Nebraska, Lincoln
Attila Németh
Eötvös Loránd University
Henry K. Njovu
College of African Wildlife Management
Andrew Osborn
Stephen F. Austin State University
Kathryn Ottaway
Indiana University
Gerhard Ernst Overbeck
Universidade Federal do Rio Grande do Sul
Maria Gabriela Palomo
Museo Argentino de Ciencias Naturales
Etienne Paradis
Université Laval
Rhama Parthasarathy
Independent Scholar
Daniel M. Pavuk
Bowling Green State University
Petina L. Pert
Ecosystem Sciences
Christopher Peters
Southern Illinois University
Jess Phelps
Historic New England
Alexandra Pineda-Guerrero
ProCAT Colombia/Internacional
E. Mark Pires
Long Island University
Renata Leite Pitman
Duke University
Robert Pitman
National Marine Fisheries Service
Gianluca Polgar
University of Malaya
Beth Polidoro
ProCAT Colombia/Internacional
Eduardo Ponce
ProCAT Colombia/Internacional
Narcisa G. Pricope
University of California, Santa Barbara
Diogo B. Provete
Federal University of Goiás
Elizabeth Rholetter Purdy
Independent Scholar
DeeVon Quirolo
Reef Relief Founders
Subekti Rahayu
World Agroforestry Centre
Alexis S. Reed
University of Kansas
Relena R. Ribbons
University of Massachusetts, Amherst
Laura Ribero
University of Malaya
Krishna Roka
Penn State University
Katherine Ruminer
Stephen F. Austin State University
Santosh Kumar Sarkar
University of Calcutta
Elsa Sattout
Notre Dame University
Jan Schipper
ProCAT/University of Hawaii
Franziska I. Schrodt
University of Minnesota, Twin Cities
Stephen T. Schroth
Knox College
Daniela Shebitz
Kean University
Thomas A. Shervinskie
Pennsylvania Fish and Boat Commission
Suman Singh
Banaras Hindu University
Joseph V. Siry
Rollins College
Kymberley A. Snarr
Laurentian University
Katherine Sparrow
Stephen F. Austin State University
Margaret J. Sporck
University of Hawaii at Manoa
Elizabeth Stellrecht
D'Youville College
John Richard Stepp
University of Florida
Kerry Bohl Stricker
University of South Florida

Lindsey Noele Swierk
Pennsylvania State University
Hesti Lestari Tata
Research and Development Centre for Conservation and Rehabilitation
Yasmin M. Tayag
New York University
Brian W. Teasdale
Kean University
Lida Teneva
Stanford University
Janelle Wen Hui Teng
Stanford University
Patricia K. Townsend
University of Buffalo
Clay Trauernicht
University of Tasmania
Marcella Bush Trevino
Independent Scholar
Melanie L. Truan
University of California, Davis
Krishna Prasad Vadrevu
University of Maryland
Francisco Javier Gómez Vargas
University of Barcelona
Monika Vashistha
Guru Gobind Singh Indraprastha University, Delhi
Karthikeyan Vasudevan
Wildlife Institute of India
Ivan Mauricio Vela-Vargus
Sierra to Sea Institute
Troy Vettese
St. Andrews University
Christian Vincenot
Kyoto University
Yvonne N. Vizina
Métis National Council
George Vrtis
Carleton College
John Walsh
Shinawatra University
Sarah M. Wandersee
San Diego State University
Jennifer Rhode Ward
University of North Carolina, Asheville
Judith S. Weis
Rutgers University
Ken Whalen
Universiti Brunei Darussalam
Robert C. Whitmore
West Virginia University
Jeffrey Wielgus
Ocean Conservancy
Benjamin Theodore Wilder
University of Arizona
Stephen Wood
Columbia University
Jennifer Stoudt Woodson
Gordon College
Sarah Wyatt
Yale University
Kien-Thai Yong
University of Malaya
Diego Zárrate-Charry
ProCAT Colombia/Internacional
Jian Zhang
University of Alberta

Illustrations

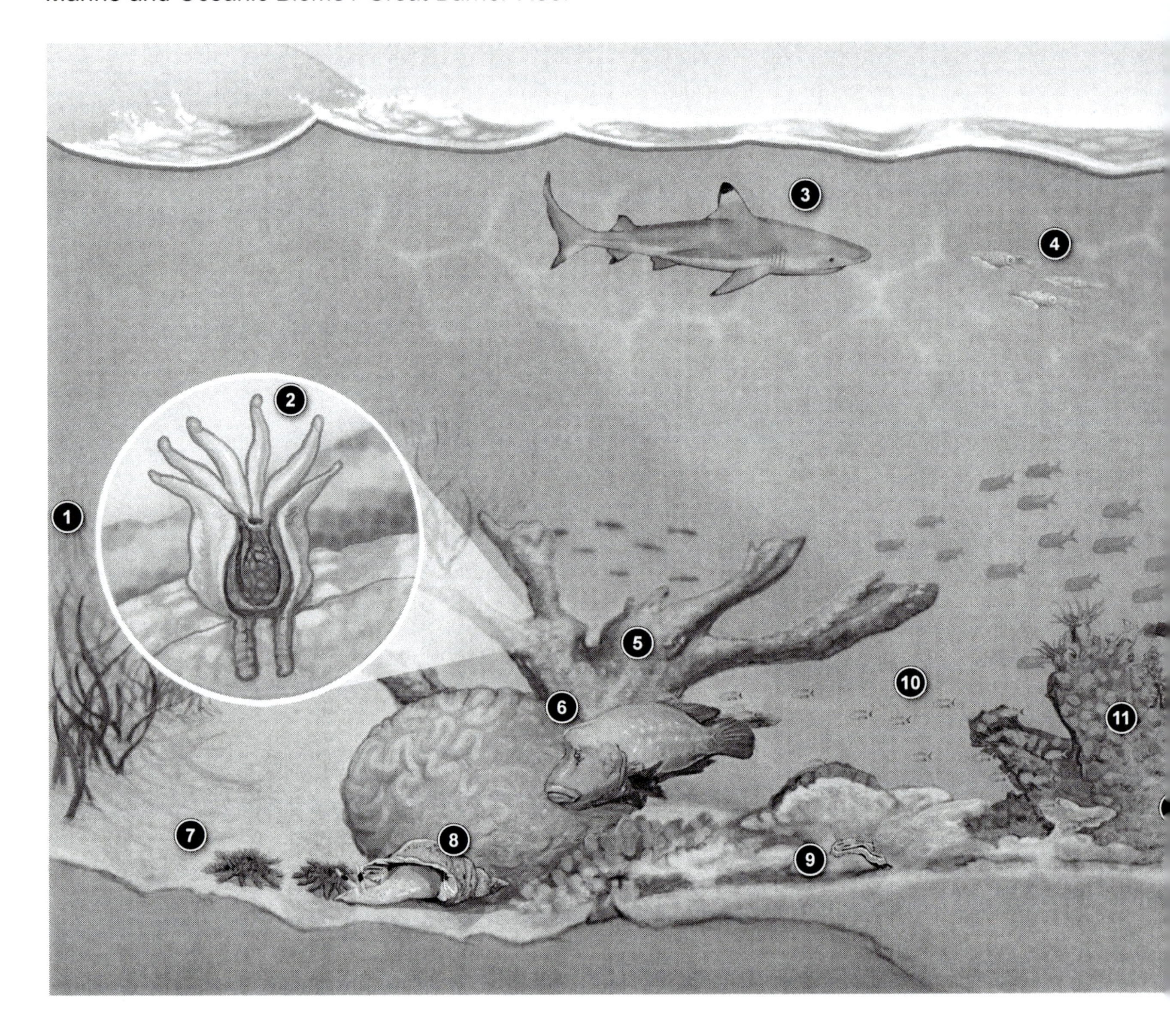

1. Seagrass *(Halodule uninervis)*
2. Bubble-tip Anemone *(Entacmaea quadricolor)*
3. Blacktip Reef Shark *(Charcharhinus melanopterus)*
4. Bigfin Reef Squid *(Sepioteuthis lessoniana)*
5. Elkhorn Coral *(Acropora palmata)*
6. Hump Head Wrasse *(Cheilinus undulatus)*
7. Crown-of-thorns Seastar *(Acanthaster planci)*
8. Triton's Trumpet *(Charonia tritonis)*
9. Nudibranchs *(Chromodoris daphne)*
10. Orange Basslet *(Pseudanthias squamipinnis)*
11. Moon Coral *(Favia lizardensis)*
12. Crustose Coralline Algae *(Porolithon onkodes)*

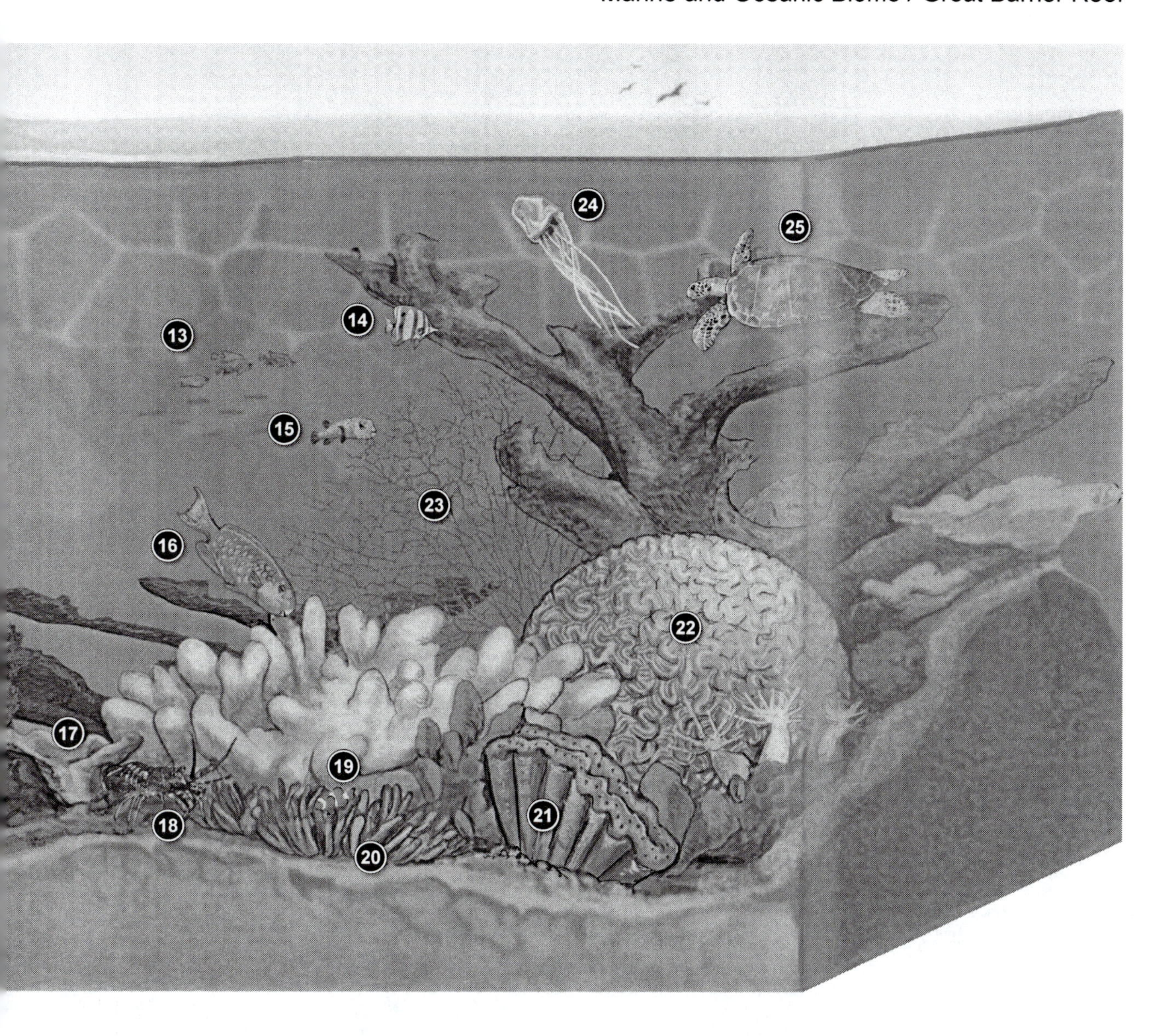

13. Yellow-tail Blue Damselfish *(Chrysiptera cyanea)*
14. Copperband Butterflyfish *(Chelmon rostratus)*
15. White-spotted Pufferfish *(Arothron hispidus)*
16. Yellow-head Parrotfish *(Scarus spinus)*
17. Toadstool Mushroom Leather Coral *(Sarcophyton* spp.*)*
18. Rock Lobster *(Jasus edwardsii)*
19. Clownfish *(Amphiprion ocellaris)*
20. Tube Sponge *(Cribrochalina olemda)*
21. Giant Clam *(Tridacna gigas)*
22. Brain Coral *(Leptoria phrygia)*
23. Gorgonian Fan Coral *(Muricella* spp.*)*
24. Box Jellyfish *(Chironex fleckeri)*
25. Hawksbill Turtle *(Eretmochelys imbricata)*

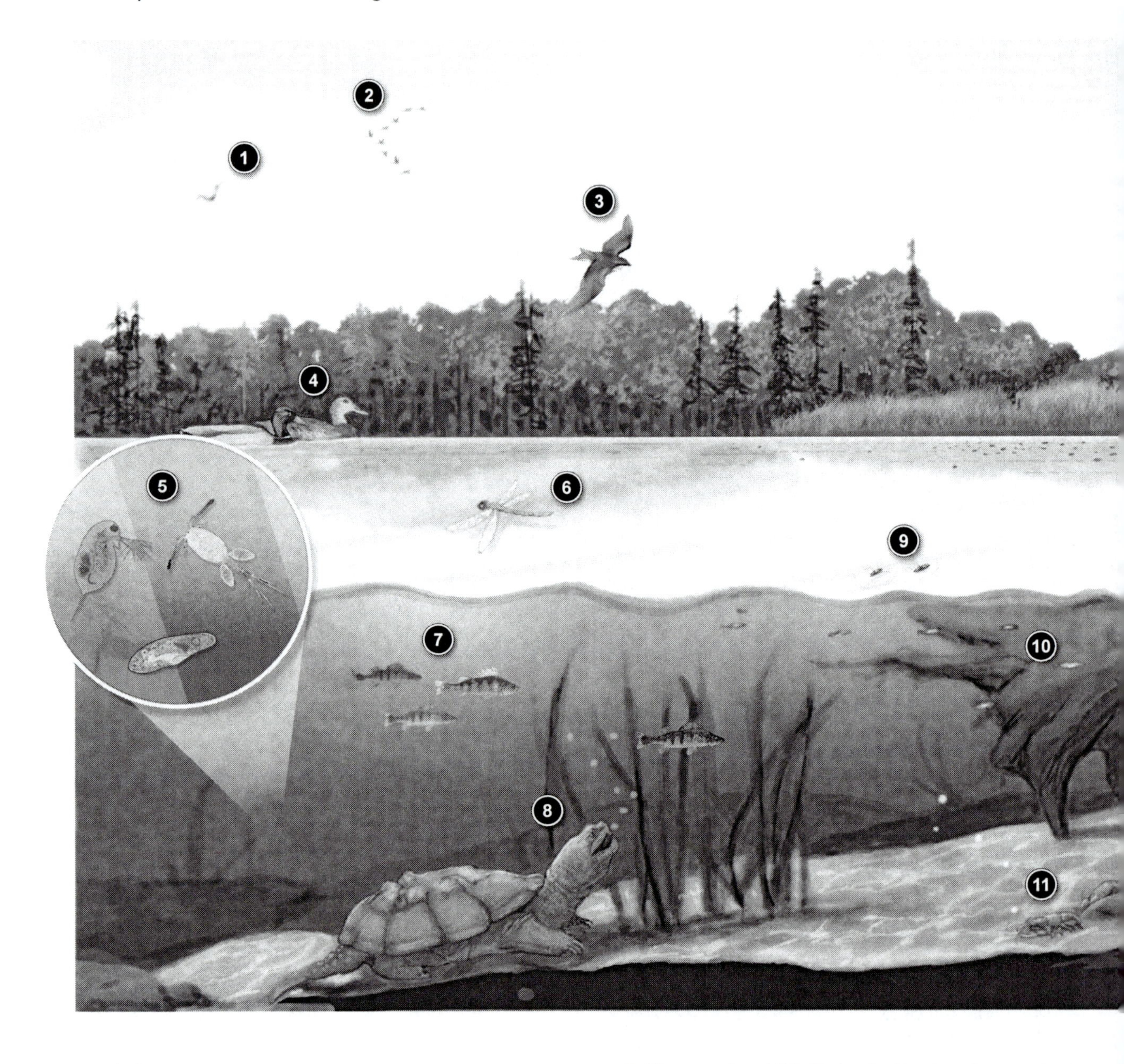

1. Osprey *(Pandion haliaetus)*
2. Canada Goose *(Branta canadensis)*
3. Tree Swallow *(Tachycineta bicolor)*
4. Mallard *(Anas platyrhynchos)*
5. Zooplankton (multiple species)
6. Northern Hawker Dragonfly *(Aeshna juncea)*
7. Three-spined Stickleback *(Gasterosteus aculeatus)*
8. Snapping Turtle *(Chelydra serpentina)*
9. Water Boatman *(Hesperocorixa castanea)*
10. Common Minnows *(Phoxinus phoxinus)*
11. Crayfish *(Cambarus* spp.*)*

12. Great Blue Heron *(Ardea herodias)*
13. Fragrant Water Lily *(Nymphaea odorata)*
14. Water Strider *(Gerris remigis)*
15. Chain Pickerel *(Esox niger)*
16. Pumpkinseed Sunfish *(Lepomis gibbosus)*
17. American Bullfrog *(Rana catesbieana)*
18. Raccoon *(Procyon lotor)*
19. Common Cattail *(Typha latifolia)*
20. Pickerelweed *(Pontederia cordata)*

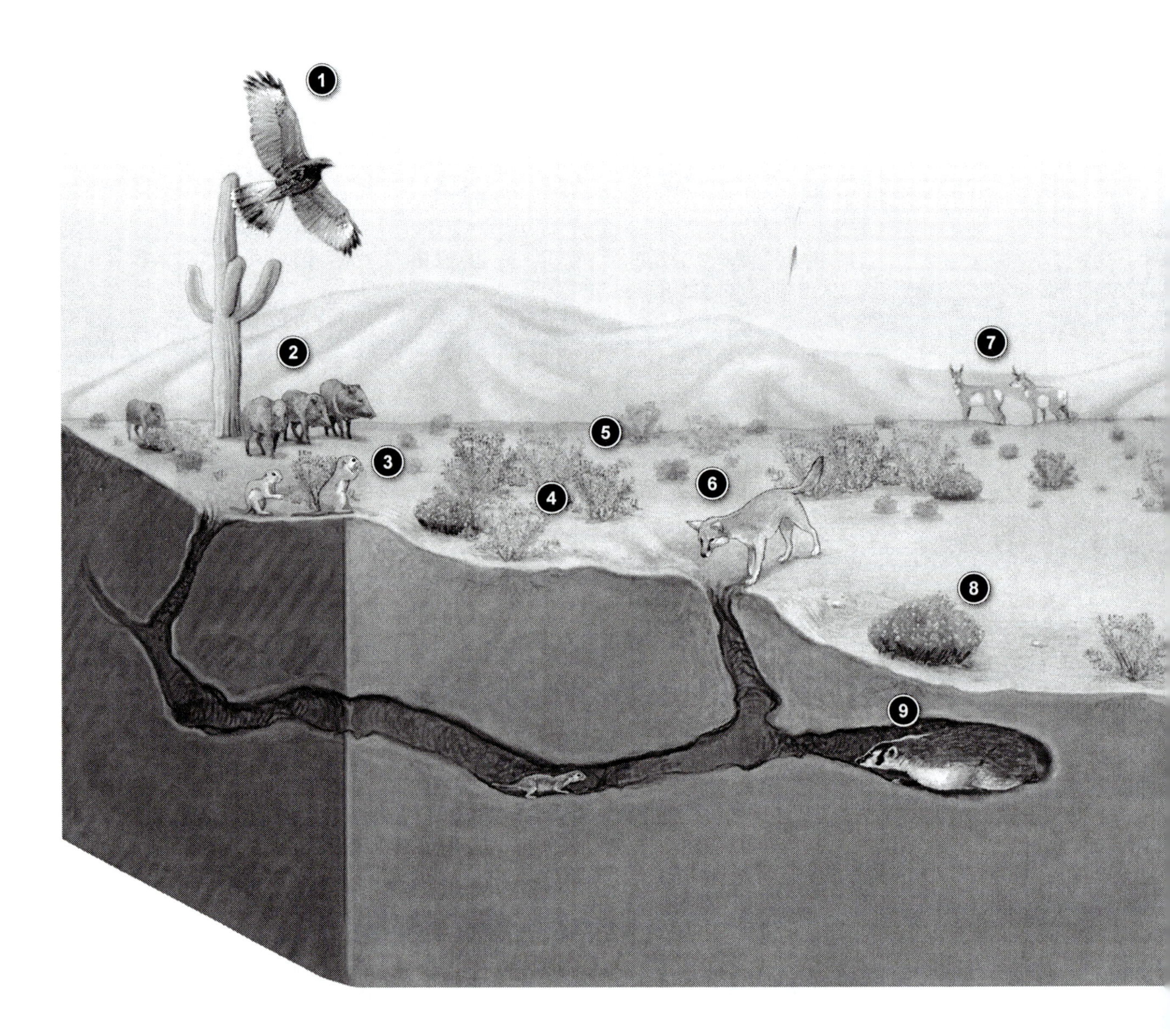

1. Harris's Hawk *(Parabuteo unicinctus)*
2. Collared Peccary *(Pecari tajacu)*
3. Round-tailed Ground Squirrel *(Spermaphilus tereticaudus)*
4. Western Ornate Box Turtle *(Terrapene ornata)*
5. Sagebrush *(Artemisia tridentata)*
6. Kit Fox *(Vulpes macrotis)*
7. Pronghorn Antelope *(Antilocapra americana)*
8. Brittle Brush *(Encelia farinosa)*
9. American Badger *(Taxidea taxus)*

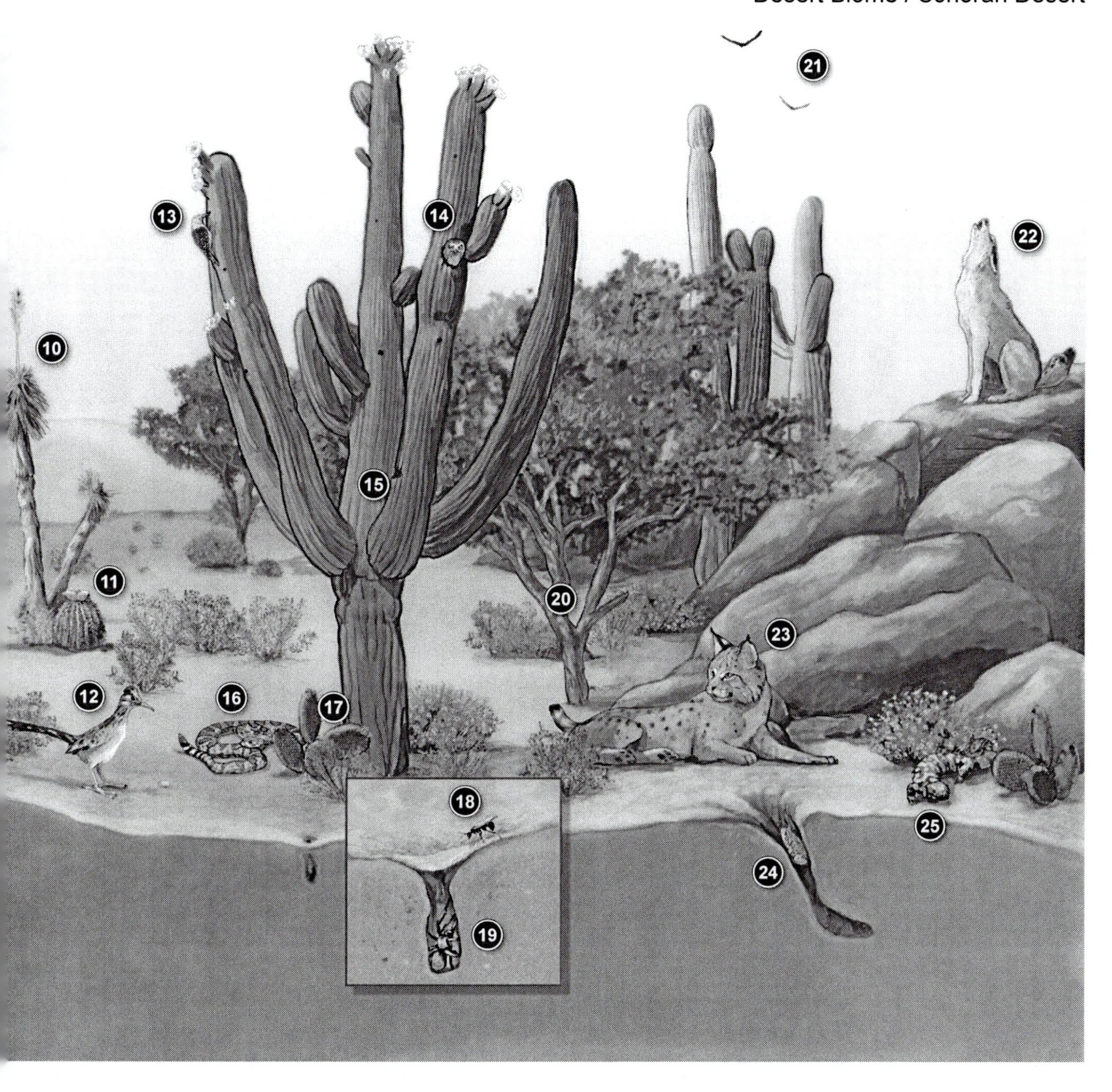

10. Soaptree Yucca *(Yucca elata)*
11. Fishhook Barrel Cactus *(Ferocactus wislezeni)*
12. Greater Roadrunner *(Geococcyx californianus)*
13. Gila Woodpecker *(Melanerpes uropygialis)*
14. Elf Owl *(Micrathene whitneyi)*
15. Saguaro Cactus *(Carnegiea gigantea)*
16. Western Diamondback *(Crotalus atrox)*
17. Pancake Prickly Pear *(Opuntia chlorotica)*
18. Paper Wasp *(Polistes* spp.*)*
19. Trapdoor Spider *(Ummidia* spp.*)*
20. Honey Mesquite *(Prosopis glandulosa)*
21. Black Vultures *(Coragyps atratus)*
22. Coyote *(Canis latrans)*
23. Bobcat *(Lynx rufus escuinipae)*
24. Sonoran Desert Toad *(Bufo alvarius)*
25. Gila Monster *(Heloderma suspectum)*

1. Soprano Pipestrelle *(Pipestrellus pygmaeus)*
2. European Stag Beetle *(Lacanus cervus)*
3. Red Deer *(Cervus elaphus)*
4. Bohemian Waxwing *(Bombycilla garrulus)*
5. Common Raven *(Corvus corax)*
6. European Mole *(Talpa europaea)*
7. Norway Spruce *(Picea abies)*
8. Wild Boar *(Sus scrofa scrofa)*
9. European Aspen *(Populus tremula)*

10. Eurasian Red Squirrel *(Sciurus vulgaris)*
11. Diamond Willow Fungus *(Haploporus odorus)*
12. Buckler Fern *(Dryopteris carthusiana)*
13. Red Campion *(Silene dioica)*
14. European Blueberry/Bilberry *(Vaccinium myrtillus)*
15. Brown Bear *(Ursus arctos)*
16. Red Fox *(Vulpes vulpes)*
17. Schreber's Moss *(Pleurozium schreberi)*
18. Scotch Pine *(Pinus sylvestris)*
19. Great Grey Owl *(Strix nebulosa)*

1. Great Egret *(Casmerodius albus)*
2. Flatwhiskered Catfish *(Pirinampus pirinampu)*
3. Capybara *(Hydrochoerus hydrochaeris)*
4. Arawana *(Osteoglossum bicirrhosum)*
5. Red-bellied Piranha *(Pygocentrus nattereri)*
6. Swan Orchid *(Cycnoches loddigesii)*
7. Jaguar *(Panthera onca)*
8. Brazil Nut Tree *(Bertholletia excelsa)*
9. Liana Vines *(Memora schomburgkii)*
10. Toco Toucan *(Ramphastos toco)*
11. Fig Tree *(Ficus carica)*
12. Common Squirrel Monkey *(Saimiri sciureus)*
13. Giant Philodendron *(Philodendron giganteum)*

14. Harpy Eagle *(Harpia harpyja)*
15. Amazon Tree Boa *(Corallus hortulanus)*
16. Blue Clitocybulas Mushroom *(Clitocybula azurea)*
17. False Bird-of-paradise *(Heliconia bihai)*
18. Leaf Cutter Ants *(Atta cephalotes)*
19. Kapok Tree *(Ceiba pentandra)*
20. Brown-throated Three-toed Sloth *(Bradypus variegatus)*
21. Scarlet Macaw *(Ara macao)*
22. Flaming Star Bromeliad *(Aechmea recurvata)*
23. Stag Horn Fern *(Platycerium andinum)*
24. Giant Blue Morpho Butterfly *(Morpho didius)*
25. Poison Dart Frog *(Ranitomeya amazonica)*
26. Brazilian Tapir *(Tapirus terrestris)*

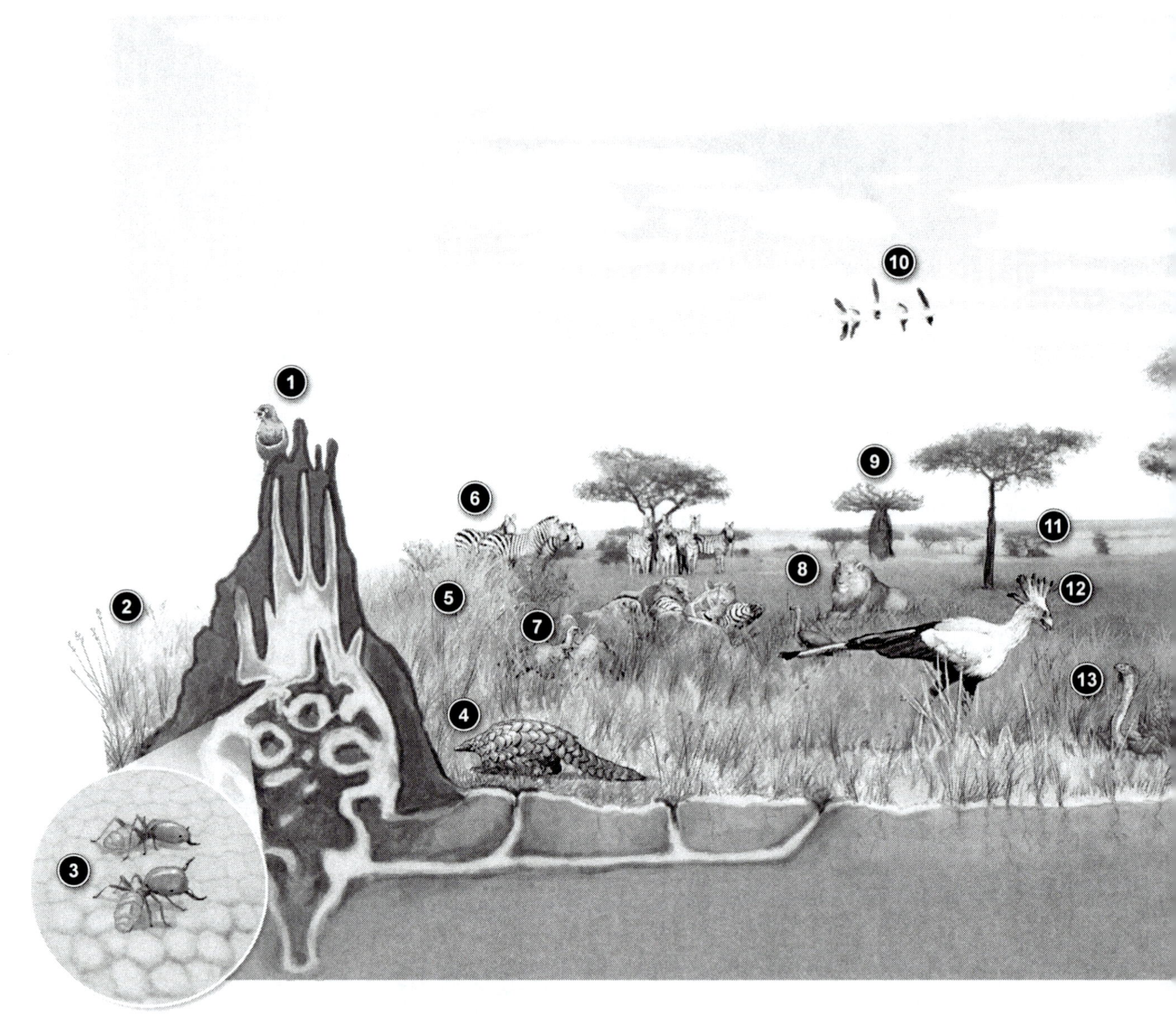

1. Golden Breasted Starling *(Cosmopsarus regius)*
2. Nine Awned Grass *(Enneapogon cenchroides)*
3. Termite *(Trinervitermes geminatus)*
4. Giant Pangolin *(Manis gigantea)*
5. Foxtail Buffalo Grass *(Cenchrus ciliaris)*
6. Zebra *(Equus zebra)*
7. White-backed Vulture *(Gyps africanus)*
8. Lion *(Panthera leo)*
9. Baobab Tree *(Adansonia digitata)*
10. Lesser Flamingo *(Phoenicopterus minor)*
11. African Boxthorn *(Lycium ferocissimum)*
12. Secretary Bird *(Sagittarius serpentarius)*
13. Egyptian Cobra *(Naja haje)*

14. White Rhino *(Ceratotherium simum)*
15. African Wormwood *(Artemisia afra)*
16. Striped Ground Squirrel *(Xerus erythropus)*
17. Honey Badger *(Mellivora capensis)*
18. Curly Leaf Grass *(Eragrostis rigidior)*
19. Giraffe *(Giraffa camelopardalis)*
20. Umbrella Thorn Acacia *(Acacia tortillis)*
21. Candelabra Tree *(Euphorbia ingens)*
22. African Wild Dog *(Lycaon pictus)*
23. Savannah Monitor *(Varanus exanthematicus)*
24. Common Finger Grass *(Digitaria eriantha)*

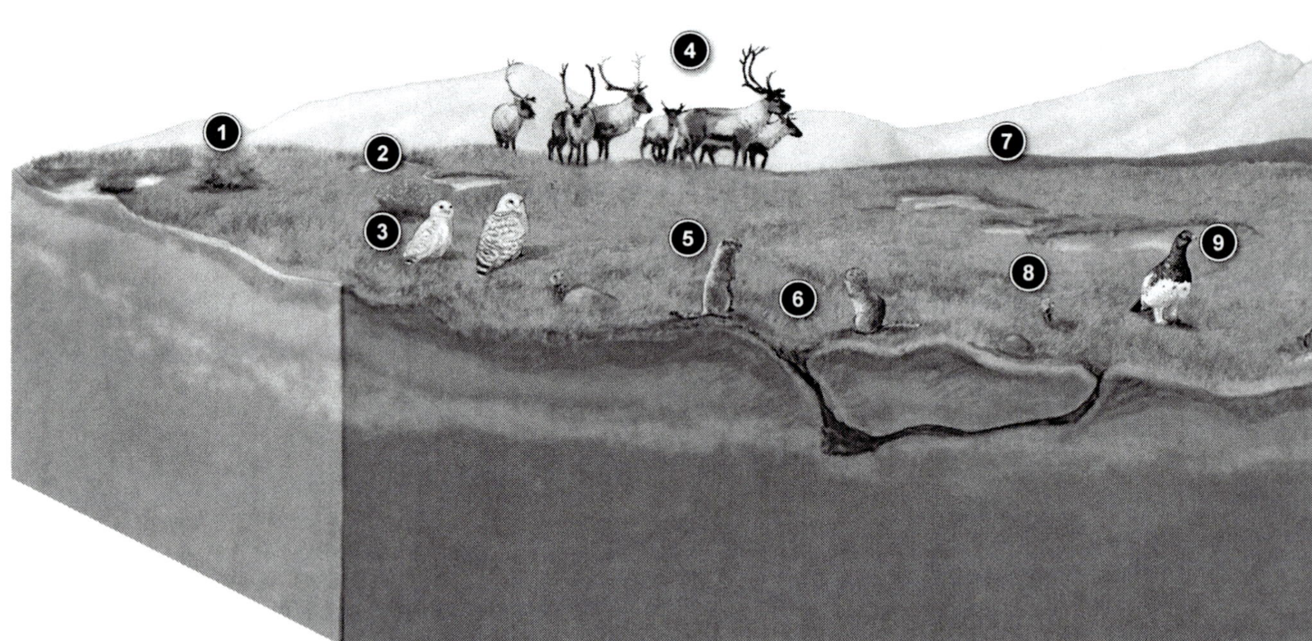

1. Bear Berry *(Arctostaphylos uva-ursi)*
2. Shrubby Cinquefoil *(Dasiphora fruticosa)*
3. Snowy Owl *(Nyctea scandiaca)*
4. Caribou *(Rangifer tarandus)*
5. Arctic Ground Squirrel *(Spermophilus paryii)*
6. Arctic Bluegrass *(Poa arctica)*
7. Gmelin's Sedge *(Carex gmelinii)*
8. Pasque Flower *(Anemone patens)*
9. Rock Ptarmigan *(Lagopus muta)*

10. Caribou Moss *(Cladonia rangiferina)*
11. Purple Saxifrage *(Saxifrage oppositifolia)*
12. Eurasian Wolf *(Canis lupus)*
13. Artic Hare *(Lepus arcticus)*
14. Snow Mosquitoes *(Culiseta alaskaensis)*
15. Arctic Willow *(Salix arctica)*
16. Ermine *(Mustela erminea)*
17. Brown Bear *(Ursus arctos horribilis)*
18. Snow Goose *(Chen caerulescens)*
19. Arctic Fox *(Lagopus alopex)*
20. Alaskan Blackfish *(Dallia pectoralis)*
21. Arctic Moss *(Calliergon giganteum)*
22. Brown Lemming *(Lemmus sibiricus)*

Human (Anthropogenic) Biome / American Suburb

1. Coyote *(Canis latrans)*
2. Grey Squirrel *(Sciurus carolinensis)*
3. Black-capped Chickadee *(Poecile atricapillus)*
4. Blue Jay *(Cyanocitta cristata)*
5. American Robin *(Turdus migratorius)*
6. Turk's Cap Lily *(Lilium superbum)*
7. Dandelion *(Taraxacum officinale)*
8. White-footed Mouse *(Peromyscus leucopus)*
9. Silver Maple *(Acer saccharum)*

10. Harvestman *(Phalangium opilio)*
11. Carpenter Ant *(Camponotus pennsylvanicus)*
12. House Cricket *(Acheta domesticus)*
13. American Raccoon *(Procyon lotor)*
14. Domestic Cat *(Felis catus)*
15. Eastern White Pine *(Pinus strobus)*
16. Rock Dove/Pigeon *(Columba livia)*
17. Herring Gulls *(Larus smithsonianus)*
18. Canadian Yew *(Taxus canadensis)*
19. Kentucky Bluegrass *(Poa pratensis)*
20. Garter Snake *(Thamnophis sirtalis)*
21. Virginia Opossum *(Didelphis virginiana)*

Part 1: Overviews

Marine and Oceanic Biomes

Coastal Seas

Coastal seas are extremely important for biodiversity and ecosystems. Coastal seas can be defined in terms of the continental shelves, as understood geologically, and by reference to international conventions. In geological terms, the continental shelf is the extended perimeter of the continent and the associated coastal plain. The continental margin, which is located between the continental shelf and the abyssal plain, consists of a steep continental slope, followed by a flatter continental rise. Sediment from the continent descends down the slope, and accumulates at the continental rise located at the base of the slope.

Under the United Nations Convention for the Law of the Sea (UNCLOS), the continental shelf is the seabed adjacent to the shores of a country. The waters are also known as territorial waters. UNCLOS states that a country's continental shelf extends to the limit of the continental margin, but no less than 200 nautical miles (370 kilometers) from the baseline. The waters above the continental shelf are also known as territorial waters, and are subject to the sovereignty and control of coastal nations. The legal definition of a continental shelf under UNCLOS differs significantly from the geological definition. Inhabited volcanic islands, such as the Azores and the Canaries, which have no continental shelf geologically, have a legal continental shelf.

Understanding Types of Coastal Seas

Some coastal seas are pelagic seas that are located along an open ocean. Examples of pelagic seas include the North Sea and the South China Sea. Coastal seas may be enclosed seas or semi-enclosed seas. Examples of enclosed seas include the Black, Baltic, and Mediterranean Seas, while the Gulf of Mexico, Gulf of Carpentaria, and Hudson Bay are semi-closed seas. Some seas may have multiple characteristics over time, such as the Wadden Sea, which was a pelagic sea that became enclosed over time through human action. Some coastal seas can be considered estuarine, because they are influenced by river flows and sediment inputs. Examples of these estuarine-influenced seas include the Gulf of Mexico, Mississippi River, Beaufort Sea, Mackenzie River, Adriatic Sea, and Po River. Last, polar seas that are adjacent to the Arctic and Antarctic coasts can be considered a unique type of coastal sea because of their extreme environmental conditions and predominantly pristine nature. Coastal seas, particularly estuaries and adjacent coasts, have played a critical role in human development for most of human history, and are most affected by human activities. Historically, coastal seas were primary commercial fishing grounds. Based on their depth and seafloor, they can provide protection from moderate storms and extreme events. Coastal seas contain abundant life because of sunlight diffusing through the shallow waters over the continental shelves, in contrast to the sparser life of the deeper abyssal plains. Most commercial exploitation from the sea, such as mining and hydrocarbon development, also takes place on continental shelves.

Increasing numbers of the world's people live in coastal regions adjacent to coastal seas. Many of the world's major cities have been built near natural harbors and have port facilities. The coast is also a frontier for military invaders, smugglers, and migrants. Coastal beaches and warm water can be important tourist attractions. As a result of these increasing populations and uses, coastal seas face many environmental challenges from human-induced impacts. While human activity and adverse impacts have been occurring in coastal seas since early times, the decline of coastal seas has significantly accelerated in the last two to three centuries. For example, records from all coastal seas illustrate that seagrass habitats have been destroyed, water quality has decreased significantly, and marine species and diversity have lessened.

A perfect storm of overexploitation, nutrient pollution, and climate change is creating an uncertain future for coastal seas. Land-based and marine pollution includes petrochemicals and biological waste, while climate change contributes to sea-level rise, which threatens coastal ecosystems. Jeremy Jackson of the Scripps Institution of Oceanography refers to the synergistic effects of habitat destruction, overfishing, ocean warming, increased acidification, and massive nutrient runoff that are transforming complex ocean ecosystems. Areas that previously had intricate marine food webs with large fish and mammals are being converted into simplistic ecosystems dominated by microbes, algal blooms, jellyfish, and disease. Knowledge of coastal seas and their original ecosystems is important for understanding the current and possible future ecosystems. Managing and conserving the world's coastal seas and their ecosystems now and in the future will require science-based approaches, commitment, and resources for managing these seas and human activities.

Pelagic Seas

The North Sea is located on the European continental shelf, and connects to the Atlantic Ocean through the English Channel in the south and the Norwegian Sea in the north. The North Sea is an important shipping lane and supports major fisheries. The sea is popular for tourism and recreation, has significant hydrocarbon production, and wind and wave energy generation. The coasts adjacent to the North Sea are subject to high populations and industrialization, and intense uses of the sea and coasts. Environmental impacts on the sea include overfishing, industrial and agricultural pollution, dredging, and dumping. Climate change is affecting its food web and encouraging its fisheries to migrate northward, while new species enter from southern waters.

The European Union has been implementing coastal zone management and marine spatial planning in the North Sea, and there is extensive regional cooperation. The Wadden Sea is adjacent to the North Sea and borders Denmark, Germany, and the Netherlands. It is a pelagic sea that has been modified through a system of dikes and causeways to become a semi-enclosed sea. The Wadden Sea has long been an important source of food and transport for adjacent communities. Eutrophication and nutrient pollution extend over 100 years. Given its significance, a Joint Declaration on the Protection of the Wadden Sea was agreed to by the three adjacent countries in 1982, and a Trilateral Wadden Sea Plan was adopted in 1997.

The South China Sea is bordered by China to the north; the Philippines to the east; Malaysia, Singapore, Indonesia, and Brunei to the south; and Thailand, Cambodia, and Vietnam to the west. The sea lies above a drowned continental shelf that was above the sea during the prior ice ages, when the global sea level was hundreds of meters lower. The South China Sea contains many small islands, atolls, cays, shoals, reefs, and sand-

The Wadden Sea was a pelagic sea bordering Denmark, Germany, and the Netherlands that was modified using dikes and causeways to reclaim land, creating one of the world's most modified coastlines. Now one of the largest coastal wetlands in the world, it is home to more than 10,000 species of plants and animals. (Thinkstock)

bars, many of which are uninhabited and below water at high tide. One of these archipelagos is the Spratly Islands. Jurisdiction over the South China Sea is disputed by adjacent countries, who advocate different management regimes. The sea includes the second-most used sea lane in the world, extensive proven oil reserves, and significant fishing reserves.

The sea contains an estimated one-third of the world's marine biodiversity, making it an important ecosystem that includes mangrove forests, seagrass beds, and coral reefs. Environmental impacts include land-based pollution, overfishing, destructive fishing practices, the loss of mangrove forests, coral reef degradation, and damage to seagrasses and wetlands. For example, most mangrove forests have been lost to shrimp farms, industrialization, or tourism.

Enclosed Seas

The Baltic Sea is a brackish inland sea. The sea resembles a riverbed with two tributaries, the Gulf of Finland and the Gulf of Bothnia, and can be understood as the estuary of all rivers flowing into it. As a result of these inflows, its salinity is lower than that of the Atlantic Ocean. Baltic Sea water flows are complex, with brackish surface waters that discharge into the North Sea, while heavier saline water enters from the North Sea and moves in the opposite direction. The Baltic Sea is stratified, with the denser saline waters remaining near the bottom, resulting in a shifting dead zone along up to a quarter of the seafloor, with only bacteria flourishing in the dead zone. Freshwater and marine species reside in the Baltic Sea, but there is limited diversity. Algal blooms have occurred every summer for decades. Agricultural pollution from adjacent areas has encouraged these blooms and increased eutrophication. Climate change is likely to increase temperature and vertical stratification of the Baltic Sea, and the extent and duration of dead zones and algal blooms. Since the demise of the Soviet Union, all countries around the Baltic Sea are collaborating more extensively in the management of the sea and its watershed and the use of its resources.

The Mediterranean Sea is connected to the Atlantic Ocean and almost entirely enclosed by the Mediterranean region, which is characterized by hot dry summers and wet cool winters. Being enclosed affects many aspects of the Mediterranean Sea. For example, tides are very limited as a result of the narrow connection with the Atlantic Ocean. Evaporation exceeds precipitation and river flows, driving water flows. Evaporation is especially high in the eastern half, causing the water level to decrease and salinity to increase eastward. The sea has been very important throughout human history, with extensive agriculture and industrial development. The Mediterranean Sea and adjacent region is very vulnerable to climate change. Under the changing climate regime, sea surface temperatures and salinity will increase. Depending on local characteristics, erosion, sediment deposition, drought, desertification, and flooding may intensify or shift. Coastal and beach tourism is also an important source of income in the Mediterranean and south Atlantic regions. The ongoing economic viability of these regions and local communities may hinge on the maintenance of the coastal and marine ecosystems on which tourism activity and other activities such as fisheries depend.

"Coastal seas are economically and environmentally important, but are vulnerable to human activities and global and climate changes."

Semi-Enclosed Seas

The Gulf of Mexico is a coastal sea, largely surrounded by North America and Cuba. It is connected to the Atlantic Ocean through the Florida Straits between the United States and Florida, and with the Caribbean Sea through the Yucatán Channel between Mexico and Cuba. The outer margins of the continental shelves of Yucatán and Florida receive cooler, nutrient-rich waters from the deep by a process known as upwelling, which stimulates plankton growth and attracts fish, shrimp, and squid. The Gulf Stream is a strong, warm Atlantic current that originates in the gulf, and warms the coasts of eastern North America and western Europe. A number of rivers flow into the gulf, including the Mississippi and Rio Grande Rivers in the north, and the Grijalva and Usumacinta Rivers in the south. The land around the gulf is mostly low lying, and includes barrier islands, marshes, swamps, and sandy beaches.

Human activities along the coast include fishing, hydrocarbon development, shipping, petrochemical processing and storage, military use, paper manufacture, and tourism. Major environmental threats are agricultural pollution and hydrocarbon activities. There are frequent red tide algal blooms that kill fish and marine mammals and cause health problems in humans. The gulf also contains a hypoxic dead zone that is aggravated by agricultural pollutants. There are also 27,000 abandoned oil and gas wells beneath the gulf that are not consistently monitored.

The 2010 Deepwater Horizon oil spill in the Gulf of Mexico was the largest accidental marine oil spill to date and caused extensive damage to marine and wildlife species, ecosystems, fishing, and tourism throughout the gulf. The oil spill illustrated the vulnerability of coastal seas to offshore activities when those activities are not properly regulated and implemented. Since that time, there have been efforts to improve the regulation of offshore drilling.

Hudson Bay is a coastal sea in northern Canada that drains more than a third of North America. The bay is shallow, with an average depth of 0.06 mile (100 meters), being 851 miles (1,370 kilometers) long and 656 miles (1,050 kilometers) wide. It connects to the Atlantic Ocean through the Hudson Strait, and with the Arctic Ocean through the Foxe Basin. Hudson Bay is part of the north

Atlantic Ocean, but is also viewed as the southern extension of the Arctic Ocean. Hudson Bay has a lower average salinity than ocean water because of its low rate of evaporation, the large volume of freshwater entering the bay, and the limited circulation of this brackish water with the Atlantic Ocean. This lower salinity layer also decreases the overall time that the bay is free of ice.

Because of its large dimensions, the Hudson Bay marine ecosystem has many coastal ecozones and varied habitats that are used year-round by local peoples and Arctic and sub-Arctic species, and seasonally by migratory fishes, marine mammals, and birds. There are few settlements and limited industrial activity in the bay, so it is mostly affected by global environmental and climate change.

The Gulf of Carpentaria is a warm and shallow sea in northern Australia, which measures 115,830 square miles (300,000 square kilometers) and is fed by the flows of over 20 rivers. The gulf is enclosed on the west by Arnhem Land and on the east by Cape York Peninsula. The gulf floor is the continental shelf common to Australia and New Guinea, with a very low gradient. Prawn fishing developed rapidly in the gulf in the late 1960s. Sport fishing is also prevalent. Manganese and bauxite deposits are exploited around the gulf. As a result of these economic developments, settlement on the shores and islands of the gulf has increased from traditional aboriginal populations to several thousand people. The Gulf of Carpentaria Commonwealth Marine Reserve has just been proposed for portion of the sea to protect important turtle nesting and bird breeding areas, submerged coral reefs, and high concentrations of marine life.

Polar Seas

The Beaufort Sea is a complex marine ecosystem within the Arctic. The nearshore productivity of the sea has been an important resource, supporting human occupation for thousands of years. The circulation system is dominated by the Beaufort Gyre, a clockwise rotating surface current that results in large-scale movements of sea ice and surface waters. Ice usually begins to form in October. Most of the marine system is covered by ice until spring breakup, limiting biological production. The Mackenzie River significantly influences the Beaufort Sea. Its freshwaters mix with marine waters, and a relatively fresh, mixed layer forms along the coastal areas and stratifies the ocean waters, allowing numerous species to use the area. Marine mammals, fish, and birds aggregate and use the adjacent area of the sea as migratory routes and seasonal breeding and feeding areas. Polynyas, or open water areas such as the Cape Bathurst polynya, are most important habitats and attract high densities of birds, benthic organisms, and marine mammals. The Beaufort Sea is relatively pristine. There are large hydrocarbon reserves underlying the sea, but they have not yet been extensively developed. Renewable resources, such as fish and whales, are important for both subsistence uses by local peoples and tourism, but there are no commercial fisheries. Ecotourism is increasing in popularity. In Canada, the Beaufort Sea Large Ocean Management Area has been identified for integrated ocean management planning. This is located in the northwestern corner of Canada, and includes the marine portion of the Inuvialuit Settlement Region, established under an indigenous land claim agreement.

The Ross Sea is one the most pristine marine ecosystems on Earth. Located adjacent to Antarctica and without human occupation, it has not been subject to pollution, invasive species, mining, or overfishing. The Ross Sea is the most productive stretch of water in the Southern Ocean, and has high concentrations of wildlife, including large predatory fish, whales, seals, and penguins. Because of its pristine nature and abundance, the Ross Sea provides a unique opportunity to understand the functioning of healthy marine ecosystems. The Commission for the Conservation of Antarctic Marine Living Resources is the international body that manages living marine resources around Antarctica. The commission has made a commitment to designate a network of marine protected areas by 2012, with the Ross Sea identified as one of these areas. The Ross Sea ecosystem may be protected if fishing is eliminated and the Ross Sea, including the continental shelf and slope, are designated as a marine protected area.

Magdalena Ariadne Kim Muir

Coastal Wetlands

Coastal wetlands are flooded periodically by tides, so that resident plants and animals spend part of the time in the water and part of the time in the air. Some of the plants and animals that live in these areas have origins as land species, such as grasses, insects, birds, and mammals. Others, such as mollusks, crabs, and fishes, come from the sea. The plants must be able to live in saltwater and in soil that tends to be waterlogged and low in oxygen. Twice a day, many of the marine animals must be able to survive exposure to the air, sun, wind, and possibly rain at low tide. These are harsh conditions, in which drastic changes occur regularly. Bays, inlets, harbors, and sounds are all places where freshwater from rivers and streams mixes with saltwater from the ocean; these areas are called *estuaries*. Salt marshes are found on protected shorelines and on the edges of estuaries. In the United States, they span the entire east coast, and are extensive along the Gulf of Mexico, but are less common on the west coast.

On the Pacific Coast, most of the shoreline tends to be rocky, so salt marshes are relatively scarce, except in the major estuary/marsh systems of California's San Francisco Bay and Washington's Puget Sound. They are also found in other coastal regions, from the Arctic tundra to the tropics, and on every continent, except Antarctica. Salt marshes are extensive between New Jersey and northern Florida, particularly along the coasts of the Carolinas and Georgia. Further south, in Florida, they are replaced by mangrove swamps, which are the tropical equivalents of salt marshes. Mangroves are not grasses, but trees, and the aerial prop roots of red mangroves provide special habitat for juvenile fishes, some of which, as they mature, move to coral reefs.

The lower portions of marshes on the east coast of the United States are alternately flooded and drained twice a day by the tides, caused by the gravitational pull of the sun and the moon on the water. The term *flood tide* refers to the incoming tide, while *ebb tide* indicates the falling, receding tide. The exact height and timing of the tides at any given place is determined by the phase of the moon and many other factors, such as the shape of the shoreline and the strength and direction of the wind. One extreme case is the Bay of Fundy in Nova Scotia, where there is a difference of almost 50 feet between high and low tides. In some places in the Caribbean, the difference between high and low water is less than one foot. The environmental conditions in a salt marsh are highly variable because the incoming freshwater and ocean water are always mixing in the estuary. If a marsh is located right on the ocean, its water will be more saline than the water in a marsh located further upstream in an estuary that receives much freshwater. The ratio of saltwater to fresh predicts the habitat and resident species. A brackish marsh, for example, occurs in the portion of an estuary where the saltwater is more diluted with freshwater. Further up the estuary, there is a transition to freshwater marshes that are still affected by the tides, although the concentration of salt is low. Still further upstream are freshwater marshes that are untouched by the tides.

There is a parallel transition to less salt-tolerant plants and animals as the water in the marsh becomes fresher. Salinity also fluctuates, depending on the phase in the tidal cycle and the amount of recent rainfall. The same salt marsh may experience salinities ranging from almost freshwater to full strength seawater, and anything that lives there must be able to tolerate these wide swings in environmental conditions. The amount of dissolved oxygen undergoes similar swings as water alternately covers and uncovers the marsh. Temperature also varies widely across the seasons. The air in summer is much warmer than the water, but in winter, the water is warmer than the air. Plants and animals must cope with these variations on a daily basis, but despite these inhospitable and inconsistent conditions, salt marshes are thriving ecosystems.

Coastal Wetland Plant and Animal Species

Certain species of grasses found only in the shallow intertidal areas are highly specialized to deal with saltwater and salty soil. They are called halophytes (Greek *hals,* meaning salt, and *phyton*, meaning plant). Because they have to deal with being submerged in water part of the time, they are also considered hydrophytes (*hydro*, meaning water). In addition to having to tolerate salt and being immersed in water, the lack of oxygen in the waterlogged marsh soil also makes the environment stressful to plants. In east coast marshes, this zone is occupied by cord grass, *Spartina alterniflora*. The low marsh zone in some other regions frequently lacks vegetation altogether; these areas are mudflats. The diversity of plant species increases as one moves higher in the marsh to areas that are only affected by the saltwater during very high tides. In the high marsh, there are grasses, rushes, herbs, and shrubs, with different species adapted to different zones. Zonation is a typical feature of salt marshes, with different plant species found at different elevations, depending on their ability to tolerate immersion and salt, and their ability to compete for space.

Marine animals must find a way to keep moist during low tide, and when the tide is out, there is no food for those that obtain their food directly from the water. Most marsh animals are inactive during low tide, with a few exceptions. Fiddler crabs (genus *Uca*) forage at low tide and remain in their burrows at high tide, safe from predatory fish. Smaller crustaceans, called amphipods and isopods, may also remain active on the marsh surface during low tide, and the salt marsh mussel (*Geukensia demissa*) frequently "air gapes" (gape their shells open for air) at low tide. Virtually all residents of the salt marsh time their lives and their reproductive cycles according to the tides and the phases of the moon. On the highest tides (at the full and new moon), fiddler crabs release their young where tidal currents can carry them out to the ocean. Horseshoe crabs come up onto the beach to lay their eggs at the high tides before the full moon.

"Bays, inlets, harbors, and sounds are all places where freshwater from rivers and streams mixes with saltwater from the ocean; these areas are called estuaries."

Marsh Accretion

Salt marshes develop gradually when tides deposit sediments (soil, fine-grained clay, sand, and silt from rivers and shorelines) across low-lying land, creating wet mudflats or sand flats. Salt marshes develop in sheltered areas that are flat and drain slowly, where mats of microscopic algae and bacteria form and stabilize the sediment so that its elevation can increase. This process speeds up when the plants take root as the marsh ecosystem develops. Salt-tolerant grasses arrive when plants raft in on the tides and seeds are carried in by the air or water. The plants slowly take hold and spread, usually by means of underground horizontal stems or rhizomes, which further stabilize the sediment through the growth of dense root systems and add volume, helping to raise elevations. Once the grasses are in place, the force of wave action is reduced, thus permitting additional layering and settling of the sediment to occur. As the plants decay, fine sediment builds up, causing marsh accretion. As sea levels rise, the marsh pushes landward and seaward, and freshwater plants on the land are replaced by marsh grasses that can tolerate the salt. Underneath the marsh, peat accumulates from the belowground (root and rhizome) plant debris. Peat formation eventually raises the elevation of the sediment surface so that it is flooded less of the time. At that point, other somewhat less salt- and flood-tolerant plant species are able to survive. Eventually, this leads to the development of high marsh areas that are flooded less often by the incoming tides. The bulk of the roots and rhizomes contribute to both sediment deposition and marsh accretion. The rate that matches marshland growth to changing sea level has been about 4 to 10 inches (10.16 to 25.4 centimeters) per century. With climate change causing thermal expansion of water and melting of glaciers, sea level is rising faster than the sedimentation rate in many areas. Unless marsh accretion can keep pace with sea-level rise, marshes may be replaced by open water.

The result of the natural sedimentation is a general spreading of the marsh system and development of the pattern of zonation of plants, in which particular species are located in bands at different elevations. A fully developed marsh includes creeks that form a network over the marsh and are routes used by tidal water as it enters and leaves the marsh, the primary link of the estuary to the land. The general ecology of salt marshes and the roles played by different species of animals and plants are seen worldwide. Salt marshes are important for many reasons. Many species of migratory shorebirds depend on tidal wetlands as stop-over points during their migration between summer and winter habitats, and some birds overwinter in the marsh. Wading birds such as egrets and great blue herons feed in the productive salt marshes during summer months.

About 85 percent of waterfowl and migratory birds use coastal areas for resting, feeding, and breeding habitat. Estuarine animals in various stages of life can be found among the salt marsh plants. At high tide, juveniles of small fishes move up and swim among the stems of the marsh grasses, where they are protected from larger predators that cannot penetrate the shallow waters of the creeks. It is estimated that over two-thirds of the commercially harvested fish on the east coast spend some portion of their life cycle in the marsh-estuary ecosystem. Salt marshes are thus considered nurseries, where many species of fish and shellfish live during the early stages of their lives, depending on the marsh for food and shelter. The plants provide shelter for spawning and protect the juveniles of many different species of commercial and ecological importance, including blue crabs, croakers, flounders, and spot. Bluefish spawn out in the ocean, for example, but the juveniles (snappers) move into estuaries in the spring and remain there through the summer, feasting on the smaller fish that live in the marsh/estuarine system. Some fish, like the mummichog or common killifish (*Fundulus heteroclitus*), spend their entire lives in the marsh/estuary system.

Functional Nature of Marshes

Salt marshes also serve to protect coastal areas from storms and floods. They can stabilize shorelines because they can take the brunt of storm waves, buffering the shore from flood and storm damage. One common source of flood damage occurs following storms when runoff from the land hits the coastal plain. Marshes are natural sponges that can absorb much of this runoff, reducing its impact on coastal environments and real estate. Established marsh grasses are also very effective against erosion. The roots and rhizomes of marsh plants help the sediments cohere and consolidate, resulting in less erosion in vegetated areas. When marshes are removed, the effects of storms on coastal communities are much more severe and devastating, and therefore more costly. In places where marshes are strong, functional systems, storm damage to the land and coastal communities is much less severe. In Sri Lanka, for example, the Indian Ocean tsunami in December 2004 caused much less devastation in areas where mangroves were present than in areas where they had been removed. Hurricane Katrina in 2005 produced a catastrophe in Louisiana in part because of the loss of tidal wetlands.

Marshes also purify water, filtering out pollutants, sediments, and nutrients from the water that washes down from the land. Marshes along the shore slow the movement of runoff from land into the estuary while they increase sedimentation

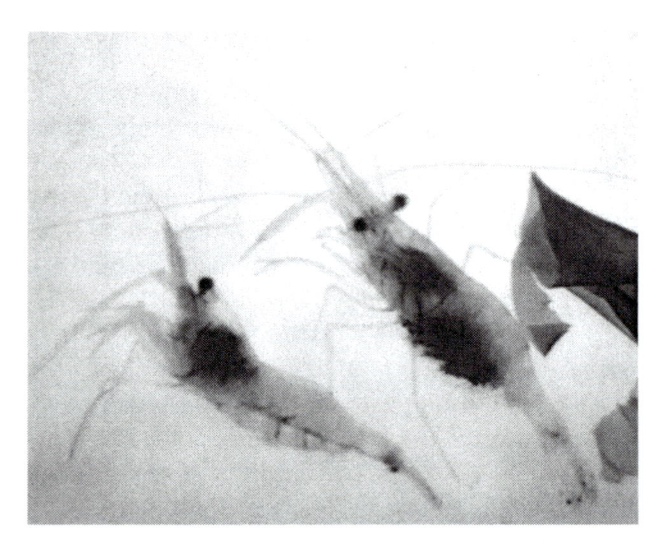

Grass shrimp swim in Winyah Bay National Estuarine Research Reserve in the North Inlet near Georgetown, South Carolina. These creatures, about an inch in size as adults, are important links in the marsh food web. (NOAA)

and the removal of wastes. Salt marsh sediments retain or sequester many kinds of toxic contaminants, helping to reduce the degree of contamination of coastal waters, and intercept nutrients that arrive in runoff from the land, protecting adjacent estuarine areas from excess nutrients that can be harmful. Resident microbes in marshes can remove wastes from the water. They break down and process pollutants in their normal metabolism. In some areas, engineers have constructed wetlands specifically to treat wastewater, a process that capitalizes on this natural ability.

Marshes are highly productive, rivaling tropical rainforests in producing the most basic food energy per acre. They are highly productive because they are rich in nutrients and cycle the nutrients very efficiently. They also have many different types of plants that perform photosynthesis: rooted grasses, seaweeds, and microscopic algae that are found in the water (plankton) and living on the surface of the mud. There have been some attempts to attach a dollar value to the services provided by salt marshes. For example, it has been estimated that the amount of annual storm surge protection services provided to areas most vulnerable to hurricane and tropical storms is $23 billion. However, monetizing their benefits and then making decisions about the conservation or destruction of that region in purely economic terms is an overly simplistic approach. Many of the important functions of a marsh are intangible and its benefits accrue to everyone, so fixing its value is unrealistic, if not impossible. Unfortunately, U.S. coastal wetlands are in decline, and are further threatened by increasing coastal development and rising sea levels.

Coastal Disturbances

Marshes can be disturbed by both natural factors and human activities. Natural disturbance by ice and floating debris can damage temperate marshes, particularly in the colder northern areas. Tidal action can lift up pieces of marsh and deposit them on higher areas, where they literally suffocate the underlying vegetation. Floating masses of dead plant material (*wrack*) are produced in the fall, when some plants die and collapse, and then are swept away by the tides. Winter ice can amass and destroy large chunks of low marsh vegetation. Fire and overgrazing by animals like geese can also destroy large areas of marshes. However, these effects are trivial compared with the damage that results from human activities. Early settlers in America found marshes important for raising livestock, but when agriculture moved inland because of expanded settlement, salt marshes were regarded as useless wastelands that were extensively filled for expanding towns and cities.

Since the 1950s, an average of over 25 square miles (65 square kilometers) per year of Louisiana's coastal wetlands has been lost because of numerous factors, including development and oil and gas exploration and production. The engineering of channels and the system of levees along the Mississippi River prevents regular tidal flooding, stopping the flow and process of salt marsh sedimentation and vertical growth. The normal accumulation of sediments washing down into the marsh cannot occur, and damage from hurricanes is exacerbated. If the area had been undisturbed, accumulated sediments would have elevated the marshes and reduced the effects of the storm surge. The 2005 hurricanes caused the additional loss of 215 square miles (557 square kilometers) of salt marsh, an order of magnitude greater than the average annual loss. According to the Army Corps of Engineers, for every 3 square miles of wetlands, storm surge can be reduced by a foot. Among other lessons, Hurricane Katrina and its aftermath demonstrate the importance that wetlands play in protecting communities from floods.

As more and more people choose to live in coastal areas, impacts on tidal marshes have increased. Today, larger coastal cities and towns hug the shores of estuaries, continuing to encroach on these vital habitats. Cities developed in these locations because they were ideal for trade when goods were carried by ships. Forest and agricultural products could be easily moved downriver to a coastal port. Ports were often dredged out of marshes, and networks of roadways and railroads were built that crossed the marshes in every direction. This construction disrupted the flow of water, impeded tidal exchange, and slowly strangled the marshes.

Marshes have been filled, drained, diked, ditched, grazed, and harvested. They have become depositories for pollutants that wash into streams and rivers from the land, as well as for contaminants introduced from coastal waters. They have been sprayed with insecticides for mosquito control, chemicals that have a wide range of negative effects on desirable estuarine animals and plants. A range of nonnative species have been introduced or arrived accidentally and have altered their ecology. East coast marshes have been affected by a strain of the common reed, *Phragmites australis*, which has taken over many lower salinity marshes, reducing the diversity of native plant species, while the native *Spartina alterniflora* of east coast marshes has become an undesired invasive species on the west coast and in China.

Salt marshes have been treated as something devalued and "reclaimed"—that is, converted to a use that is considered to have more value. There was previously a widespread perception that wetlands were wastelands and pestilent swamps. They were drained, dredged, and filled with dredge materials for urban development, fenced in for livestock grazing, or polluted through use as garbage dumps. In the past several decades, however, people have begun to understand wetlands, to appreciate the "ecosystem services" they provide, free of charge.

Mudflats and salt marshes lack the public appeal and mystique of tropical rainforests and coral reefs, although they are interesting and ecologically valuable. While protective legislation in the 1970s slowed the rate of loss, marshes are still degraded by the invasion of exotic species, poor management practices, development of urban and suburban areas, and the rise in sea level concomitant with climate change. It is estimated that 65 percent of all coastal marshlands and swamps in the contiguous United States would be inundated with 3 feet of sea-level rise. As marshes are destroyed and altered, even if at a slower rate than in the past, the vulnerability to floods and storms, the purity of water systems, the vitality of communities, and the health of populations are all at risk. A myriad of plant and animal species face dire consequences if coastal marshlands continue to be lost or exploited.

In the late 1980s, it was estimated that about 60,000 acres (249 square kilometers) of wetlands were lost per year. Since President George H. W. Bush established a national policy of no-net-loss-of-wetlands in 1989, the overall rate of loss has slowed down. However, from 1996 to 2006, 431 square miles (1,116 square kilometers) of coastal marshes were lost from the United States. Numerous conservation and restoration projects have taken place, with varying degrees of success. The practice of creating new salt marsh habitats to replace ones that have been destroyed elsewhere is widespread. But in many cases, there has been little, if any, follow-up to measure the success of the project. Despite increased public awareness of the value of wetlands, they are still losing ground to development, and some people still view them merely as obstacles in the way of urban expansion. Salt marshes and mangroves also provide unique opportunities for fishing, kayaking, and educational field trips. They are fascinating places to study, and should be valued for their beauty, wilderness, and the sense of peace and tranquility that they provide.

Judith S. Weis

Continental Shelves

Continental shelves are the submerged margins of land around landmasses that extend outward from the intertidal zone. These shelves eventually slope away into much deeper waters, the true ocean, known as the abyssal plane. At their widest, continental shelves can underlie entire seas, such as the North Sea. While continental shelves compose just 8 percent of the entire seabed, they are extremely important marine biomes because of their plant and animal life, as well as nonliving resources.

For most humans, the experience of the marine realm is completely limited to continental shelf habitats, and the value attributed to the ocean as a whole is based on knowledge and perception of these regions. While the geological mineral composition of continental shelves varies considerably throughout their worldwide distribution, usually similar to that of the exposed terrain, they share several common features. Continental shelves are relatively shallow, less than 650 feet (100–200 meters), whereas the open ocean is typically greater than 1,312 feet (400 meters) deep. Waters overlying continental shelves are often termed *shelf seas* to distinguish them from the *high seas,* or the ocean proper. The uses made of continental shelves are also similar the world over. Not only are most marine recreational activities focused in these areas, so too are aquaculture, mining, oil extraction, and other important economic activities. Natural influences on shelf waters are also similar and include deposition, pollution, ice scour (particularly in shallower areas), currents, stratification, plumes, turbulence, and upwelling.

The Earth's landmasses are generally aligned north–south, meaning that major coastlines tend to be eastward or westward facing. There are distinct differences that arise as a result of these individual aspects, caused in part by the rotation of the Earth. Eastern continental shelves, for example, are exposed to some of the Earth's strongest currents, taking the form of gyres, such as the Pacific Gyre and the North Atlantic Gyre.

Further, sea and air interactions contrive in a complex manner, resulting in a greater variation in sea temperature, which in turn has repercussions for the communities of animals and plants that live there. Western continental shelves are notably productive, with four of the world's major upwellings occurring on western shelves; they are generally narrow in their extent, and adjacent landmasses are typically high and mountainous or fjordic (higher latitudes). Desert conditions are common on these shorelines, and riverine inputs are fewer than on eastern shelves. Western shelves of the Americas are particularly vulnerable to El Niño Southern Oscillation (ENSO) effects. These can manifest in periodically higher sea level, elevated water temperatures, and a depressed thermocline.

Habitats

Continental shelf habitats are diverse and heterogeneous, with benthic (seafloor) habitats ranging from silty sands, muds, and clays to coarser gravel, stones, and boulders, and even bare rock. Different substrata support particular types of plants and animals, with organisms specifically adapted to a certain set of conditions. Creatures that live on or in sediment thrive in muddy and sandy areas, and typical animals that might be found in such environments include worms, clams, prawns, lobsters, and flatfish. Rocky areas provide more complex physical habitats and support organisms, including lush algae, anemones, sponges, bryozoans, urchins, starfish, octopus, and more agile predatory fish. The substratum is not the only habitat in continental seas; the water column is also an important habitat for a plethora of pelagic species of fish and invertebrates such as squid, jellyfish, sardines, haddock, whiting, cod, and even larger animals such as basking shark, porpoises, dolphins, and whales.

The marine habitat that humans are most familiar with is the intertidal zone, because it is accessible and readily witnessed. Even this single marine ecotope can offer extreme ambient conditions. At their simplest, intertidal shores may be exposed or sheltered, with levels of exposure rated along a gradient. Substrates may be rocky, stony, gravelly, sandy, or muddy. Salinity changes drastically according to proximity to constant freshwater inputs and the effects of meteorological conditions. Exposure to solar radiation, wind, and predation by terrestrial organisms is periodical, according to the tides, and can change drastically in its intensity according to the season. Fringing systems range from tidal mudflats and sand flats to mangroves, wetlands, marshes, and barrier reefs.

At its oceanward extreme, where the continental shelf becomes the continental slope, the seabed declines steeply. The seafloor of the abyssal plain curves upward at the continental shelf edge at a gentler gradient, to form the deeper continental rise. During this transition from shelf to seafloor, the geology of the benthos also changes from continental crust to oceanic crust. As a habitat, continental slopes and rises tend to be rich in mineral deposits dumped there by oceanic currents, or terrestrial runoff and glacial deposition over geological time when sea level was periodically low. Conditions are much colder in these waters compared to shelf seas, light is limited, and barometric pressure is much higher. Species of the deeper slope and rise area tend to be long-lived, with slower metabolism and growth rates.

Continental Shelf Communities

The relatively shallow waters of the world's continental shelves are some of the richest regions of the ocean with respect to marine life. This is partly because of the nutrient-rich runoff from rivers and streams on land. Shallow waters also allow sunlight to penetrate to the seafloor, which makes photosynthesis possible. Marine flora then sustain invertebrates, such as crustaceans, mollusks, cnidarians (urchins, starfish, and sand dollars), and worms, which in turn sustain fish, sharks, marine mammals, and seabirds. Nature is rarely simple, however, and in the marine realm, food webs are probably more complex than anywhere else on the planet.

Phytoplankton (plant) and zooplankton (animal) form the basis of life on the continental shelf. The world of plankton consists of minute, drifting, single-celled plants and tiny animals, fish larvae and eggs, foraminifera, copepods, radiolaria, and a whole host of other strange and astonishing creatures. Among these, perhaps krill is one of the best known: Although one of the smallest animals on the planet, this crustacean is the main food source of a number of marine mammals and predatory fish by virtue of its abundance, particularly in upwelling zones. Some organisms spend their entire life as plankton (holoplankton), while others are planktonic for just the early stages of their life cycle (meroplankton). Plankton, however, is not limited to continental shelf waters. Seagrass meadows are blankets of lush green foliage that are found on continental shelves throughout the world. These delicate, habitat-forming plants are hugely important in the marine biome, both as a source of food for larger animals such as sea turtles and dugongs, and as a refuge for very rare and endangered species such as seahorses. Seagrass meadows also provide ecological services through the stabilization of sediments and the absorption of excess nutrients in eutrophic runoffs.

Kelp forests are perhaps one of the lesser known assets in coastal waters. Not the most beautiful plant, kelp is rubbery-brown seaweed with a thick stalk that attaches to the seabed, topped with thick, wavy, herbaceous fronds. Despite their lack of aesthetic appeal, kelp stands are one of the most important living habitats in higher latitude waters, both a source of shelter and nutrition for everything from marine invertebrates to otters and seals. The jewel in the crown of coastal waters is coral reefs. These living ecosystems abound with diverse and prolific species, and are typical in the cleanest, warmest waters of the world. It has been suggested that coral reefs are home to about a third of all described fish species in the world. Coral reefs are abundant along western Pacific continental shelves, with Australia's Great Barrier Reef the largest-known coral reef in the world, which is in fact a collection of close to 3,000 reefs, rather than one continuous structure. Toward its northern extreme, the Great Barrier Reef lies only a few miles from the coast, reaching several hundred miles offshore in its southern stretches.

As well as resident species, the transient vertebrate community is important on continental shelves. Marine vertebrates are numerous and diverse, ranging from fish and sharks to turtles, porpoises, whales, and dolphins to seabirds. Fish are among the most prominent life forms in continental shelf waters. They are important for a number of reasons, as prey and predators within ecosystems, as well as a valuable natural resource to humans. Continental shelf habitats are crucial to fish in a number of ways: as nursery habitats, feeding grounds and spawning grounds, and as a highway for diadromous species that spend part of their life cycle in freshwater and part in the sea. Continental shelves are also home to a plethora of marine mammal species, from sea otters to seals, sea lions, walruses, porpoises, dolphins, and whales.

The nutrient-rich waters of the shelf attract both the plankton-feeding baleen whales, as well as toothed predators such as sperm whales and orcas. Larger marine mammals are found throughout a wide range of pelagic conditions, from the shelf to slope and deeper oceanic waters, many will utilize shelf areas at least temporarily. In the northeastern region of the U.S. Continental Shelf, for example, right, humpback, and minke whale species prefer shelf environments, whereas sei and blue whales show preference for the outer slope and oceanic depths.

Living Resources

Most of the world's fisheries are concentrated along continental shelves. Fishing comes in many forms: subsistence, artisanal, commercial, and large-scale commercial. For many thousands of years, marine fish and invertebrates have been exploited for subsistence and trade. Nowadays, fish products are collected not only for human sustenance, but also in the production of several secondary products such as animal feed and fertilizer, alternative medical treatments, and for the aquarium trade. Fishing is carried out with an array of gear and technologies. Gear used in continental shelf fisheries ranges from a handheld rod or line, to pelagic and bottom longlines, gillnets, purse seines, trawls, pots, traps, and dredges. Spear fishing and hand collection through free diving are also common practices in shallower, warmer waters. In U.S. coastal waters, groundfish have long been the most heavily exploited stocks, including fish such as cod, haddock, flounder, and sole.

For many decades, cod was the currency behind North American prosperity; it fed millions and found a fond place at the heart of Western culture. The crash of cod stocks in the northeastern U.S. and Canadian continental shelves was unexpected. For at least 500 years, cod fishing along the Newfoundland and Labrador shelves was highly profitable. It is believed that Portuguese fishermen were crossing the Atlantic to catch these fish, drying and salting them, and shipping them back to Europe even before the long-held discovery of the New World in 1492. From the Georges Bank offshore of New England to the Grand Banks off Newfoundland and further still to the Gulf of St. Lawrence, cod aggregated in huge numbers to spawn.

Fishing these shelves remained sustainable for hundreds of years, but the post–World War II arrival of hi-tech factory ships marked the end of these cod stocks. These new ships could trawl and haul huge nets in virtually any weather conditions, 24 hours a day, remaining at sea for weeks at a time, as well as process and freeze fish onboard, and were equipped with sonar, radar, and "fish-finders." To make matters worse, at the time, these banks were considered international waters, and therefore did not fall under the jurisdiction of any one nation.

Growing mainly on the Pacific Coast along rocky coastlines, giant kelp can grow from around 10 inches to 2 feet a day. This kelp forest at the Monterey Bay National Marine Sanctuary in California provides an important habitat to invertebrates, birds, marine mammals, and fish, such as the Senorita fish above. (NOAA).

Although the creation of the 200 exclusive economic zones (EEZs) in the early 1980s diminished the power of these large vessels, the damage to the stocks was already done. Catches continued to rise until 1968, and then took a turn for the worst, falling slowly until by the early 1990s, the great northeastern cod stocks were depleted.

Oil Extraction and Mining

The outer continental shelf (OCS) is the region of most interest with regard to extracting mineral resources. OCS regions are the seaward extents of the shelf that in the United States are managed under federal jurisdiction. Four regions have been demarcated in U.S. shelf waters: the Atlantic OCS, Pacific OCS, Alaskan OCS, and Gulf of Mexico OCS. The most recent data from the Bureau of Ocean Energy Management reported that there were 3,409 functioning extraction facilities in the Gulf of Mexico OCS in 2010, and 23 facilities on the Pacific OCS, with no exploitation of the Alaskan or Atlantic shelf regions. Following a flurry of activity in the early 1980s, the overall trend in recent years has been downward, with fewer and fewer installations established. In 2010, approximately 30 percent of U.S. oil production was from OCS reserves, a contribution that has increased steadily over time. Natural gas accounted for around 10 percent of the national total. Marine sands and minerals are also highly prized. Since the 1990s, the United States has seen mining of marine sediments along the Atlantic and Gulf coasts for the replenishment of beaches where erosion has taken a toll. Mining activities in the United States within three nautical miles (5.56 nautical kilometers) of the shoreline are generally regulated by individual states, while those beyond fall under federal jurisdiction and are regulated by the Bureau of Ocean Energy Management (formerly the Marine Mineral Service [MMS]).

"Continental shelves are underwater extensions of coastal terrain, rich in valuable living and nonliving resources."

Plate Tectonics and Sea-Level Change

Evidence suggests that over geological time, sea levels have risen and fallen, both exposing and concealing areas of the continental shelf during this process. Sea level can change by two principal mechanisms, eustatic and isostatic change. Eustatic change is climate-driven and linked with ice formation and melt, while isostatic change is driven by the rise and fall of tectonic plates. As little as 21,000 years ago, sea levels were 410 feet (125 meters) lower than at present, and in places the sea would have receded to the edge of the continental slope, exposing the entire continental shelf. Evidence of such changes in sea level and the geological history of continental shelves has been demonstrated by the discovery of fossilized tree remains in deep shelf waters. Continental crust is less dense than oceanic crust, therefore, where these plates meet, the oceanic crust tends to be pushed under, or subducted by, the continental plate.

Often, this causes an oceanic trench to form along the edge of the continental shelf plate edge. The impact causes the terrestrial plate to buckle and lift up, eventually forming mountain ranges, often associated with very narrow continental shelves with steep continental slopes. Although sediment runoff can be considerable from mountain streams, these are usually carried directly offshore to quite deep water. This type of continental shelf is typical along the Pacific coast of North and South America. Earthquakes are common where plates meet, as is changing sea level. This is not because of changes in the water mass, but rather because of movements of the tectonic plates. These movements are characterized by gradual uplifting, interspersed with sudden dropping of the coastline when energy is released during seismic events. Rifts form where plates move apart; these trailing coastlines are typical of much of the African continent, as well as the eastern seaboard of North America. Here, continental shelves tend to be more extensive and more prone to sedimentation.

International Law of the Sea

Throughout history, the sea was seen as a common resource that could be navigated and exploited by all. In the early 17th century, however, the principle of Freedom of the Seas, *mare liberum*, was born. This was a doctrine wherein sovereign nations were given jurisdiction over continental shelf waters within 3 nautical miles (6 nautical kilometers) of shore, while the rest of the sea remained a common resource. By the mid-20th century, concerns about exploitation of fish stocks, pollution, and the potential profits available from both living and nonliving marine resources incited a number of coastal nations to unilaterally extend their claim to govern and manage adjacent waters. In 1945, U.S. President Harry Truman declared national jurisdiction over all continental shelf waters, mineral resources, and underlying geology, in a move to secure valuable oil, gas, and mineral reserves, and in doing so began a revolution in the way that shelf resources are managed. Several South American countries soon laid claim to waters within 200 nautical miles (370 nautical kilometers) of shore, while north African and Arabian Gulf nations extended their territorial claims from 3 to 12 nautical miles (6 to 22 nautical kilometers). The world over, nations were disputing claims to marine resources, from tin to diamonds, gas and oil. The 1958 United Nations Convention on the Law of the Sea (UNCLOS I), which entered into force in 1964, defined "continental shelf" in terms of its use, rather than its geological extent. Here, the continental shelf was considered the seabed and subsoil, beyond territorial seas, extending to a depth of 656 feet (200 meters), and coastal states were given exclusive sovereign rights over its exploitation.

Responding to continued international unrest, the United Nations convened the Third United Nations Conference on the Law of the Sea in 1973, a process that continued for nine years, finally ending in 1982 with the publication of the UN Convention on the Law of the Sea (UNCLOS III). The convention provided for the setting of limits, resolution of disputes, inauguration of EEZs, as well as exploitation, environmental protection, and research. Under UNCLOS III, territorial waters extended to 12 nautical miles (22 nautical kilometers) from the mean low water mark, with a further contiguous zone extending to 24 nautical miles (44 nautical kilometers). EEZs were waters extended up to 200 nautical miles (370 nautical kilometers). The degree and nature of regulation that coastal nations can impose in each of these jurisdictions varies, but EEZs demarcate the waters within which a state has sole rights to exploitation of all natural resources, both living and nonliving. In the United States, management of the continental shelf is shared by state and federal governments. In most cases, these limits are delineated by the 3-mile limit (9-mile limit in Florida), in nautical miles.

The legal definition of *continental shelf* was modified from that of UNCLOS I, and now refers to the natural prolongation of territory to the continental margin or 200 nautical miles (370 nautical kilometers), whichever is greater, satisfying nations with narrow natural continental shelves. To appease nations with a naturally wider shelf, the convention allows the extension of continental shelf designation of as much as 350 nautical miles (648 nautical kilometers) or more where specific geological criteria are met. A further development in UNCLOS III was the creation of the International Seabed Authority, which governs mineral exploitation beyond EEZ limits. The Commission on the Limits of the Continental Shelf was set up to resolve disputes about the extent of shelf regions and their sovereignty.

Ruth M. Higgins

Coral Reefs and Seagrass Beds

Coral reefs are found in the euphotic zone of continental shelves at depths up to 164 feet (50 meters), in water with a mean temperature averaging above 68 degrees F (20 degrees C), and with a salinity ranging from 32 to 42 practical salinity units. They are found in both tropical and subtropical waters, in latitudes from 25 degrees north to 25 degrees south. Coral reefs have greater taxonomic

Reef fish under a natural arch in the Rose Atoll coral reef in American Samoa. The Rose Atoll Marine National Monument was established in 2009 near Pago Pago Harbor and is one of the most pristine atolls in the world. The reef's distinctive pink hue occurs from the color of the primary reef-building species: coralline algae. (NOAA)

diversity per area than any other marine environment, although exact species counts are unknown, and many coral reef species remain undescribed. Coral reef ecosystems cover about 0.17 percent of Earth's surface, and are extremely productive, with gross primary productivity estimates of 3.3–1.1 pounds (1,500–500 grams) of coral per square meter per year; this is about 100 times greater than primary productivity in the ocean's pelagic zones.

Coral reefs are most common in the western Pacific, western Atlantic, and Indian Oceans, where waters are oligotrophic. More eutrophic waters, a consequence of upwelling, mean that coral reefs in the eastern Pacific and Atlantic have reduced species richness and are smaller. Mangrove forests and seagrass beds are often adjacent to coral reefs, and the former communities benefit coral reef ecosystems by absorbing nutrients and slowing rates of sedimentation.

Coral reefs are biogenic structures, composed primarily of calcium carbonate. Coral reefs take up over half of all the calcium ions that enter the world's oceans. Reefs are surrounded by biogenous sediments derived, in part, from the breakdown of this skeleton by bioerosion (particularly by boring sponges) and abiotic processes (particularly through wave action). The main taxonomic groups that create reef structures are hermatypic (reef-building) anthozoans, including stony corals and stony hydrozoans, though crustose coralline algae and mollusks are also important contributors to the reef's infrastructure. Other common organisms found in coral reef communities include sponges, gorgonians (sea fans), echinoderms, and fishes.

Hermatypic corals are cnidarians which grow clonally; about 75 percent of them can also reproduce sexually, and release of sperm and eggs is triggered in response to changes in temperature, day length, or lunar cycles. Corals can feed on plankton using their tentacles, which contain pneumatocysts. They can also acquire planktonic food with their mesenterial filaments, which release digestive enzymes and can be everted to absorb food particles that have not entered the gastrovascular cavity. All hermatypic coral polyps also contain endosymbiotic, mutualistic dinoflagellates, termed *zooxanthellae*, which live between the cells of the polyp's gastrodermis.

The zooxanthellae, which are members of multiple taxonomic groups, are autotrophs that exchange photosynthate and amino acids with their animal hosts in return for protection and nitrogenous waste. Zooxanthellae provide up to 90 percent of their primary productivity to the polyp; this allows the polyp to produce and deposit calcium carbonate, which it embeds in a carbohydrate matrix to form a calyx into which it anchors. Polyps continue to deposit calcium carbonate throughout their lifetimes, allowing the reef's skeleton to continue to grow in size.

Unfavorable environmental conditions can trigger polyps to expel their endosymbionts, a process called bleaching. Such conditions include water temperatures in excess of 95 degrees F (35 degrees C), high light conditions (particularly ultraviolet radiation), or eutrophication. As the mutualism is obligate for many coral species, the polyps often die in the days or months after a bleaching event. Some polyps are able to regain zooxanthellae, and recent studies have shown that zooxanthellae species and genotypes differ in their resistance to stress and likelihood to abandon their host polyps.

Since coral polyps are small (usually less than .12 inch, or 3 millimeters in diameter), and calcium carbonate deposition rates are relatively low, a coral colony grows less than 3.9 inches (10 centimeters) per year. Thus, formation of an entire reef structure takes up to 10,000 years. There are three main types of coral reefs, each of which has its topographic features determined by abiotic factors including light, temperature, the direction and strength of waves and currents, and water depth. Coral reef classification is based on both reef morphology and reef origin.

Types of Reefs

The most common reef type worldwide is the fringing reef, found in narrow bands parallel to tropical shorelines. Fringing reefs are found in nearshore habitats with existing hard substrate to support the settlement of a coral's planula larvae, which can metamorphose into polyps after settlement. Regions of a fringing reef include the reef flat (inshore), which extends to the reef crest and drops off to the reef slope, which interfaces with the open ocean. The world's longest coral reef, in the Red Sea, is a fringing reef. Barrier reefs like Australia's Great Barrier Reef are farther offshore, and a lagoon separates the reef from the shoreline. The reef begins with a back-reef slope, which transitions to a reef flat, then reef crest, which absorbs much of the energy from wave impacts. The reef crest ends in the forereef slopes, which trails off to the deep ocean. Atolls, found mostly in the Pacific and Indian Oceans, consist of a lagoon ringed by a coral reef. The edge of the reef is the reef flat, and the outer slope moves into the open ocean. Atolls are formed around sunken volcanoes, a theory first proposed by Charles Darwin. Fringing reefs first form around islands. As islands slowly subside, reefs gradually gain height through calcium carbonate accretion by hermatypic corals and others. Over long periods of geologic time, the island is replaced by a lagoon.

Interspecific interactions like competition are the primary factors determining community assemblages in coral reefs. Space is at a premium on reefs, and is the primary resource for which organisms struggle. Some scleractinian species actually digest neighboring species with their mesenterial filaments to acquire more space. Bottom-down interactions mediated by herbivory also affect the species composition of reefs. Mutualisms are also common in coral reef systems; these include cleaner symbioses and mutualisms between zooxanthellae and sponges, mollusks, or other cnidarians. Herbivores are important grazers of macroalgae, and reductions in macroalgae densities allow coral polyps and their symbionts to acquire sufficient light.

Robert Costanza et al. valued coral reef ecosystems at $6,075 per hectare per year, with a global value of over $375 million per year. Coral reefs contribute to local economics by attracting tourism dollars, and they are important sources of organisms for commercially important, as well as sustenance, fisheries. Coral reefs also prevent shoreline erosion and dampen the energy of waves associated with storm activity.

Threats to Coral Reefs

Natural threats to coral reefs include damage from wind and waves associated with hurricanes and cyclones. The slow growth rates of the reef skeleton means that recovery times for reefs can be long. Predation from fishes and invertebrates (polychaetes, mollusks, crustaceans, and echinoderms) is also a threat to coral polyps. Extreme low tides or drops in water level can cause physiological stress to polyps and trigger bleaching. Though coral polyps secrete mucus, this mucus provides only limited protection from desiccation.

Nearly 60 percent of the world's coral reefs are at immediate risk because of human activity, and these threat levels have increased over 30 percent from 1997 to 2007. The most threatened ecosystems are found in southeast Asia. Anthropogenic threats to coral reef ecosystems include overfishing, which can cause trophic cascade effects that allow macroalgae to overgrow coral reefs and trigger bleaching. Fishing practices that use toxins like cyanide can kill the corals' sessile polyps, and dynamite-based fishing or trawling can destroy reef structures. Pollutants can cause eutrophication, can be toxic to the polyps, can trigger bleaching, can reduce water clarity so that zooxanthellae are below their photosynthetic compensation points, or can interfere with episodic sexual reproduction events.

Sedimentation or siltation is particularly harmful to coral polyps, because it can reduce rates of photosynthesis or actually suffocate polyps. Although viral, bacterial, and fungal diseases occur naturally in coral reef ecosystems, the incidence of disease is heightened by other stressors to corals, including increased ultraviolet radiation or temperature. Finally, the effects of global climate change are particularly problematic for sensitive coral reef communities. Increased oceanic temperatures can contribute to increased bleaching rates and resultant coral reef mortality. Rising levels of atmospheric carbon dioxide diffuse into oceans, where they decrease water pH in a process known as ocean acidification. Carbonate ions combine with carbon dioxide and water, producing bicarbonate while decreasing the relative availability of carbonate. Since hermatypic corals deposit carbonate in their exoskeletons, the ability of reefs to maintain or grow their structures has been impeded.

When threats of climate change and direct human impacts are considered together, the World Resources Institute estimates that over 75 percent of the world's coral reef ecosystems are threatened. Although 27 percent of the world's coral reef ecosystems are in Marine Protected Areas, enforcement of protection laws is uneven, and many of the world's most threatened reefs completely lack protection.

Seagrass Beds

Seagrasses are submerged aquatic angiosperms that live in relatively low-current regions within the euphotic zone of estuarine or saline waters. They have a broad thermal tolerance, living in waters with average temperatures ranging from 21 to over 104 degrees F (minus 6 to over 40 degrees C), and can tolerate salinities from 0 to 50 practical salinity units. All reproduce asexually via rhizomatous growth, and this process contributes to most of the growth rate of seagrass populations. Seagrasses, which comprise less than 0.02 percent of the described flowering plant flora, are distributed across two families (Potamogetonaceae and Hydrocharitaceae) and a total of 12 genera. Of these genera, seven are tropical and five are temperate. All but one seagrass species (*Enhalus acoroides*, the most recent to evolve) utilize hydrophilous pollination from simplified and reduced flowers. The number of seagrass species is estimated to range from 50 to 60. The precise number is still in dispute, as molecular and morphological taxonomists struggle to resolve their data. Other factors contributing to the confusion surrounding seagrass taxonomy is the difficulty in finding reproductive organs, most useful in anatomical classification schemes, for many species, and the fact that many vegetative traits used in classification are highly plastic.

Seagrasses evolved from either terrestrial (marsh) species or freshwater macrophytes, and the sparse fossil record indicates that they invaded coastal marine habitats about 100 million years ago. Biochemical and molecular evidence supports the polyphyletic origin of this evolutionary group. Current levels of population genetic diversity vary widely among species and regions, and are affected greatly by seagrass species' extensive rhizomatous (clonal) reproduction.

Habitat requirements restrict the global distribution of seagrass-dominated biomes. Although some abiotic needs are species-specific, all seagrasses require saline waters, substrate in or on which to root, and high levels of light. Salinity levels ideal for seagrass growth range from 10 to 45 practical salinity units, with values above or below this resulting in solute stress, necrosis, and/or greater disease susceptibility, particular to fungal pathogens. Most species root in substrates with high sand or organic contents, where they absorb nutrients primarily through their root systems. Areas where particulate matter is fine are less likely to be colonized due to increased sediment turnover rates and decreased light penetration.

Rocky intertidal and subtidal regions are frequently inhabited by members of the genera *Phyllospadix*, *Posidonia*, and *Thalassodendron*, which can attach their rhizomes and roots to rocks in intertidal or subtidal regions. Seagrasses require from 4 to 29 percent of incident radiation to stay above their compensation point, the light level at which the glucose generated via photosynthesis and that used by cellular respiration are in balance. These light levels are higher than those estimated for many shade-adapted terrestrial plants and macroalgae.

Because of their high light requirements, seagrass beds are restricted to the shallow, oligotrophic, littoral portions of continental shelves. They are found along the margins of all Earth's land masses, except Antarctica. Seagrasses are often separated into nine different flora based on their distribution: temperate east Pacific, temperate west Pacific, temperate north Atlantic, temperate south Atlantic, Caribbean, Indo-Pacific, Mediterranean, New Zealand, and South Australia. The species diversity of seagrasses is highest in the waters surrounding the Indo-Pacific region. Biogeographic boundaries between such flora are often sudden, and such shifts are usually determined by changes in water density or nutrient availability. Within a single region, species evenness is usually depressed compared to even the most depauperate of temperate regions, and many stands of seagrasses are composed of only a single species. As in temperate flora and other marine communities (including mangals and coral reef ecosystems), species diversity and evenness declines along a latitudinal gradient from the tropics to the poles.

Seagrass Ecosystems

Seagrass ecosystems are rich in macro- and microalgae, which can be benthic, or use the seagrass organs as a substrate. Epiphytic diversity within seagrass communities can be quite rich, composing up to 45 percent of all algae in that ecosystem, and algal primary productivity can meet or exceed that of the submerged aquatic macrophytes. All major phyla of animals can be found residing in seagrass beds, in infaunal, epifaunal, or motile forms. Experimental manipulations as well as mensurative studies have shown that community composition of seagrasses can be controlled by resource limitation, bottom-down forces, or a combination of these.

Although many animals use seagrass blades as a substrate, or graze on the seagrasses' epifloral, epifaunal, and periphyton communities, few consume seagrasses directly as a major component of their diets. This is because of the high cellulose content and high carbon to nitrogen ratio in seagrasses, and the fact that many are chemically defended.

Exceptions to the rule of direct consumption are found in communities which lack keystone predators, where large populations of isopods, amphipods, parrotfishes, or sea urchins may overuse their preferred algal food source and prey switch onto seagrasses. Some invertebrates selectively graze the inflorescences or seeds of seagrasses, with potential impacts on their sexual fitness. Grazing of periphyton, which lack structural or biochemical deterrents to herbivory, actually increases seagrass fitness, as it increases the amount of incident light available for photosynthesis and reduced hydrodynamic drag on seagrass leaves. Larger herbivores, like West Indian manatees (*Trichechus manatus*), have diets consisting mostly of seagrass species, and other Sirenia like dugongs (*Dugong dugon*) eat only seagrasses.

"Coral reef ecosystems cover about 0.17 percent of Earth's surface, and are extremely productive, with gross primary productivity estimates of 3.3–1.1 pounds of coral per square meter per year."

Aquatic birds like American widgeons or brant geese might rely on seagrasses for a major part of their diet during migration. After spending their first year of life in the pelagic zone, where isotopic studies have shown them to be omnivorous, green sea turtles (*Chelonia mydas*) have diets that consist exclusively of seagrass species. Grazing at intermediate levels is also beneficial to seagrass productivity and population growth, because the physiological integration among seagrass clones allows the plant to produce increased standing crops of shoots. Isotopic analyses have revealed that seagrasses are also an important component of detrital food chains in both nearshore (adjacent) and deeper oceanic environments. In fact, more than half of the primary productivity of a seagrass bed is connected to other trophic levels in the food web via detritus. Seagrass leaves themselves are rich in aerenchyma, can float over long distances, and can directly export carbon-rich leaves to adjacent ecosystems.

Seagrass beds are both ecologically and economically important. Costanza et al. valued seagrass ecosystems at $22,832 per hectare per year, with a global value of over $4 trillion per year. The extensive root and rhizome systems of seagrasses, as well as their baffle-like blades, allow them to stabilize sediments and improve water quality and clarity. They also do significant carbon and nutrient cycling, sequestering up to 12 percent of the ocean's total carbon and supporting primary productivity levels of as much as 8 grams of coral per square meter per day. Seagrass beds are vital nursery habitats for many organisms, though the number of species that they support and the commercial importance of those species varies by latitude. The hydrodynamic properties of seagrass beds help them entrap planktonic larvae and small nekton.

Seagrass beds also provide shelter from predation for a number of animals, particularly small invertebrates, and are higher than most ecosystems in their levels of secondary productivity. In addition, they are important feeding grounds and nursery habitats for coral reef and mangrove-dwelling fishes; these species often migrate diurnally from reefs to seagrass beds to feed, or may recruit their offspring into the more quiescent seagrass habitat. Seagrass beds house more species than habitats with similar abiotic conditions that lack seagrasses, even those with extensive macroalgal communities. Rates of predation on both fish and invertebrate species are lower in seagrass beds than on bare substrate. Common taxonomic groups found as juveniles or adults as members of seagrass communities include fishes, crustaceans, mollusks, and polychaetes.

Seagrass species are sensitive to disturbance, especially those that reduce light availability, such as sediment loading and dock building. Eutrophication, in addition to creating resource regimes more conducive to the growth of macroalgae, increases the abundance of epiphytes and phytoplankton, which can reduce light availability. Declines in seagrass can occur in response to ice scouring, storm activity, grazing by waterfowl, or infection (particularly wasting disease, a fungal infection). Recent decades have seen dramatic and worldwide declines in seagrass cover, and Frederick T. Short and Sandy Wyllie-Echeverria estimated that most of these declines can be attributed directly to anthropogenic disturbances. Oil spills can reduce seagrass production, as can changes in salinity because of changes in freshwater input. One example of this occurs because of modifications of water flow through the Everglades, leading to hypersaline conditions in Florida Bay. Mechanical damage from prop scarring or dredging activities harms seagrass beds.

Climate change and associated warming of waters might harm temperate species near the tropical edges of their biogeographic distributions. Sea-level rise might also harm seagrasses by increasing water depth and thus decreasing light availability. Current legislation in the United States requires that seagrass losses be mitigated, under provisions of the Clean Water Act. However, effective restoration of complete seagrass communities that mimic the function of undisturbed communities has proven elusive.

JENNIFER RHODE WARD

Estuaries, Bays, Lagoons, and Coastal Rivers

Estuaries are partially enclosed bodies of water with a connection to open ocean in which seawater enters according to the rhythm of the tides. Seawater entering the estuary is diluted with freshwater flowing from rivers and streams, and the pattern of dilution depends on the volume of freshwater, tidal amplitude range, and evaporation rate. These ecosystems are known to thrive where the inflow of both seawater and freshwater provide high levels of nutrients in both the water column and sediments. The high concentration of nutrients supports phytoplankton, which is the primary producer in the ecosystem. Each estuary is unique with respect to its physical, chemical, and biological characteristics. Classification into general groups is made according to the geological processes that formed their embayment or, alternatively, according to their water circulation and mixing characteristics. Dynamics inside an estuary depend on its geographic location, the shape of the coastline and ocean floor, the depth of the water, tidal patterns and currents, local winds, freshwater input, and any restrictions to water flow. Additionally, there is a constant fluctuation in dissolved oxygen, salinity, temperature, and sediment load. At any point, salinity and temperature can vary considerably over time and season. Sediment often settles in intertidal mudflats, and dissolved oxygen variations can limit livability for many organisms. The wide range of salinity, temperature, and other chemical conditions, coupled with high primary productivity, creates environments that support many species of fishes and invertebrates.

Estuary Classification Based on Geomorphology

Four types of estuaries are recognized: coastal-plain estuaries, bar-built estuaries, tectonic estuaries, and fjords. In coastal-plain estuaries, formation occurred at the end of the last ice age, between 10,000 and 18,000 years ago as sea level rose to flood river valleys. These estuaries, often called *drowned river valleys*, are especially abundant on passive margins, such as the east coast of the United States. Their width-to-depth ratio is typically large, appearing wedge-shaped in the inner part, broadening and deepening seaward. Water depths rarely exceed 98 feet (30 meters). Chesapeake Bay, Delaware Bay, and New York Harbor are examples of coastal-plain estuaries. The Thames River in England, the Ems River in Germany, the Seine River in France, the Si-Kiang River in Hong Kong, and the Murray River in Australia are other examples. The Mississippi River delta is an example of a former coastal-plain estuary that has been filled with river-borne sediment.

Bar-built estuaries are formed when a sandbar is constructed parallel to the coastline by wave action and long-shore drift, and the bar separates the ocean from a shallow lagoon. Bar-built estuaries typically develop on gently sloping plains located along tectonically stable edges of continents and marginal sea coasts. They are extensive along the Atlantic and Gulf coasts of the United States in areas with active coastal deposition of sediments and where tidal ranges are less than 13 feet (4 meters).

The average water depth is usually less than 16 feet (5 meters) and rarely exceeds 33 feet (10 meters). Examples include Albemarle Sound and Pamlico Sound in North Carolina, Barnegat Bay in New Jersey, and Laguna Madre in Texas. Some estuaries have mixed characteristics. For example, the Hudson River estuary that passes through New York Harbor and Raritan Bay is primarily a drowned river valley, but the Sandy Hoop spit gives Raritan Bay certain bar-built estuary characteristics.

The formation of tectonic estuaries occurs when a section of land drops or tilts below sea level as a result of vertical movement along a fault, volcanoes, or landslides. Such estuaries are found on coasts along a subduction zone or on a transform fault plate boundary. There are only a small number of tectonically produced estuaries in the world. One example is the San Francisco Bay, which was formed by the movements of the San Andreas Fault system, causing the inundation of the lower reaches of the Sacramento and San Joaquin Rivers.

Fjord-type estuaries are found in drowned valleys that were cut by glaciers when sea level was lower. The U-shaped estuaries are generally steep-sided, both above and below sea level, and are often deep. Depth can exceed 980 feet (300 meters). Many fjords have a shallow sill near the mouth that was formed by sediment deposited at the lower end of the glacier when it flowed through the valley. Fjords are common along the coast of Alaska, the Puget Sound region of western Washington State, eastern Canada, Greenland, Iceland, New Zealand, and Norway.

Estuary Classification Based on Water Circulation

Movement and mixing of freshwater and seawater in estuaries are affected by many factors, including tidal currents and mixing; wind-driven wave mixing; shape and depth of the estuary; rate of freshwater discharge; friction between moving freshwater and seawater layers, and between the water and the seafloor; and the Coriolis effect. For this, estuary circulation is extremely important because major cities are located on estuaries. These cities discharge large quantities of waste, particularly sewage treatment plant effluents and storm water runoff, into the estuaries. Additionally, circulation in many estuaries is altered by piers and other port structures, dredging, filling of wetlands, and construction of levees and other structures. These alterations can affect the life cycles of marine species that inhabit and transit estuaries. Based on the major characteristics in water circulation, five types of estuaries are recognized: salt wedge estuaries, partially mixed estuaries, well-mixed estuaries, fjord estuaries, and inverse estuaries.

In salt wedge estuaries, river output exceeds marine input, and freshwater flows down the estuary as a surface layer separated by a steep density interface (halocline) from seawater flowing up the estuary. The differences in density form a wedge-shaped intrusion. Vertical mixing is slow, but a velocity difference between seaward-flowing river water and the underlying seawater generates internal waves mixing the seawater upward with the freshwater. Almost no freshwater is transferred into the lower seawater layer. Examples of salt wedge estuaries are the Mississippi, Columbia, and Hudson Rivers.

In partially mixed estuaries, freshwater and seawater layers are separated by a relatively weak halocline, and therefore the density gradient between the two layers is much less pronounced. Partially mixed estuaries are most common where tidal currents are relatively fast, river flow rate is moderate, and river current speed does not greatly exceed tidal current speed. Examples include the Chesapeake Bay and Narragansett Bay, the Puget Sound, and San Francisco Bay.

In well-mixed estuaries, tidal mixing forces exceed river output, and no halocline is present. Salinity decreases progressively toward the head of the estuary, although it is uniform from surface to bottom. Some variation may occur during

The Bay of Kotor on the Adriatic Sea in southwestern Montenegro is often confused with a fjord, which occurs when the surrounding valley has been cut by glaciers. The bay is instead a submerged river canyon of the disintegrated Bokelj River and its extremely wet environment has been inhabited since ancient times. (Thinkstock)

the tidal cycle, increasing as the tide floods and decreases as the tide ebbs. This type of estuary tends to be wide and relatively shallow, with limited freshwater inputs and strong tidal currents. Some examples include the Delaware Bay and the Raritan River in New Jersey.

Most fjords have a deep, elongated basin where vertical mixing does not reach the bottom waters, even if tidal currents are strong. Consequently, almost all fjords have a halocline separating high-salinity bottom water from lower-salinity surface water. Fjords are almost never well-mixed. Fjord-type estuaries are found along glaciated coasts such as British Columbia, Alaska, Chile, New Zealand, and the Scandinavian countries.

In arid regions, shallow estuaries can sustain high evaporation rates where salinity is higher than the ocean water outside. High-salinity estuarine water flows seaward in the bottom layer, and ocean water flows landward as a relatively lower-salinity surface layer. Thus, estuarine circulation is inverted. Inverse estuaries generally occur at in subtropical regions (30 degrees north and south), where evaporation is strong and rainfall is low. Some examples include the Red Sea, San Diego Bay in California, and Laguna Madre in Texas.

Estuaries are stressful environments for organisms because of variable temperature, salinity, turbidity, currents, and other environmental factors. Salinity can induce stressful changes and organisms adapted to live in an estuary must be able to cope with widely varying osmotic pressures. Because of the high-stress environment, estuaries support fewer species than the adjacent ocean or freshwater environment. However, there is an abundant supply of nutrients and sunlight, resulting in large biomass.

Despite being an ecosystem with a high stress environment, estuaries are ideal habitats for many larvae and juvenile stages of marine animals because they provide abundant food and substantial protection. Many marine fish and invertebrates use estuaries as nursery rounds and other species briefly visit the estuary to spawn. Along the southeast coast of the United States, more than half of the commercially important marine fish species are known to use estuarine wetlands as nursery areas or for breeding. Because of the abundant supply of detritus, estuaries also support huge populations of commercially important shellfish, including clams, oysters, mussels, and crabs.

Estuaries are also among the most heavily populated areas throughout the world, with about 60 percent of the world's population living along estuaries and the coast. As a result, these ecosystems are suffering degradation by many factors, including sedimentation from soil erosion from deforestation, overgrazing, and poor farming practices; overfishing; drainage and filling of wetlands; eutrophication; and pollution with heavy metals and polychlorinated biphenyls (PCBs).

Bays

A bay is a small area of water or broad inlet set aside from a larger body of water where the land curves inward. Bays have calmer waters than the surrounding sea, since land is blocking waves and often reducing winds. A large bay may be called a gulf, a sea, a sound, or a bight. A cove is a circular or oval coastal inlet with a narrow entrance, and some coves may be referred to as bays. Bays are essential for many bird, fish, and marine mammal species. For example, the Bay of Fundy located on the Atlantic (east) coast of North America, on the northeast end of the Gulf of Maine, has been compared in marine biodiversity to the Amazon Rainforest. Other examples include San Francisco Bay, off the coast in northern California. Other bays include the Bay of Pigs in Cuba, Hudson Bay in Canada, Chesapeake Bay in Maryland and Virginia, and the Bay of Bengal, near India. When large and deep enough, bays become natural harbors that are often of great economic and strategic importance. Many of the great cities around the world are located on natural harbors.

"Ecosystems like coral reefs and seagrass beds are particularly susceptible to environmental changes in coastal river flows."

Shark Bay is located in Gascoyne, the most western point of Australia, in the federal territory of Western Australia, 497 miles (800 kilometers) north of Perth. Shark Bay is currently composed of two bays, several peninsulas, and islands with a coastline that extends around 932 miles (1,500 kilometers) and covers an area of 3,861 square miles (10,000 square kilometers). Shark Bay gathers three of the most important climatic regions of the world, and has great ecological importance. This bay is home to more than 10,000 dugongos, a large community of bottlenose dolphins (*Turciops truncatus*), and is an important reproduction zone for hundreds of species of fishes, cnidarios, and crustacean, including sharks and rays that give its name to the bay. Additionally, Shark Bay has also the largest and most varied area of seagrass in the world, with a cover of about 1,544 square miles (4,000 square kilometers). The bay also hosts a unique community of microbes (located in Hamelin Pool) that are building colonies of algae around 3,000 years old (stromatolites).

The Bay of Fundy is located on the Atlantic coast of North America, on the northeast end of the Gulf of Maine between the Canadian provinces of New Brunswick and Nova Scotia, with a small portion in the state of Maine. It extends 170 miles (270 meters) and it is known for having the highest tides in the world (54 feet, or 16 meters), with the potential of becoming one of the world's greatest producers of tidal energy. The marine biodiversity in the Bay of Fundy is vast; at least 12 species of whales can be found here. An abundant number of dolphins, porpoises, fish, seals, and seabirds populate its waters.

The Bay of Islands is one of the most popular holiday destinations in New Zealand, located in the Northland Region of the North Island, 37 miles (60 kilometers) northwest of Whangarei. The bay is an irregular, 10-mile (16-kilometer) wide inlet and contains 144 islands, many secluded bays, and sandy beaches. This bay has abundant marine life including whales, penguins, dolphins, and marlins. It is also a popular gathering place for sailing yachts and international sport fishermen. The bay is also historically significant because it was the first part of New Zealand settled by Europeans.

The Bay of Kotor is a winding bay on the Adriatic Sea, located in southwestern Montenegro.

This bay is often mistaken for a fjord, but is in fact a submerged river canyon of the disintegrated Bokelj River. Because of its location and peculiar topography, this bay is the second-wettest place after Japan's Kii Peninsula, north of the Himalayas. The Bay of Kotor is considered a picturesque place to visit, with numerous beaches.

Phang Nga Bay is located over 60 miles (95 kilometers) from the island of Phuket and the Malay Peninsula in the Andaman Sea. It is an extensive bay, and since 1981 an extensive section was protected as the Ao Phang Nga National Park. In 2002, this bay was declared a Ramsar Site of international ecological significance, comprising shallow marine waters and intertidal forested wetlands, with 28 species of mangroves, seagrass beds, and coral reefs. Species like the dugong and the black finless porpoise (*Neophocaena phocaenoides*) are also found in the bay.

San Francisco Bay is situated on the Californian coast, and it includes an entire group of interconnected bays. The area around it is considered the second-largest urban area, with approximately 8 million residents. Water drains from the Sacramento and San Joaquin Rivers from the Sierra Nevada Mountains, where both rivers then flow into the Suisun Bay. This water then flows through the Carquinez Strait, which meets with the Napa River at the entrance of the San Pablo Bay. This group of rivers finally drains into the Pacific Ocean through the San Francisco Bay. The bay occupies between 400 and 1,600 square miles (1,040 to 4,160 square kilometers), and it is considered the largest estuary in the Americas. The San Francisco Bay and the Sacramento–San Joaquin Delta are considered California's most important ecological habitats, where organisms like the Dungeness crab, California halibut, and the Pacific salmon fisheries rely on the bay as a nursery. This bay is protected by the California Bays and Estuarine Policy.

Lagoons

Coastal lagoons are among the most common coastal environments, occupying around 13 percent of the world's coast. A lagoon can be defined as a shallow body of water, usually oriented parallel to the coast and separated from the ocean by a barrier, connected to the ocean by one or more restricted inlets. These ecosystems are found on all continents, from arctic to equatorial, and from arid to humid, but are less common on emergent high-latitude coasts. They may have one or more natural entrances from the ocean, which can be permanent or intermittent gaps through the enclosing barriers. Lagoons are characteristically shallow, typically 3–10 feet (1–3 meters), and almost always less than 16 feet (5 meters, with the exception of inlet channels and isolated relict holes or channels), and are subject to tidal mixing. They are strongly influenced by precipitation and evaporation, which results in fluctuating water temperature and salinity.

Salinity can range from completely fresh to hypersaline conditions. Size varies substantially, with surface areas ranging up to 3,938 square miles (10,200 square kilometers), as in the case of Lagoa dos Patos in Brazil. Coastal lagoons serve as material sinks or marine filters because they trap inorganic sediment and organic matter. They often exhibit high primary and secondary productivity, and are valuable for fisheries and aquaculture, and sometimes for salt extraction. These are fragile ecosystems that experience forcing from river input, wind stress, tides, precipitation to evaporation balance, and surface heat balance. Lagoons are classified into three geomorphic types, according to water exchange with the coastal ocean. They are classified as: choked lagoons, restricted lagoons, and leaky lagoons.

Choked lagoons occur along high-energy coastlines, and have one or more long narrow channels that restrict water exchange with the ocean. Tidal oscillations are often reduced to 5 percent or less, compared to adjacent coastal tide. Circulation within choked coastal lagoons is characterized by long flushing times, dominated by wind patterns, and intermittent stratification events from intense solar radiation or runoff events. Examples of chocked coastal lagoons include Lagoa dos Patos and Lagoa de Araruama in Brazil, Lake St. Lucia in South Africa, the Coorong in Australia, and Lake Songkla in Thailand.

Restricted lagoons consist of a large and wide water body with two or more channels or inlets, and have a well-defined exchange with the ocean. Wind patterns influence circulation, and lagoons exhibit a variety of salinities, ranging from brackish to hypersaline. Examples include the Indian River Lagoon and the Pontchartrain in the United States, and Laguna de Terminos in Mexico.

Leaky lagoons are elongated shore-parallel water bodies with many ocean channels along the coast. Leaky lagoons are characterized by numerous wide tidal passes; unimpaired exchange of water with the ocean on wave, tidal, and longer time scales; strong tidal currents; and salinities close to that of the coastal ocean. Examples include the Mississippi Sound in the United States and the Wadden Zee in the Netherlands.

Coastal lagoons are extensive in low-lying coasts and are poorly developed on coasts dominated by high retreating cliffs, such as the Great Australian Bight, and the steep and rocky fjord coasts of Norway, Chile, and southern New Zealand. They are also rare on macro-tidal coasts such as the Bay of Fundy between Canada and the United States, and the Baie de Mont Saint Michel in France. Other examples include the Albemarle Sound in North Carolina; Great Sound Bay between Long Island and the barrier beaches of Fire Island in New York; Isle of Wight Bay, which separates Ocean City, Maryland, from the rest of Worcester County, Maryland; Banana River in Florida; and Lake Illawarra in New South Wales. In India, the Chilika Lake in Orissa, near Puri, and the Vembanad Lake in Kerala are also considered lagoons. Some other well known lagoons include the Lagos Lagoon and the Keta Lagoon in Ghana, Africa.

Coastal Rivers

Rivers are the result of precipitation falling across the land, coalescing into even larger streams and rivers. These ecosystems are one of the most dramatic features of a continent, delivering sediment and nutrients downstream, eroding valleys, and eventually reaching the sea or an inland lake. This water movement shapes terrain, creates a variety of freshwater environments, and allows the evolution of several species of plants, animals, and microbes. Rivers vary in latitude, topography, and size, contributing to a great variation in biodiversity and ecological characteristics. Climate, including temperature and precipitation patterns, further characterizes the landscape and is influenced greatly by global air circulation patterns, ocean currents, shape of the continent, and mountain ranges.

Coriolis forces have a predominant effect in river discharge patterns, and significantly influence the living marine resources of coastal areas. Gravity causes the higher-density oceanic water to sink beneath the less-dense surface water, and mixing processes produce water of intermediate density. Buoyancy-driven flows that are supplied by this mixed water formation cause positively buoyant materials and vertically migrating planktonic organisms to accumulate in the slowly sinking waters of the convergent frontal zone. In this scenario, the distribution of food particles becomes highly concentrated, and feeding organisms may dramatically increase caloric intake per unit. This process tends to work its way upward through the trophic levels.

For this reason, coastal rivers are extremely important to biological population dynamics. Major factors that threaten river biodiversity, and consequently the marine environment, include agriculture, deforestation, freshwater extraction, and human and industrial waste discharge, plus the effects of artificial changes to river hydraulic characteristics. Nutrient pollution, also known as overenrichment of nitrogen and phosphorous from coastal rivers, can lead to phytoplankton blooms, causing estuaries to receive more nutrients per surface area than any other ecosystem. In the United States, for example, more than 60 percent of coastal rivers and bays are moderately to severely degraded by nutrient pollution. Although such problems occur in all coastal states, the situation is particularly acute in the mid-Atlantic states, southeast, and Gulf of Mexico.

High sediment loads and stagnant nutrient-rich organic matter flows, plus excessive extraction of freshwater in coastal rivers, can also lead to high turbidity rates and bottom anoxia in many coastal waters. These changes in nutrients, light, and oxygen can favor some species over others and cause shifts in the structure of phytoplankton, zooplankton, and bottom-dwelling (benthic) communities. For example, blooms of harmful algae such as red and brown tide organisms can become more frequent and extensive, sometimes resulting in human shellfish poisonings and even marine mammal deaths. Other factors, such as how often a bay or estuary is flushed and its nutrients diluted by open ocean water, make some coastal ecosystems more susceptible to nutrient overenrichment than others. Ecosystems like coral reefs and seagrass beds are particularly susceptible to environmental changes in coastal river flows.

The largest coastal system affected by eutrophication in the United States is the dead zone in the Gulf of Mexico, an extensive area of reduced oxygen levels. In the early 1990s, the zone covered an estimated 3,668 square miles (9,500 square kilometers) of the gulf, extending out from the mouth of the Mississippi River. By the summer of 1999, this hypoxic area had doubled to 7,722 square miles (20,000 square kilometers), an area the size of Lake Ontario or New Jersey. Other severely impacted coastal systems in the United States include Chesapeake Bay, Long Island Sound, and the Florida Keys. In Europe, the Baltic, North, Adriatic, and Black Seas have all experienced problems from nutrient overenrichment. In the Black Sea, eutrophication was partially reversed during the early 1990s as nutrient inputs decreased following the collapse of the Soviet Union and fertilizer use in eastern Europe dropped sharply. This decrease was temporary, and the Black Sea has recently reached high levels of nutrient overenrichment.

Maria Jose González-Bernat

High Latitude Oceanic Regions

High latitude oceanic regions, also known as oceanic polar regions, refer to sections of the oceans near either pole, especially the regions within either the Arctic Circle or the Antarctic Circle. The Northern Hemisphere's high latitude area is between the Arctic Circle, which is at 66 degrees 33 minutes north latitude, and the North Pole, sitting at 90 degrees north. Parts of Alaska, Canada, Europe, Russia, and Asia are within the Arctic Circle in this northern high latitude area. The Southern Hemisphere's high latitude area is located between the Antarctic Circle, at 66 degrees 33 minutes south latitude, and the South Pole, at 90 degrees south latitude. Antarctica is located at the South Pole.

The difference between these two high latitude oceanic regions is that Antarctica has a land mass that has ocean currents circulating around it, while the Arctic has no land mass, but a year-round existing ice cover. High winds, high sea state, extreme cold temperatures, seasonal sea ice, and the remoteness of the regions make them unique, harsh places. There is now an increased awareness of the importance of polar regions in the Earth system, as well as their vulnerability to anthropogenically derived change, including global climate change. Both high latitude oceans have sets of currents, each of which influences weather, upwelling—and therefore food webs—and seasons, among many other factors in their corresponding hemisphere. Although these two regions share many commonalities, they are also very different.

The Northern Hemisphere's high latitude Arctic region is a landlocked ocean, covered by pack ice that can persist for several years. The Arctic has large areas of tundra and permafrost and several very large river systems. It contains Greenland, covered by the massive Greenland ice sheet, which is on average 1.24 miles (2 kilometers) thick. In contrast, the Antarctic is made up of a land mass almost entirely covered by the huge east and west Antarctic ice sheets 1.24–2.48 miles (2–4 kilometers thick) that are separated by the Transantarctic Mountains.

The Antarctic continent is completely surrounded by the Southern Ocean, of which 6,177,634 million square miles (16 million square kilometers) freeze over every year, effectively doubling the area of the frozen Antarctic continent. Whereas the Arctic has land connections with other climate zones, the Antarctic is effectively cut off from the rest of the world because of the barrier of the Southern Ocean (the shortest distance is the 503-mile (810-kilometer) wide Drake Passage between South America and the Antarctic Peninsula). Together, the Greenland and Antarctic ice sheets account for more than 90 percent of the Earth's freshwater, and if they were to melt, sea level would increase by about 223 feet (68 meters), 200 feet (61 meters) from the Antarctic ice sheets and 23 feet (7 meters) from the Greenland ice sheet. These differences in land and ice also account for the varied biodiversity in each region.

Despite their very different characteristics and the vast expanses covered by the oceans of the world, they are all interconnected by a large-scale movement of water that is referred to as the *meridional overturning circulation* (MOC), *global thermohaline circulation,* or *global ocean conveyor belt.* The basis of themohaline circulation is that a kilogram of water that sinks from the surface into a deeper part of the ocean displaces a kilogram of water from the deeper waters. As seawater freezes in the Arctic and Antarctic and ice sheets consolidate, cold, highly saline brines are expelled from the growing ice sheet, increasing the density of the water and making it sink. In the conveyor-belt circulation, warm surface and intermediate waters (0–3,281 feet, or 1,000 meters) are transported toward the northern North Atlantic, where they are cooled and sink to form the North Atlantic Deep Water that then flows southward. In southern latitudes, rapid freezing of seawater during ice formation also produces cold high-density water that sinks down the continental slope of Antarctica to form Antarctic bottom water. These deepwater masses move into the South Indian and Pacific Oceans, where they rise toward the surface. The return leg of the conveyor belt begins with surface waters from the northeastern Pacific Ocean flowing into the Indian Ocean, and then into the Atlantic Ocean.

It is not just the temperature and salinity of the deepwater formation in the polar regions that is crucial to the ocean circulation. These water masses are rich in oxygen, and so are fundamental

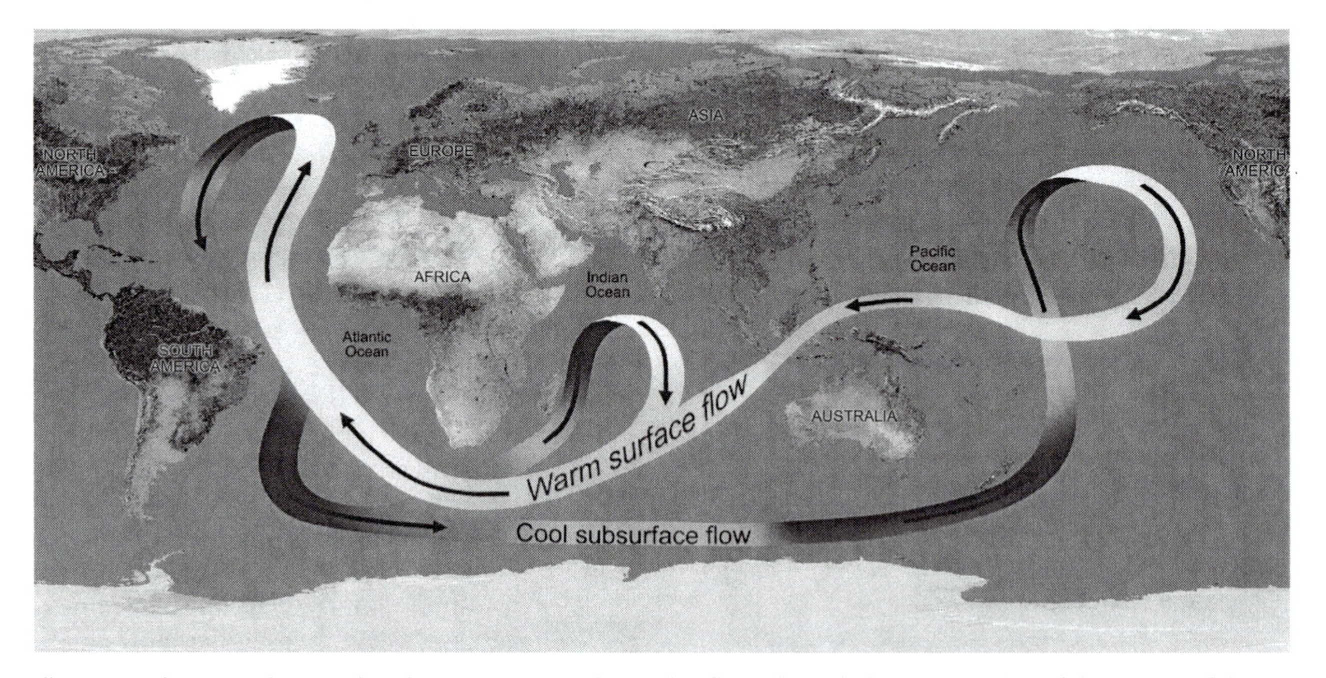

Illustration depicting the meridional overturning circulation that flows through the vast expanse of the oceans of the world. The circulation carries warm waters (band in the foreground labeled "warm surface flow") northward near the surface and cold deepwaters (band in the background labeled "cool subsurface flow") southward. (NASA/JPL)

for transporting oxygen to the ocean depths, where respiration by deep-sea organisms consumes oxygen. The transport of dissolved organic matter and inorganic nutrients is also governed fundamentally by this transportation system, increasing the nutrients that are remineralized during the transfer of the deepwater masses. Therefore, water rising at the end of the conveyor belt in the northeastern Pacific has higher nutrient loading and lower oxygen concentrations than North Atlantic waters at the beginning of the conveyor belt. The ocean circulation in high latitude waters has been determined only relatively recently, using a somewhat different method from that employed for the more temperate ocean currents.

The principal difficulty in studying the polar currents is that large portions of the oceans are covered by ice for a large part or all of the year. There is still a very small database. Whereas the positions of the Gulf Stream and the Agulhas Current have historically been determined by trading ships during their passage through these areas, the general circulation of the Weddell Sea and the Arctic Ocean were initially determined using more dramatic means. Unlike other oceans, the general circulation of polar waters was determined by temperature and salinity analysis carried out with relatively few measurements.

Northern High Latitude Ocean Currents

The Arctic Ocean is a deep basin, almost completely surrounded by continental land masses, with just one deep entry and exit point called the Fram Strait. The Arctic Ocean is permanently covered in ice up to 16.4 feet (5 meters) thick, and this ice cover has made its circulation difficult to determine. Toward the end of the 19th century, several renowned scientists began to believe that there might be a current from the Bering Strait across the North Pole to the Fram Strait, based on the discovery of items from a shipwreck, as well as Siberian fir trees on the coast of Greenland. This hypothesis was not tested until Fridtjof Nansen led the carefully designed *Fram* (Norwegian for "forward") to the Arctic Ocean, where she drifted across the basin from 1893 until 1896, in what we now know as the Transpolar Drift.

Exploration of the Arctic Basin really began in earnest in the 1930s, with the drift of the Sedov and the start of the Russian North Pole annual drifting ice stations, which began in 1937 and continued until 1990. At any one time, there were up to three North Pole camps on the ice, each studying various physical and biological processes. Position data from these drifting ice camps and the addition of modern satellite-tracked drifting buoys deployed on the surface of the ice revealed that the speed of the Transpolar Drift can vary between .39 and 1.57 inches (10 and 40 millimeters). The other main feature of the Arctic Ocean circulation is a large, slow anticyclonic (referring to a large-scale circulation of winds around a central region of high atmospheric pressure, clockwise in the Northern Hemisphere, counterclockwise in the Southern Hemisphere) circulation over the Canadian Basin called the Beaufort Gyre.

At cold temperatures, salinity has a strong effect on controlling the density of seawater. Coupled with the Arctic halocline, this means that most of the transport in the Beaufort Gyre (approximately 80 percent) is in the upper 984 feet (300 meters) of the gyre. The two ocean currents have a significant effect on ice conditions within the Arctic Ocean. The Transpolar Drift moves a large volume of ice from the Siberian Coast toward the Fram Strait. The open water caused by the removal of the ice allows significant ice growth and salt rejection, and may contribute to the maintenance of the halocline. The Beaufort Gyre is responsible for the piling up (ridging) of large volumes of ice on the north Greenland and Canadian coasts, making the mean ice thickness in these places up to as much as 26 feet (8 meters). There is still much we do not know about the circulation within the Arctic Ocean; satellite-tracked drifting buoys have revealed that the Beaufort Gyre has been known to slow down and stop for periods of several months. There are many large research programs planned in order to investigate these features.

As the transpolar drift exits through the Fram Strait, it is responsible for transporting large quantities of sea ice into the Nordic Basin. This ice travels down the east coast of Greenland within the East Greenland Current, which is one of the major sources of water entering the Nordic Basin. Nansen first correctly suggested, and then proved with data from the *Fram*, that the origin of the flow was from the rotation of the Earth and the operation of buoyancy forces along the coast. The other major current of the Nordic Basin is the remnant of the North Atlantic drift, which first becomes the Norwegian Current along the coast of Norway, and then becomes the West Spitsbergen Current at roughly 78 degrees north. The flow of the warm, salty, Atlantic-derived waters is complex at the Fram Strait. A certain fraction enters the Arctic Ocean, while another fraction of this water turns westward and then southward at approximately 79 degrees north to join the flow of the East Greenland Current in the Return Atlantic Flow.

"High latitude oceanic regions, also known as oceanic polar regions, refer to sections of the oceans near either pole, especially the regions within either the Arctic Circle or the Antarctic Circle."

The source waters of the cold fresh Arctic water and the warm salty Atlantic water form a cyclonic gyre, which is closed in its southern section at approximately 72 degrees north by the eastward-flowing Jan Mayen Current. This current is cold and relatively fresh. In winter, it is covered by a famous ice feature called the Odden, which may be an important factor in the driving of the ocean conveyor belt. The cold East Greenland Current leaves the Nordic Basin through the Denmark Strait and turns north at the southern tip of Greenland to become the West Greenland Current. As this current heads north, it collects icebergs from some of the most active glaciers in the world. At the Nares Strait, the current then turns south again to become the Labrador Current. Past Newfoundland, the current, now iceberg-laden, can be a major hazard to shipping.

Southern High Latitude Ocean Currents

The current structure in the southern high latitudes is less complex than in the northern high latitudes because of the absence of land. Antarctica is bounded on all sides by deepwater, and the most significant current in the high latitudes is the Antarctic Circumpolar Current (ACC), or West Wind Drift, which lies between 40 and 60 degrees south. The current flows around the Antarctic continent in a clockwise direction, uninterrupted by land, with surface speeds that can range between 1.64 and 4.92 feet (0.5 and 1.5 meters) per second, and can be considered the ocean analogue of the atmospheric jet stream.

The passage of the ACC around Antarctica is heavily determined by topography, but this is not well defined in the high latitudes because of sea ice cover. The ACC is especially strong below the Agulhas Current and in the Drake Passage, where recent measurements have determined the transport as 130 × 106 cubic meters per second, with an uncertainty factor of roughly 10 percent; the instantaneous flow may, however, differ by as much as 20 percent from this figure. The ACC derives from the Ekman transport, which is to the left of the prevailing wind (the Roaring Forties) and raises the sea surface to the north. The surface slope generates a current that is stable and balanced by the Coriolis force.

The ACC delineates two major frontal regions: at the northern boundary, the Sub-Antarctic Front; and at the southern boundary, the Antarctic Front. The fronts are highly variable, and changes in position of 62 miles (100 kilometers) in 10 days have been observed. Larger meanders in the fronts can form eddies similar to those formed in the Gulf Stream. The greater of the two fronts is the Antarctic Front (southern), which separates the warmer surface waters of northern origin from the cold waters of southern origin. This front is historically identified as the Antarctic Convergence, given that water is converging in this region, but recent measurements have shown that the ACC between the two fronts is zonal and complex in structure throughout its north–south extent, and there is a series of convergences.

Consequently, the region is now termed the *Antarctic Polar Frontal Zone.* It is important for biological processes because deepwater rich in nutrients is brought to the surface, resulting in an area of high localized primary production. The primary production is then grazed by zooplankton, and in particular by krill, which is a staple food for many species of birds, seals, fish, squid, and whales. Close to the Antarctic continent, there is a narrow and westward-flowing current at approximately 65 degrees south, which is called the Antarctic Coastal Current, or occasionally the East Wind Drift. This current has not been well studied, but it has a speed typically of .33 feet (0.1 meter) per second. The current is not, however, continuous around Antarctica, and it is absorbed in the two large gyre systems of the Weddell Sea and the Ross Sea.

The Weddell Sea is one of the few areas of the Antarctic to retain a permanent ice cover. A gyre was first suspected from the drift of the *Deutschland*, which was trapped in the southeast Weddell Sea in 1911, and drifted for nine months before escaping from the pack ice close to the Antarctic Peninsula. A worse fate struck Ernest Shackleton's Imperial Trans-Antarctic Expedition in the Weddell Sea when the *Endurance* was beset just off the coast of the Filchner Ice Shelf in January 1915, and sank in November of the same year. Shackleton and his men drifted within the northward portion of the Weddell Gyre until April 1916, when they landed on Elephant Island at the edge of the peninsula and went on to complete one of the most famous self-rescues in polar history. The existence and continuity of both the Weddell Sea and Ross Sea gyres have been subsequently confirmed from hydrographic measurements and satellite-tracked drifting buoys. The region at the north edge of the Antarctic Peninsula marks the Weddell–Scotia Confluence, a region that has been identified as extremely biologically important.

Alexandra M. Avila

Intertidal Zones

Intertidal zones, which are fundamentally marine, are common in coastal ecosystems worldwide and are subject to the influence of high and low tides on a daily basis, and organisms living on these shorelines are exposed to air during periods of low tides. Many factors that are either directly or indirectly related to the shoreline influence the type of habitat available for colonization of organisms. The effect of heterogeneity in topography has a subtle influence on intertidal shores, with localized irregular features such as cracks, crevices, undersides of boulders, gullies, overhangs, caves, tide pools, damp areas, and areas of freshwater drainage, all of which can serve as refuges from harsh environmental conditions and predators.

The type of substratum, sandstone or basalt, rock type, and rock texture determine how well water drains from the shore. Exposure to wave action varies greatly between intertidal shores because prevailing winds and the slope of the shore influence the type of waves and the point at which they break. Waves break far out from the shoreline on a relatively flat shoreline, then "spill" over the shore. On steep shorelines, waves come close before "surging" up the rock face.

Surging waves on steep shores may increase the effective intertidal zone to many times the height predicted purely from tidal rise and fall. Gentle sloping shores will drain more slowly than steep slopes, allowing organisms that are characteristic of lower zones to expand their ranges into the upper intertidal area. Shores with heavy wave action increase the splash area on the upper intertidal sections that would normally be covered by the tides. Waves can also exert a destructive mechanical effect on the shoreline, encourage scour by sand and shingle, circulate the water, disturb or deposit sediment, renew oxygen, and reduce dissolved carbon dioxide.

They also affect the movement of organisms, limiting feeding raids or keeping away predators. Conditions during emersion are not only affected by the position of the shore with respect to water currents, climate, and latitude, but also by the direction in which the shore faces and the amount of sunlight that it receives. The time at which the greatest low tides occur may be a critical factor, as evaporation and dessication may be intense during the middle of the day, but in places where it

This intertidal zone in Estonia features a tide pool. These habitats can also contain irregular features like cracks, crevices, gullies, overhangs, caves, damp areas, and areas where freshwater drains. Because they are affected by high and low tides, marine organisms dwelling here are exposed to the air on a daily basis. (Thinkstock)

occurs in early morning or late afternoon, there will be less physical stress.

Tidal and wave exposure, in addition to substratum type (among other factors), have a great influence on distributions of organisms on the intertidal zones, and play a role in the myriad of adaptations of form, function, and life history that are observed among the organisms of the shore to counteract the effects of the changing physical conditions of the environment in which they live. On rocky headlands, organisms must withstand the hydrodynamic force generated by the water and tremendous impact of tons of water breaking over them, and here tough or hard bodies and strong attachment mechanisms are an adaptation. Organisms abundant in relatively protected locations are simply not equipped to take the battering and the shearing force of heavy surf. In addition to considering the physical conditions on the shore, vertical zonation is also controlled by the biological interactions that occur between species, such as larval settlement, competition between species and among individuals of the same species, predation, grazing, and behavior. Many mobile organisms have been shown to possess endogenous activity rhythms that trigger activity at each high tide, and may help to maintain the organisms in the appropriate zone, reducing the possibility of being stranded above the normal resting level on shore.

Tides

Periodic emergence in air, or emersion, is an overwhelming feature affecting life on intertidal shores, and the rise and fall of tides over the shore can be referred to as the *emersion-submersion cycle.* The tidal cycle is produced by the interrelation of the gravitational and centrifugal forces between the Earth, moon, and sun on the ocean waters of a rotating Earth. The Earth and moon rotate about each other in space and are held in position relative to one another by the balance of the gravitational forces that tend to attract the two masses, and centrifugal forces that keep them apart. The centrifugal force from this rotation on Earth is everywhere equal in magnitude and direction, with the direction being away from the moon. The moon's gravitational pull is directed toward the moon and is strongest on the side of the Earth closest to the moon, but is directly opposing the centrifugal force of the Earth. Depending on the location on the Earth relative to the moon position, the net effect of these two opposing forces will vary. On the side of the Earth closest to the moon, strong gravitational forces override the centrifugal forces and result in a pull away from Earth (toward the moon). On the opposite side of the Earth, centrifugal forces override the weaker gravitational forces and again result in a net force directed away from the Earth.

Tidal fluctuation is then related to the interplay of gravitational and centrifugal forces as these combined forces would create two bulges of water on opposite sides, one closest to the moon and one farthest from it. Every 24 hours, the Earth rotating on its axis would then seem to rotate under these bulges of water. A point on the Earth's surface thus would experience alternately a high tide, a low tide, another high tide, and another low tide almost every 24 hours. During the time taken for the Earth to complete one revolution on its axis, the moon has advanced somewhat on its orbit, requiring an additional 50 minutes of the Earth's rotation for a point to catch up to its previous position directly opposite the moon. This means that the tides would occur approximately 50 minutes later each day as a complete tidal cycle of two high tides and two low tides takes about 24 hours and 50 minutes, or one lunar day. Tide tables show predictions for the time and height of tides in a number of standard ports around the world.

Differences in height between tides are observed in one day, but larger changes are usually seen over periods of weeks because of changes in the relative tide-raising forces generated by the sun and the moon. The influence of the sun on producing tides on the Earth is only half that of the moon because the sun is much farther away, even though it is much larger. During a new or full moon, when the sun and the moon are nearly in a line with the Earth, their forces combine to produce spring tides, which are of much greater range than average because the tide-raising forces are at a maximum. The word *spring* has to do with the Anglo-Saxon words *to rise*, not from the season.

A few days after these occurrences, high tides are very high, while low tides are very low, thus tidal amplitude is very large. During a first or fourth quarter moon, when the sun and moon are at right angles to each other, the gravitational attraction of sun and moon act in opposition and the tidal forces of the sun partially cancel those of the moon. This produces high and low tides of minimum range in which high tides are not very high, and low tides are not very low. In these cases, tidal amplitude is very small and the tides are called *neap* tides. The monthly cycle contains two neap tides and two spring tides. The highest spring tides occur at the equinoxes (September 21 and March 21), whereas the spring tides are at their lowest amplitude at the solstices (June 21 and December 21).

Global Variations

Tidal patterns and amplitudes vary immensely around the world, as well as times of tidal occurrence. A common one is the semi-diurnal tidal pattern, when there are two high tides and two low tides a day. A simple pattern of two equal tides per day (with a delay of approximately 50 minutes from day to day) is also common, but not universal. These tides can be of different sizes. The overall variation of tidal patterns runs from fully semi-diurnal through semi-diurnal with unequal tides, to diurnal, where there is only one high and one low tide each day. Many factors are involved in determining the type and magnitude of the tidal changes characteristic of a particular location at a particular time, like position of land masses, and local conditions such as weather, barometric pressure, and onshore winds.

As a result, organisms on shore in different parts of the world experience quite different patterns in the times when they are forced to tolerate life in air. Intertidal zones are characterized by many different organisms, which in one way or another have adapted to life in this zone. On shores with similar exposure to wave action, the distribution patterns of organisms are remarkably similar around the world. Almost everywhere, organisms are found in distinct vertical bands or zones at particular heights, and are not distributed randomly; this is known as vertical or tidal zonation. Organisms occur in certain vertical zones, but do not always occur uniformly in those zones. Patches of organisms, or even bare space, frequently occur. One boulder in a boulder field may be covered with mussels, and next to it, another may be covered with algae. The mosaic in the horizontal dimension is a vital contributor to the richness of intertidal life.

Shores that are neither extremely sheltered nor pounded by extreme wave action usually have three distinct vertical zones and are appropriately called the *littoral fringe, eulittoral zone,* and the *sublittoral zone.* The high shore zone, or littoral fringe, often extends well above the levels reached by tidal cover, and is dominated by small snails and black lichen with blue-green algae. The midshore, or eulittoral zone, is characterized by barnacles and limpets, mussels, and algae. The sublittoral zone at the bottom of the shore extends gradually downward to regions well below those that are ever uncovered by the tides, and contains red algae and kelps (e.g., laminarian algae) or large tunicates (as in parts of the Southern Hemisphere). Even though the zones have been defined entirely in biological terms as they are related to tidal levels, zonation patterns on the shore are caused by the relative influence of multiple biological and physical factors.

In many cases, intertidal species are controlled by some physical factor at their upper limit of distribution, while they tend to be controlled by sets of biological factors at their lower limit. A good example is the distribution of the barnacle *Chthamalus stellatus* because its upper limit is set by tolerance to dessication, while the lower limit is set by competition with the barnacle *Semibalanus balanoides* and predation by the dog whelk *Nucella lapillus.* Second, the zones are defined using conspicuous organisms as indicators, but these organisms are far from evenly distributed across the zone. Some eulittoral zones may have clumps of algae forming an irregular and changing mosaic among a background of barnacles. Third, the boundaries between zones are not necessarily sudden, but can occupy a transition belt. For example, barnacles and black lichens might overlap considerably.

Inhabitants of Intertidal Zones

Organisms on intertidal zones can be divided into four large subgroups: algae, grazers, suspension feeders, and predators. Algal growth is a major factor influencing shore ecology because they provide a primary energy source and thus form much of the basis for intertidal food webs. Great biomass of brown, green, and red algae, collectively known as macroalgae or macrophytes, can be abundantly present in some shorelines. Shores that do not have abundant macroalgae usually have a film of microalgae, which is a collective term that includes the blue-green algae, diatoms, and other protists such as the euglenoids, and the spores and sporelings of macro-algae. The microalgae are often more important than the macrophytes in providing a food supply for grazers because, in spite of their low biomass, they have very high rates of production. On most exposed shores, there is a rich assemblage of algae at the bottom of the littoral and sublittoral zones, with brown kelps and often a belt of short red algal turf above them. On sheltered shores, the diversity of algae increases and the whole shore may show extensive algal cover, and in many cases this may be because of growth of brown algae. On all shores, the variety of species increases toward the sublittoral, and red algae tend to take over the browns at the bottom of the shore. Green algae do not usually occur in large masses, but some of their genera are found worldwide.

The grazers are herbivorous gastropods that are specialized for intertidal zones, such as limpets, winkles, topshells, neritids, chitons, and urchins. Some of these organisms can be found worldwide. Herbivores rasp microfilms off rock surfaces, taking in microalgae, sporelings, fungi, and probably algae detritus. Urchins that feed sublittorally may take both live and dead macroalgae. Suspension feeders, which are sessile, such as barnacles and mussels, usually outnumber mobile organisms on rocky shores. On top of rocks, barnacles and mussels may entirely cover rocks, while the underside of rocks and macroalgae are covered with bryozoan colonies, tunicates, sponges, hydrozoans, and polychaetes. Predators such as sea stars, crabs, whelks, and nemertine worms are common on all vertical zones of the intertidal.

The slope of the shore will affect the foraging ranges of the many predators that move up in the intertidal to capture prey during high tide, and then retreat to lower zones at low tide. On steep slopes, such predators can cover greater vertical distances in a given amount of time, and as result, the vertical distribution of both predator and prey may be affected. Despite their mobility, gastropods such as winkles and crustaceans such as crabs usually maintain a well-defined zonation pattern. When covered by high tide, many species roam widely. Some crabs move up-shore as the tide rises, but retreat again as the tide falls. Some species of snails begin to move down-shore as soon as the tide covers them, but return to their usual zone when the tide rises again. Some species, especially grazing gastropods, have been shown to exhibit movements that are random in direction and extent, at least in the short term. But since these animals usually remain within a relatively restricted zone, there must in the long term be some limits to movements in the vertical plane. Very mobile species, such as crabs, show directional movements. Grazing gastropods displaced from their normal position are often capable of returning to it.

"Intertidal zones, areas where land and sea meet, can serve as refuges from harsh environmental conditions and predators."

Wave exposure may influence the distribution of organisms indirectly through the effect it can have on the foraging ability of predators and distribution of sessile organisms. On exposed areas to full wave impact, predators such as crabs or predatory snails may be largely restricted to protective crevices from which they can only make quick foraging raids on prey populations. On the other hand, for other organisms such as barnacles, the vertical distribution may become broader along a vertical gradient of increasing exposure to waves. This is often observed with sessile, upper intertidal organisms as the more frequent wetting of the upper zones by splash and surf in exposed areas lessens the restrictive influence of desiccation and allows these organisms to live higher on the shore than they would under conditions of quieter water.

One of the structural features of intertidal zones are tide pools, which are dynamic and often constantly submerged, undergoing changes in temperature of the water, carbon dioxide concentrations, and pH, depending on the time for which they are isolated from the sea. The amazing diversity of fauna and flora in tide pool systems vary greatly in relation to height on the shore, shelter, shading, size and depth of the tide pool, and also from random disturbance events. Pools high on the shore undergo the consequences of prolonged separation from the main body of the sea, and conditions within them fluctuate dramatically. For example, in tide pools on the Atlantic coast of Cape Peninsula, South Africa, the maximum temperature during the day increases from about 16 degrees C in pools near the mean low water in spring tide, to near 86 degrees F (30 degrees C) near mean high water in spring tides.

The situation with oxygen concentration, carbon dioxide concentrations, and pH is less simple because here organisms directly affect the variables, as well as being affected by them. Salinity may increase from evaporation, or decrease from rainfall and water seepage, and these changes can be overwhelming in high shore pools. Algae in tide pools produce oxygen during the day by photosynthesis, and oxygen concentrations may rise above saturation value, so that bubbles rise from dense beds of algae. Carbon dioxide, absorbed by algae, is followed by a rise in pH, which can be extreme. Tide pools can exhibit equally drastic changes in the opposite direction during the night, when photosynthesis stops.

LUCIA M. GUTIERREZ

Tropical and Subtropical Oceanic Regions

The Earth is tilted by 23.5 degrees and it rotates on its axis. This results in seasons because of the exposed sun angle. Some regions remain less affected and are exposed to the sunlight in a more constant fashion. The regions between the Tropics of Cancer and Capricorn (23.5 degrees north and south latitude) make up the tropics. Its center is the equator, with the Intertropical Convergence Zone (ITCZ) at its core. In concert with the polar regions, it is a global weather engine. Many ocean and wind currents are also found in this region. The tropics are bound in the north and in the south by the subtropical region, where much of the rain falls that is created by the ITCZ. Both regions, tropical and subtropical, make for areas where a rather high biodiversity can be found. A diversity of human life can also be found there.

Easter Island is a remote island in the southeastern Pacific Ocean at the southeastern end of the Polynesian Triangle, which is the triangle of ocean between Hawaii, Easter Island, and New Zealand. (Thinkstock)

Humans fared in this ocean region very well because of the warm temperatures and the superabundance of food and resources, including plants and fish. Approximately 20 percent of the tropical and subtropical region is covered by land, including Polynesia, Greece, Papua New Guinea, Melanesia, and Japan. The historic use of ocean vessels was commonly found in these regions, for example, off Chile, in Polynesia, Portugal, Spain, and in ancient Greece.

Coastal fisheries have been a mainstay of human life in the tropical and subtropical ocean regions. The coastal wetlands of this region, including estuaries and mangroves, play a major role for global well-being. The mangroves of this region are a major contributor when it comes to global carbon cycling. Mangroves are also nurseries for fish, and host congs (shells that were harvested) and marine mammals like dogongs and manatees (many of these species are now endangered, or even extinct). Further, the Amazon, Nile, Congo basin, Ganges, Jangtsekiang, and Mekong Rivers all drain into tropical and subtropical oceans, and are major global hot spots because of sediment inflow, mixing, and fish nurseries. Marine mammals can be found there (indicative of rich food chains), as well as many human population centers (now often expressed as modern port locations).

The tropical seas are the hot spots of mankind. Most people live in the tropics, and there, most of them actually live near the coast. Coastal fisheries play a major role. Famous interfaces between islands and oceans are found, for instance, in the southeast Pacific with its millions of islands, the Florida Keys, the Caribbean islands, and the Seychelles. Such ocean–land interfaces include entire cultures and nations like the Philippines, Indonesia, Papua New Guinea, Jamaica, Madagascar, Japan, and northern Australia. Subtropical examples are found in New Zealand, the Azores, the Baleares, and Easter Island.

Examples of tropical and subtropical oceans can be found in the Yellow Sea, Indian Ocean, Caribbean Sea, Sargasso Sea, Central Atlantic, and the Central Pacific Ocean. More landlocked seas of such water bodies are represented by the Mediterranean, Black Sea, Red Sea, Gulf of Mexico, and Sea of Japan. These ocean regions include such famous and diverse ecosystems as the coral reefs (e.g., the Great Barrier Reef, Caribbean Reefs, and Coral Reef Triangle), the deepest ocean floor (Mariana Trench), seagrass (Sargasso Sea), the Ring of Fire (a Pacific chain of tectonic plates and volcanoes), and some of the most remote islands of the world (e.g., Easter Island and Hawaii).

The tropical and subtropical oceans are a large resource of biodiversity, and many sea mammal and seabird species can be found there. Much of the marine benthos still awaits its description, as do many plankton and fish species. However, the biodiversity and habitat loss is already a major problem. Invasive species and diseases are on the rise. Marine protected areas (MPA) are planned, but these initiatives will hardly protect relevant areas. The well-being of these oceans and their sophisticated and evolved land- and seascapes, ecosystems, and human cultures have declined rapidly in less than 50 years. Overfishing is occurring on a global level. Illegal fishing and piracy are also big unresolved issues. The freedom of the international seas cannot be upheld.

The approach of single-species management fisheries has been criticized for decades. Top predators disappear (e.g., marine mammals such as sharks and tuna) in the study area. Dynamite fishing is still practiced, destroying entire ecosystems. Coral reefs are bleaching on a massive scale. Hypoxias occur, and so does ocean acidification. Plastic pollution has been raised as another major issue. Shipping, ports, marinas, oil pollution, and offshore gas and oil development are further unresolved problems. Many of the destructions are directly linked with the promotion of economic growth, poverty increase, political heritages, and the rise of neocolonialism schemes. Policies set by the Rio Convention or the Convention on International Trade in Endangered Species of Wild Fauna and Flora (CITES) will not be able to stop these destructions alone, judged by the results thus far. The situation has become worse in most areas, including missed biodiversity targets, widening poverty gaps, economic crises, and increased human populations. The tropic and subtropic ocean regions and their web of life will not disappear any time soon, but with the increase of climate change, sea-level rise, a population of more than 9 billion people projected by 2050, and an increase of consumption of products and goods, the outlook for the tropical and subtropical ocean regions, and thus the future for the world, seem rather grim.

Falk Huettmann

Upwelling Ecosystems

Upwelling ecosystems occur where cold, often nutrient-rich waters from the ocean depths rise to the surface. They can also occur in freshwater bodies such as lakes. Upwelling ecosystems are considered some of the most productive ecosystems in the world. Upwelling is the process in which surface currents move away from one another, or when winds push surface waters away from the shoreline. This draws deeper water upward to replace the surface water. This process of the rising of cold nutrient-rich waters is essential for productivity in marine and freshwater ecosystems. The importance of the nutrient-rich waters has a direct tie to the food web. This is because most primary productivity in the oceans occurs in surface waters, but most of the organic matter can be found at the bottom of the sea. The reason for this is because almost all food chains begin with photosynthesis, and photosynthesis occurs in sunlight. Marine plants, therefore, will only live near the surface where light can penetrate, keeping herbivores and their predators up near the surface. But once these organisms die or shed scales, eggs, leaves, shells, or feces, they sink down toward the seafloor and decompose.

Upwelling thus brings all these organic nutrients back up to the surface water. With these nutrients, the phytoplankton and seaweed population grow, forming the basis of many food webs in marine and freshwater ecosystems. These blooms form the ultimate energy base for large animal populations higher in the food chain, including marine and freshwater fish, mammals, and seabirds. The raising of benthic nutrients to the surface waters occurs in regions where the flow of water brings currents of differing temperatures together. There are at least five types of upwellings: coastal upwelling, large-scale wind-driven upwelling in the ocean interior, upwelling associated with eddies, topographically associated upwelling, and broad-diffusive upwelling in the ocean interior.

Coastal Upwelling

Winds affected by the rotation of the Earth create a phenomenon known as the Coriolis effect. Along a coastline oriented north–south, like much of the west coast of the United States, winds that blow from the north tend to drive ocean surface currents to the right of the wind direction, thus pushing surface waters offshore. As surface waters are pushed offshore, water is drawn from below to replace them; these are coastal upwellings. Coastal upwelling usually occurs in the subtropics along the continental coasts, where prevailing trade winds drive the surface water away from shore, drawing deeper water upward to take its place. Because of the abundance of krill and other nutrients in the colder waters, these regions are rich feeding grounds for a variety of marine species, including whales, sharks, and many other fish species, as well as avian species (coastal birds). Some of the most well-known examples of this type of upwelling include: the Canary Current in Northern Africa, the Benguela Current in Southern Africa, the California Current off of the west coast of the United States, the Humboldt Current in South America, and the Somali Current off the western side of India. All of these currents support major fisheries. Along the west coast of southern Africa, the Benguela upwelling distinguishes itself from other upwelling systems by the intrusion of warm waters at its two boundaries: to the north, via the Angola Benguela front, and to the south, via the Agulhas Current that terminates the western boundary current of the Indian Ocean.

To spawn, sardines (*Sardinops sagax*) and anchovies (*Engraulis encrasicolus*) migrate toward the Agulhas Bank, where the warm waters transported by the Agulhas Current create a highly stratified environment. Eggs and larvae are then rapidly transported northward by a coastal current. After several days, they reach the upwelling region of the east coast of Africa, where certain numbers are transported out to the open sea by wind-driven currents, a source of mortality. Other sources of early life cycle mortality must not be overlooked, in particular, those linked to starving larvae and predation of eggs and larvae. A week after hatching, larvae must feed. As their swimming abilities are limited, they need not only a high density of suitably sized prey (mainly microzooplankton), but also optimal turbulence conditions. Excessive turbulence disperses plankton aggregations and reduces capture. Using more precise modeling of currents and production should shed new light on these issues. Although primary productivity of the Benguela system seems globally in excess, deficits

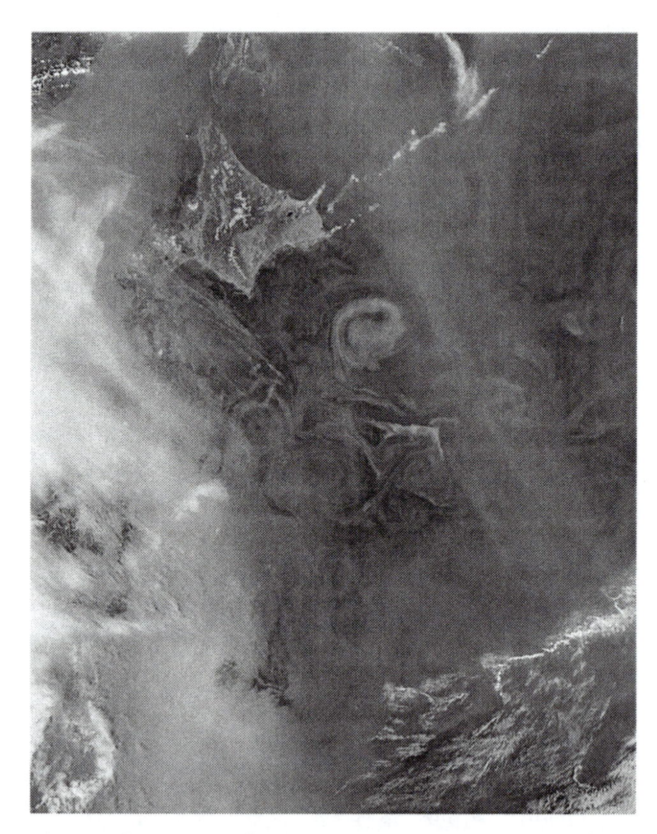

This satellite image illustrates how the convergence of the Oyashio and Kuroshio Currents affect phytoplankton. When currents with different temperatures and densities collide, they create eddies. Phytoplankton growing there become concentrated and create visible swirls consistent with the water movement, seen above. (NASA)

may occur in certain spatiotemporal strata, such as that of the Agulhas Bank or the offshore feeding area during the reproductive season. In these conditions, feeding oases may be found within the large offshore eddies caused by the meeting of different currents and water masses. The low productivity of the Agulhas Bank creates intense food competition among the millions of tons of spawners that congregate there during the spawning season. This encourages parental cannibalism and the predation of eggs and larvae by other species. Because of the absence of monthly spatialized data on fish distribution, modeling of this phenomenon remains difficult but efforts at quantification are possible.

The Benguela ecosystem can be divided into two subsystems, the northern Benguela (southern Angola and Namibia) and the southern Benguela (South Africa and Namibia), separated by the permanent upwelling cell of Luderitz, the strongest in the world. In South Africa, historical fisheries data suggest a stock collapse of sardines in the late 1960s, followed by slow recovery, whereas the first level of collapse was noted in Namibia during the same period, but worsened in the late 1970s. The biomass of the Namibian stock has remained at a very low level. Scientists are not sure if these collapses are because of an environmental process (upwelling) or overfishing.

The structure and ecosystem dynamics of the southern Benguela have been undergoing progressive changes for over 30 years, whereas the northern Benguela ecosystem has been completely "reorganized" following a "regime shift." Today, the northern Benguela is dominated by jellyfish and fish feeding on detritus (gobies). Pelagic fishes have become low in abundance, and fisheries for most commercial species (hake and horse mackerel) are threatened.

Upwelling in the Ocean Interior

Upwelling can also occur in the middle of the oceans where cyclonic circulation is relatively permanent or where southern trade winds cross the equator. The churning of a cyclone eventually draws up cooler water from lower layers of the ocean, which, in turn, causes the cyclone to weaken. Another very important example of upwellings that occur in very special habitats are upwellings that occur off the shore of islands, ridges, or seamounts because of a deflection of deeper currents. Just like other upwelling habitats, this provides a nutrient-rich area in otherwise low productivity ocean areas. Upwelling in seamounts has made them a very unique habitat with much endemism, and can be considered a biodiversity hot spot. A specific example includes upwellings around the Galapagos Islands with the Cromwell Current, and the Seychelles Islands, both which have major pelagic fisheries. In the Galapagos, the Crowmwell Current is a very important feeding ground for the endemic Galapagos penguin, which heavily depends on the smaller fish that feed on the plankton caused by the upwelling. This upwelling is also very important for the marine iguana, also endemic to the Galapagos Islands. When this upwelling fails to occur during an El Niño year, there is great devastation to the populations of these endemic animals that depend so heavily on it.

Another type of powerful upwelling occurs around Antarctica. The deepwater in the global conveyer belt flows across the seafloor all the way from the North Atlantic before it reaches Antarctica, so it has plenty of time to fill up with nutrients. Antarctica has many surface currents flowing around it, which stir up the waters, including the southern parts of the Pacific, Atlantic, and Indian Ocean gyres and the Circumpolar and Subpolar Currents. Also causing upwelling in this region are the strong polar winds that usually blow offshore.

Freshwater Upwelling

Upwelling in freshwater ecosystems tends to be seasonal. It is most likely to occur in lakes in temperate regions. During the summer, the lake surface is heated with infrared radiation from the sun, causing a thermal stratification (also known as thermocline) in the lake. It is because of this stratification that upwelling does not occur. By the time fall comes around, the stratification breaks down, and the temperature of the epilimnion (surface layers) becomes the same as the temperature at the hypolimnion (deeper layers). This means that the thermocline shifts and allows upwelling to occur; also, stronger seasonal winds aid the process of upwelling. When winter comes, the surface water of the lake freezes and upwelling or mixing does not occur. As the ice begins to melt during the spring, lakes become isothermal, which means that the epilimnion and the hypolimnion have the same temperature because of the wind action that brings deeper waters to the surface as a result of mixing.

Canaries Current

In the Canaries Current, there are three major ecosystems: the northern Moroccan coast with seasonal upwellings in the summer; the south Moroccan and north Mauritanian coast with permanent upwellings (Sahara Desert); and the south Mauritanian and Senegalese coast, with boreal upwellings. The southern part of this ecosystem is characterized by high seasonal variability, alternating between an ecosystem under tropical influence in the summer and a coastal upwelling ecosystem in the winter. This alternation is accompanied by a migration of certain tropical species (tuna and tuna-like fishes) up to 20 degrees north during summer, and by a southward extension of the habitat of temperate species such as the sardine *S. pilchardus* during the winter. The Canaries Current system has wide continental shelves in the south, whereas the eastern regions are generally characterized by narrow continental shelves because of their young geological age. Principal spawning areas have been linked to regions with wide shelves. This association would be the result of physical processes developing over wide and shallow continental shelves, which would limit exchanges between the coast and pelagic areas.

The Canaries Current ecosystems were dominated by large demersal fish, which were rapidly overexploited. In the southern part of the area, the adaptability of artisanal fisheries took advantage of the seasonal migration of numerous species (a balance between seasonal upwelling and the warm season dominated by tropical influences). Over a longer period of time, West African fisheries have been spared the sudden collapses observed in other systems. A lower fishing exploitation rate until the 1980s and high seasonal fluctuations could have contributed to this relative resilience. Three countries experienced rapid and unexpected growth of octopus stocks in the 1970s.

Because of its commercial value, octopus fishing has become one of the essential components of West African fisheries. This shift in species was interpreted as the result of the absence of top-down control, following the overexploitation of demersal species that favored the development of short-lived prey species, such as the octopus (as well as shrimp and pelagic fishes). Further to the north, the reasons that have lead to the southern displacement of seiner fleets exploiting sardines remain largely unknown.

Humboldt Current

The Humboldt Current system, with its permanent upwellings off the coast of Peru and its seasonal upwelling along the coast of Chile, is by far the most productive in fish landings. With less than 1 percent of the world's ocean surface, it provides 15 to 20 percent of world marine catches (up to nearly 20 million tons per year for Peru and Chile combined). A second particularity lies in the presence of a very intense, extensive shallow zone that is very low in oxygen. A final particularity is its position under the direct influence of the El Niño Southern Oscillation (ENSO) mechanism. Alternate upwellings of nutrient-poor and nutrient-rich waters off the coast of Ecuador and Peru are associated with El Niño and La Niña episodes in the tropical Pacific. During El Niño, the pycnocline is so deep that the upwelled waters come from the nutrient-poor waters above the pycnocline. In extreme cases, nutrient-deficient waters coupled with overfishing cause fisheries to collapse, bringing about severe, extended economic impacts.

Coastal upwelling ecosystems, like the west coast of the United States, are some of the most productive ecosystems in the world and support many of the world's most important fisheries. Although coastal upwelling regions account for only 1 percent of the ocean surface, they contribute roughly 40 to 50 percent of the world's fisheries landings (this refers to the part of the fish catch that is put ashore. Frequently, landings provide the only record of total catch).

In upwelling systems, large variations in fish recruitment in fisheries appear to be from fluctuations in mortality during the early life-history stages of the fish, and thereby must be essentially associated to climatic variability. Survival through these stages is essentially linked to hydrodynamic structures which, in certain spatiotemporal strata, favor the retention, enriching, and concentration of plankton and ichthyoplankton. Fluctuations in abundance of pelagic fish stocks reflect significant changes in the structure and functioning of upwelling ecosystems.

"Upwelling is the process in which surface currents move away from one another, or when winds push surface waters away from the shoreline."

High mortality rates have been observed at upper trophic levels (birds, marine mammals, and large predatory fish) in response to the diminished abundance of prey. Effects at lower trophic levels may also bring to light evidence of reduced predation by pelagic fishes on planktonic species, bringing about in turn changes throughout the entire trophic web. An example is the change in relative abundance of two species of anchovy associated with the El Niño phenomenon. Long-term fluctuations in dominant species have been observed in most upwelling ecosystems, such as the alternating dominance of sardines and anchovies in the Humbolt, Benguela, and Kuroshio (offshore Japan) Currents.

Climate and Climate Change

Coastal upwellings also influence weather and climate. Along the northern and central California coast, upwellings lower sea-surface temperatures and increase the frequency of summer fogs. Relatively cold surface waters chill the overlying humid marine air to saturation, so that thick fog develops. Upwelling cold water inhibits formation of tropical cyclones (e.g., hurricanes) because tropical cyclones derive their energy from warm surface waters. During El Niño and La Niña, changes in sea-surface temperature patterns associated with warm- and cold-water upwelling off the northwest coast of South America and along the equator in the tropical Pacific affect the interannual distribution of precipitation around the globe. A great impact on upwelling is posed by the new installation of large wind farms.

Some upwellings are wind-driven; these new large wind farms exert a significant disturbance on the wind speed in the vicinity of the installations. A recent study shows that the size of the wind wake is an important factor for the oceanic response to the wind farm. At certain wind speeds, sufficient upwelling may be generated that the local ecosystem will most likely be strongly influenced by the presence of a wind farm.

Another factor that is currently affecting upwelling is climate change. In any one region, upwelling is intermittent; it can be strong in some years, and weak in others. The success of fishermen is greatly affected by this, since a weakening of an upwelling system can bring economic disaster. Upwelling brings nutrient-rich deepwaters to the surface, where algae can thrive in the sunlight, feeding the fish. Without nutrients, there will be no fish (as happens during El Niño conditions).

Scientists are trying to determine how climate change will affect upwelling and coastal productivity. The strong dependency of upwelling processes on the strength of trade winds contains one hint. Trade winds are zonal winds, which feed off the latitudinal temperature gradient; as this gradient weakens (computer simulations show that high latitudes become warmer than low ones), the zonal flow will weaken. Thus, the upwelling that derives its energy from trade winds will weaken. Off California, upwelling of cold water has become less common since 1975, and the productivity of the California Current has diminished accordingly. A similar decrease in productivity may be expected elsewhere in the eastern boundary currents, and in the eastern equatorial high production regions. Upwelling that depends on monsoon activity (as in the Arabian Sea) should be much less affected, or may even benefit from the change.

According to some scientists, wind-driven upwellings along the California coast have increased over the past 30 years, and some scientists postulate that the increase in wind-driven upwellings is largely because of increased greenhouse gas, but such an association has been speculative. A study on the effects of future climate change and upwellings in California showed negative results from the intensification of wind-driven upwelling events in that specific area.

According to a model scientists have made, the intensification of upwelling will likely have impacts on terrestrial and marine ecosystems. With intensified upwelling, enrichment can be increased that would be beneficial to organisms; however, concentration may decrease because of increased mixing, and retention may also be decreased by increased seaward transport of surface water. Overall, this could have a negative effect on marine ecosystems, as the current balance of these three factors will change with changes in upwelling.

ALEXANDRA M. AVILA

Inland Aquatic Biomes

Freshwater Lakes

The water present in freshwater lakes makes up approximately 0.01 percent of the total water on the surface of the Earth. By volume, this is approximately 29,989 cubic miles (125,000 cubic kilometers), in contrast to the oceans, which contain 328.68 million cubic miles (1.37 billion cubic kilometers). The freshwater lake water mass is distributed among over 920 large named lakes and countless smaller lakes over all the continents of the world.

By continent, the distributions are: Africa, 43; Antarctica, 1; Asia, 174; Europe, 147; North and Central America, 452; South America, 55; and Oceana, 48. Lake Baikal is located in the southern Siberian region of Asian Russia. Lake Baikal is considered to be the oldest lake in the world, with an estimated age of 25 to 30 million years. The lake was formed through tectonic activity that formed the Baikal Rift; it is still active and growing at a rate of 1 inch (2.5 centimeters) per year. At its present dimensions (395 miles, or 636 kilometers long; 30 miles, or 48 kilometers wide; and 0.5 miles, or 0.74 kilometers deep), Lake Baikal holds the greatest volume of freshwater of any biome on Earth, at 5,517 cubic miles (23,000 cubic kilometers), which represents one-fifth of the freshwater on Earth. Lake Baikal is also the deepest lake on Earth, with a maximum depth of greater than 1 mile (1.6 kilometers). There are roughly 300 rivers that feed Lake Baikal, but only one, the Angara River, drains the lake. The lake is home to over 2,600 known species of flora and fauna, including the Baikal seal (*Pusa sibrinica*), which feed on the deep-living Baikal oil fish (*Comephorus baicalensis* and *C. dybowskii*). The lake is also home to the salmonid omul (*Coregonus automnalis migratorius*), which is an important food source to the indigenous population of Buryat tribe people who live on the eastern side of the lake. Lake Baikal was designated a World Heritage Site by the United Nations Educational, Scientific and Cultural Organization (UNESCO) in 1996.

Lake Tanganyika, located in Africa and surrounded by Tanzania, Zambia, the Democratic Republic of the Congo, and Burundi, is considered the longest freshwater lake in the world, with a length of 420 miles (676 kilometers). Lake Tanganyika is the second-largest lake in the world, with a freshwater volume of 4,534 cubic miles (18,900 cubic kilometers). It is also the second-deepest lake in the world, following Lake Baikal at almost 1 mile (1.5 kilometers) deep. Because of its length and depth, the lake does not mix below a depth of 656 feet (200 meters), and is anoxic at these depths; therefore, all biological activity occurs above the 656 feet (200 meter) depth. Lake Tanganyika was formed tectonically as a rift lake, situated in the southwestern part of the African Great Rift Valley. It is one of seven Great Rift Valley lakes that include Lakes Victoria, Malawi, Turkana, Albert, Kivu, and Edward. Lake Tanganyika is fed by two main rivers, the Ruzizi River entering the lake from the north, and the Malagarasi River that feeds the lake from the east.

The lake drains into the Congo River drainage system via the Lukuga River on the western side of the lake. The lake is home to 250 species of cichlid fishes that have been the focus of studies on speciation and other evolutionary research areas. Numerous cichlid fish species are popular in the aquarium trade industry, and other fish species are important as a food source. Commercial fishing on Lake Tanganyika has been in operation since the 1950s. But as in other regions, the fishing industry has been in decline. However, attempts to manage the resources of the lake have been implemented since 2004 through a Water and Nature Initiative program organized by the International Union for Conservation of Nature to monitor physical, chemical, and biological conditions of the lake.

Lake Superior is the largest of the Laurentian Great Lakes of North America, which is surrounded by Ontario, Canada, from the north, and Minnesota, Wisconsin, and Michigan's upper peninsula from the south. Lake Superior is third of the volume of Lake Baikal, with a freshwater volume of 2.9 million cubic miles (12.1 million cubic kilometers), and Lake Tanganyika at 4.5 million cubic miles (18.9 million cubic kilometers), but is the largest lake in the world by surface area (31,800 square miles, or 82,414 square kilometers). The geologic formation of Lake Superior is from scouring of the land surface as the glacial ice sheets retreated at the end of the last ice age, also known as the Pleistocene Period, approximately 10,000 years ago. This makes Lake Superior one of the youngest major lakes in the world, in contrast to Lake Baikal, which is one of the oldest lakes in the world. The other Laurentian Great Lakes, as well as most of the major lakes of North America from the temperate latitudes northward, were formed in a similar manner at the end of the Pleistocene Period. Lake Superior is fed by approximately 200 rivers, and drains into Lake Huron by way of the St. Mary's River. There are approximately 80 species of fish in Lake Superior. The lake is considered oligotrophic because of the low nutrient content and low primary productivity in the water column. The shores of Lake Superior were first inhabited at the end of the Pleistocene Period, around 10,000 years ago, by the Plano, who hunted caribou in the region. They were followed by the Archaic, Ojibwe, Cree, and the Laurel peoples. The Ojibwe called Lake Superior *Gichigami*, or "big water."

Lake Superior has a freshwater volume of 2.9 million cubic miles, contains as much water as all of the other Great Lakes combined, and holds 10 percent of the Earth's fresh surface water. (Thinkstock)

Lake Titicaca is the largest freshwater lake in South America, with a surface area of 3,141 square miles (8,135 square kilometers) and a volume of 214 cubic miles (893 cubic kilometers). At an elevation of 2.37 miles (3,811 meters), Lake Titicaca is the highest navigable lake in the world, and lies between Peru and Bolivia on the eastern side of the central Andes Mountain Range. The lake has five major rivers that feed it, the Ramis, Coata, Ilave, Huancané, and Suches Rivers. There are several small tributaries that also feed into the lake from the mountains; however, there are no drainages from Lake Titicaca.

The water level has been declining since the 1980s because of a decline in rain levels in the watershed that feeds the lake, as well as receding glaciers in the surrounding mountains. The lake consists of two subbasins that are connected by the Strait of Tiquina. The large subbasin is called Lago Grande, and the smaller one is Lago Piqueño. Lake Titicaca was formed during the Oligocene Epoch of the Paleogene Period, about 23 million years ago, through strike-slip movement as the Andes Mountain Range was forming. The lake is inhabited by people living on floating reed islands or *Uros* that can be moved throughout the lake. There are about five islands that are inhabited on the lake, some by native Quechua speakers. Lake Titicaca has about 30 endemic species of fish; 28 are cyprinodont, and two catfish. The Titicaca Orestias killifish (*Orestias cuvien*) became extinct after the introduction of the rainbow trout (*Salmo gairdneri*) in 1942 and the silverside (*Basilichthyes bonariensis*) in the early 1950s for commercial purposes. Lake Titicaca is home to large populations of waterfowl, earning it status as a Ramsar Site in 1998.

Lake Ladoga is located in European Russia, just north of St. Petersburg, and is considered the largest freshwater lake in Europe, and 14th-largest lake in the world. The lake has a surface area of 6,907 square miles (17,891 square kilometers) and a volume of 200 cubic miles (837 cubic kilometers). Lake Ladoga was formed by scouring as the glaciers receded after the last Ice Age, between 12,500 and 11,500 years ago. There are five major rivers that feed Lake Ladoga, and it drains to the Gulf of Finland to the southwest via the Neva River. In the Middle Ages, Lake Ladoga was part of the trade route between Scandinavia and Greece. In recent times, overfishing in the 1950s and 1960s led to stringent regulations to help recovering resources, which rebound by the 1970s to 1990s. There are over 40 species and subspecies of fishes in the lake, including important commercial salmonids. The lake has been exposed to anthropogenically mediated eutrophication since the 1960s, from intense industrialization and agriculture within the watershed. Lake Ladoga is home to a subspecies of ringed seal known as the Ladoga seal.

Lake Murray is the largest lake in Papua New Guinea and Oceania, with a surface area of 251 square miles (650 square kilometers) and a maximum depth of 33 feet (10 meters), and can expand to five times the area during the monsoon rainy season. The lake is formed in a shallow depression in the land relief and is fed by five major rivers and drains to the Gulf of Papua via the Herbert River. Lake Murray is home to many waterfowl, designating the lake a Ramsar Site since 1987. Basalt Lake is a small freshwater lake on Byers Peninsula on Livingston Island in the South Shetland Islands, Antarctica. It is formed in a depression surrounded by volcanic outcrops, and drains to the south into Bransfield Strait.

Limnology of Freshwater Lakes

Limnology is the study of inland freshwater bodies and is derived from the Greek *limné*, which means "lake." It includes the study of rivers and streams or lotic (flowing) systems and lakes and ponds or lentic (standing) systems. George Evelyn Hutchinson (January 30, 1903–May 17, 1991) is considered the father of limnology. He published the series *A Treatise on Limnology* in four volumes from 1957 until 1993. *Volume I Geography, Physics and Chemistry* was published in 1957, *Volume II Introduction to Lake Biology and the Limnoplankton* was published in 1967, *Volume III Limnological Botany* was published in 1975, and *Volume IV The Zoobenthos* was finished posthumously in 1993. Lakes can be divided into physical zones. The nearshore or littoral zone covers areas of lakes where there is light penetration to the bottom.

The offshore or limnetic zone is the open water pelagic region of a lake. The profundal zone occurs in deep lakes and is the region below where light penetration occurs. The benthic zone is on the lake bottom and includes the region of water sediment interface. Limnological studies of freshwater lakes cover the physical, chemical, and biological nature of lakes and their interactions.

Physical studies in a lake are concerned with aspects of circulation, that is, vertical and horizontal structure and dynamics. These studies are important to determine the relationship between temperature and density that leads to determining vertical stratification within a lake. Depending on the surface area–depth ratio and the climatic regime of the geographic location, a lake may be permanently stratified, for example, the very deep Lake Tanganyika of the African Great Lakes; on the other hand, shallow lakes with large surface areas, for example, the central basin of Lake Erie of the North American Laurentian Great Lakes, is continuously mixed.

"The freshwater lake water mass is distributed among over 920 large named lakes and countless smaller lakes over all the continents of the world."

The vertically stratified sections of lakes can be divided into the upper epilimnion, middle metalimnion, and bottom hypolimnion. The epilimnion is where relatively warm water can mix vertically in the presence of wind shear on the surface, where the majority of the pelagic primary productivity takes place because of exposure to light penetration. The metalimnion is also known as the thermocline, where a sharp difference of a short vertical distance in water temperature will occur. This is the density boundary between the upper, warm, less dense lake water and the lower, cold, denser lake water. The cold, dense, lower lake water is known as the *hypolimnion*. This layer is still and stagnant, typically devoid of oxygen or anoxic, and rich in nutrients because the biological activity taking place above it in the epilimnion eventually sinks into the hypolimnion as it decomposes. If most of the microbial decomposition takes place before the organic matter reaches the hypolimnion, it oxidizes into nitrates and phosphates that can be easily recycled back into the lake biosphere; however, much of the organic material does not completely decompose to the completely oxidized inorganic form, and remains in a reduced form until exposed to oxygenated or aerobic conditions.

Monomictic lakes in temperate zone latitudes or higher, which lose heat over the winter by freezing over completely or partially, and are also considered holomictic, will mix completely from top to bottom once a year during the spring season. The mixing brings nutrient-rich bottom water to the surface and oxygen to the bottom, an important event that drives the biological activity for the lake. Dimictic lakes will mix twice a year; this typically takes place when the water temperatures are relatively low, which would be during the spring and autumn, separated by a summer stratification period. Polymictic lakes mix several times over an annual period, not necessarily based on seasonal patterns. Polymicitic lakes are typically shallow, with large surface areas that are exposed to persistent wind shear.

Chemical studies in limnology focus on biologically active chemicals dissolved in lake water. Nutrients that include nitrogen, phosphorus, and silica are important constituents in the cellular structure of aquatic plants—both vascular and single-celled organisms known as phytoplankton. The productivity of a lake is limited by the concentration of nutrients available to plants to utilize for growth. In freshwater lakes, the limiting nutrient is phosphorus. The trophic status of a lake can be defined by the concentration of phosphorus in the lake, typically measured as total phosphate, which includes both inorganic phosphate and organic phosphorus molecules in living phytoplankton cells and particulate and dissolved organic material. Oligotrophic lakes hold total phosphorus content of less than 12 milligrams per liter; these freshwater lake systems are considered pristine, healthy biomes. An example of an oligotrophic lake is Lake Superior of the North American Laurentian Great Lakes.

Mesotrophic lakes have a total phosphorus content range between 12 and 24 milligrams per liter. These freshwater lake systems are considered moderately productive and considerably healthy. Lake Ontario of the North American Laurentian Great Lakes is considered mesotrophic. Eutrophic lakes have a total phosphorus content range of 24 to 96 milligrams per liter. The process of eutrophication may occur naturally; however, in many cases these freshwater lakes are influenced by anthropogenic activity through agriculture or industry. Lake Ladoga in Russia has been eutrophicated since the 1960s, likely because of intense industrialization and agricultural efforts within the watershed. Hypereutrophic lakes have total phosphorus content greater than 96 milligrams per liter. These lakes have intense phytoplankton blooms, low water clarity, and are mostly devoid of oxygen near the bottom. The phytoplankton blooms in hypereutrophic lakes are usually cyanobacteria, such as *Aphanizomenon flos-aquae*, as found in Upper Klamath Lake in southern Oregon.

Biological studies in limnology include plankton dynamics in limnetic zones of lakes; these include phytoplankton and zooplankton, their dynamics and interactions, and their importance in the freshwater lake food web. In the littoral zone, higher aquatic plants are important constituents of the freshwater lake biome as shelter and food sources for aquatic organisms from aquatic insects to fishes, waterfowl, and aquatic mammals. Benthic organisms occupy the bottom habitat from the littoral zone to the deep dark regions of lakes. These habitats are occupied by mollusks, annelids, crustaceans, and benthic fishes.

Freshwater Lake Inhabitants

Freshwater lake biomes contain species of phytoplankton from all major groups. The Chlorophyta or green algae include desmids and flagellated *Chlamydomonas*; most are considered edible and nutritious for zooplankton. The Chrysophyta are divided into the flagellated golden Chrysophyceae and diatoms or Bacillariophyceae, which are considered highly edible by zooplankton. Members of the Chrysophyceae can be colonial such as *Dinobryon* or singular cells like *Ochromonas*. The Bacilliariophyceae or diatoms are either centric or pinnate in shape. An example of a centric freshwater diatom is *Cyclotella* and a pinnate is *Navicula*. The Pyrrophyta or dinoflagellates are flagellated unicellular organisms that are considered mixotrophic, with the ability for autotrophic or heterotrophic production, for example, *Peridinium*. The Eugleophyta can be autotrophic or phagotrophic, and some classify the group as Euglenozoa. An example of a euglenid with autotrophic capability is *Euglena*. The Cryptophyta are flagellates that are considered highly edible by zooplankton, for example, *Cryotomonas*. The Cyanobacteria are considered primitive autotrophic bacteria, without a cellular nucleus. This group can cause nuisance blooms under high phosphorus conditions and low nitrogen concentrations, where species like *Anabaena* can fix nitrogen from dissolved N^2 gas.

The zooplankton community of the freshwater lake biome consists of Protozoa, Rotifera, and Crustacea that include Cladocera and Copepoda. The Protozoa include small and large ciliates that are important in the microbial loop part of the food web, and mostly feed on bacteria and microbially processed particulate organic matter. The Rotifera are herbivores that feed on the small phytoplankton species, and are an important food source for carnivorous zooplankton. An example of a rotifer is *Asplanchna*. The dominant zooplankton group in freshwater lakes is the Crustacea. Dominant within the Crustacea are the cladoceran daphnids, which are herbivores. An example of a cladoceran daphnid is *Daphnia*. There are carnivorous clodecerans, such as *Bythotrephes longimanus*. The Copepoda include herbivore and carnivore species among this group. An example of an herbivore calanoid copepod is *Diaptomus*. An example of an omnivorous cyclopoid copepod is *Cyclops*.

The benthic fauna consists of macrobenthic organisms, including mollusks such as clams (*Corbicula*) and mussels (*Dreissena*). These are important filter feeders and have the ability to influence the trophic structure and dynamics of a lake by reducing the standing crop of the phytoplankton community. The meiobenthic community consists of organisms that range in size from 0.1 millimeter to 1 millimeter in length, and include crustaceans like harpacticoid copepods and ostracods, as well as nematodes. The meiobenthos are important in the microbial loop and contribute in the processing of organic materials into inorganic nutrients. Microbenthos organisms are key players in the microbial loop and remineralization process. These organisms are less than 0.1 millimeter in size and include bacteria, diatoms, ciliate, amoeboid, and flagellated protozoans.

Water in its freshwater form is an important resource to man and other organisms on the major global continents. The freshwater lake biome is critical to the ecology and sustainability of life within the terrestrial environment. The pattern of distribution of this resource is dependent on the major global climatic regimes. Since the global system is dynamic, the climatic regimes are constantly in flux and shifting, and these shifting patterns will cause shifts in the distribution of freshwater resources.

NASSEER IDRISI

Great Lakes

The Great Lakes are five bodies of water spanning the eastern U.S.–Canadian border. Together, Lakes Superior, Michigan, Ontario, Huron, and Erie comprise the largest group of lakes on Earth's surface, covering 80,545 square miles (208,610 square kilometers), bordered by approximately 9,500 miles (15,289 kilometers) of shoreline, and containing an estimated 35,000 islands. This one continuous drainage basin spans 750 miles (1,200 kilometers) and contains 21 percent of the world's surface freshwater, totaling 5,412 cubic miles (22,560 cubic kilometers). Their combined surface area is larger than the U.S. states of New York, New Jersey, Connecticut, Rhode Island, Massachusetts, Vermont, and New Hampshire combined. The Great Lakes are framed by the states of Minnesota, Wisconsin, Illinois, Indiana, Michigan, Ohio, New York, and Pennsylvania, and the provinces of Quebec and Ontario. In addition to their notable size and volume, several other features make the Great Lakes unique among lake ecosystems. The largest island in an inland body of water, Manitoulin Island, is in Lake Huron. This island contains a lake, called Manitou, which is the world's largest lake located on a freshwater island. The United States has more shoreline on the Great Lakes than it has on the Atlantic Ocean and Gulf of Mexico combined. Michigan's 2,232 miles (3,592 kilometers) of shoreline is second only to Alaska's. The longest suspension bridge in the world, the Mackinac Bridge, spans Lakes Michigan and Huron across the Straits of Mackinac. Finished in 1957, the bridge is 5 miles (8 kilometers) long and rises 552 feet (168 meters).

The Great Lakes formed when the Laurentide, or Wisconsin, glaciation began to retreat 14,000 years ago. As the edge of the nearly 2-miles-thick (3.29 kilometers), moving ice sheet melted, the giant basins carved by the glacier filled with water, forming huge Lake Agassiz, which held more water than all lakes on Earth today combined. The repeated drainage events of Lake Agassiz formed the first versions of the Great Lakes, which were initially much larger than they are today, and flowed east to west. The lakes continued to change shape and size as the climate shifted and the land recovered from the weight of the glaciers, a process known as *isostatic rebound.* Around 4,000 years ago, the Great Lakes took on their present configuration. They flow from west to east, starting at Lake Superior's southeastern corner through the St. Mary's River to Lakes Michigan and Huron, then down the St. Clair and Detroit Rivers to Lake Erie, then flowing north through the Niagara River, over Niagara Falls, and into Lake Ontario, and finally into the St. Lawrence River, joining the Atlantic Ocean over 2,000 miles (3,219 kilometers) from its origins.

Climate, Water Levels, and Lake Stratification

Great Lakes weather is affected by three prominent air systems. There is a very dry and cold arctic system that comes from the north; another dry, but warm Pacific system that comes from the west;

and finally, a warm, wet tropical system from the south and the Gulf of Mexico. Because of their large size and volume, the Great Lakes produce "lake effects" on regional weather. For example, the eastern shores and nearby land in Michigan, Ohio, Pennsylvania, New York, and Ontario can receive especially heavy snowfall. This happens when westerly winds blow across ice-free water in the middle of the lakes in the winter, picking up warmth and moisture. When these moisture-laden winds reach the colder landmasses on the eastern shores, heavy snowfall can occur. Another "lake effect" is the moderation of seasonal temperatures in a band around each lake. The latent summer heat in the water causes later autumn frosts, and the cold water keeps summer temperatures cooler lakeside than they are inland.

"Changes in the species composition of the Great Lakes basin in the last 200 years have been the result of human activities. Many native fish species have been lost by overfishing, habitat destruction, contaminants, or the arrival of exotic or nonindigenous species."

Like all lakes, the Great Lakes are dynamic bodies of water that undergo complex processes and have a variety of subsystems that change seasonally and on longer cycles. One such cycle is the stratification or layering of water in the lakes because of changes in water density caused by seasonal temperature changes. Colder water is denser, so it stays deep in the lake, forming a layer called the *hypolimnion* during the summer months when the sun warms the surface water, forming a layer called the *epilimnion*. A middle layer, called the *thermocline*, is an area of rapid temperature change between the cold and warm waters.

The warm waters of the epilimnion are the most productive, growing algae that form the base of the food chain. When temperatures fall in autumn, surface waters cool, becoming denser and eventually sinking, causing a mixing of layers, called *turnover*. In winter, surface waters are often colder than those below. The layering and turnover of water annually are important for water quality. Turnover is the main way in which oxygen-poor water in the deeper areas of the lakes can be mixed with surface water containing more dissolved oxygen. This prevents anoxia, or complete oxygen depletion, of the lower levels of most of the lakes. However, the process of stratification during the summer also tends to restrict dilution of pollutants from effluents and land runoff.

Water levels also go through cycles in the Great Lakes. Seasonal fluctuations of about 12 to 18 inches (30 to 46 centimeters) result from regular changes in water amounts in the lakes. In the fall and early winter, when the air above the lakes is cold and dry and the lakes are relatively warm, evaporation from the lakes is greatest. This can be seen when fog or mist forms and appears to float over the lakes. Consequently, water levels decline to their seasonal lows. As the snow melts in the spring, runoff to the lakes increases. Evaporation from the lakes is also least in the spring and summer, when the air above the lakes is warm and moist and the lakes are cold. With more water entering the lakes than leaving, water levels rise to their peak in the summer.

Changes in water levels can also be short-term. For example, a strong sustained wind from one direction can push the water level up at one end of a lake, which makes the level drop by a corresponding amount at the opposite end. These events are most common on Lake Erie because of its east–west orientation in an area of prevailing westerly winds and its generally shallow western end. Long-term fluctuations occur over periods of consecutive years. Continuous wet and cold years will cause water levels to rise. Likewise, consecutive warm and dry years will cause water levels to decline. Over the last century, the range from extreme high to extreme low water levels has been nearly 4 feet (1.2 meters) for Lake Superior, and between 6 and 7 feet (1.8 and 2.1 meters) for the other Great Lakes.

Aquatic Life

It is estimated that there were as many as 180 species of fish indigenous to the Great Lakes. Those inhabiting the nearshore areas included smallmouth (*Micropterus dolomieu)* and largemouth bass (*Micropterus salmoides)*, muskellunge (*Esox masquinongy*), northern pike (*Esox lucius*), and channel catfish (*Ictalurus punctatus).* In the open water were lake herring (*Coregonus artedi*), lake whitefish (*Coregonus clupeaformis*), walleye (*Sander vitreus,* formerly *Stizostedion vitreum),* sauger (*Sander Canadensis*), freshwater drum (*Aplodinotus grunniens*), lake trout (*Salvelinus namaycush*) and white bass (*Morone chrysops),* blue pike (*Sander vitreus glaucus*), and Atlantic salmon (*Salmo salar*). The last two species are now considered extinct from the Great Lakes. Because of their different characteristics, the lakes varied in their species composition.

Warm, shallow Lake Erie is the most productive, while deep Superior is the least productive. Currently, hatchery-reared coho salmon (*Oncorhynchus kisutch*) and chinook salmon (*Oncorhynchus tshawytscha)* are the most plentiful top predators in the open lakes, except in the western portion of Lake Erie, which is dominated by walleye. This diversity of fish formed part of a complex food web, with sunlight, dead organic matter, and bacteria at its base; plankton and small-bodied fish in the middle; and predaceous fish, birds, and humans at the top. Changes in the species composition of the Great Lakes basin in the last 200 years have been the result of human activities. Many native fish species have been lost by overfishing, habitat destruction, contaminants, or the arrival of exotic or nonindigenous species.

Early Inhabitants and Current Population

Because humans are thought to have been in North America as early as 12,000 years ago, they probably experienced some of the changes that the Great Lakes underwent. In particular, their shoreline settlements may have been affected by changes in drainage events. Over time, approximately 120 bands of native peoples have occupied the Great Lakes basin. Descendants of the first settlers were using copper from the south shore of Lake Superior 6,000 years ago. The population in the Great Lakes area is estimated to have been between 60,000 and 117,000 in the 16th century, when Europeans began their search for a passage to the Orient through the Great Lakes. The native people occupied widely scattered villages, fished and hunted, and grew corn, squash, beans, and tobacco. Present-day names for these tribes include the Chippewa, Fox, Huron, Iroquois, Ottawa, Potawatomi, and Sioux. Today, Native American fishing treaties protect the fishing rights of Great Lakes tribes.

The first written record of the Great Lakes is of Lake Huron, called *La Mer Douce*—the sweet sea—by French explorer Samuel de Champlain in 1615, and Lake Ontario appears shortly thereafter. Accounts of Lakes Superior and Michigan appeared in 1622 and 1634, respectively, and of Lake Erie in 1669. After European settlement, major cities grew up rapidly on the shores on the Great Lakes. The largest of these include Chicago, Milwaukee, and Detroit on Lake Michigan; Cleveland on Lake Erie; and Toronto on Lake Ontario. One-tenth of the population of the United States, and one-quarter of the population of Canada, currently live near the Great Lakes and depend on them for drinking water, industry, agriculture, recreation, food, transportation, and tourism.

Commercial Fishery

Fishing has been a historically important industry of the Great Lakes, increasing about 20 percent per year after 1820. Commercially important species include lake trout, salmon, lake whitefish, and yellow perch. The greatest fishing harvests were recorded in 1889 and 1899, at about 147 million pounds (67,000 metric tons). Production then decreased because of overharvesting, pollution, habitat degradation, and the introduction of exotic species. The value of the fishery also decreased when it shifted to smaller, low-value species. Catches of Lake Michigan whitefish declined from over 5,000 tons in 1930 to 25 tons in 1957. Even more dramatic, lake trout catch declined from almost 7,000 tons in 1943 to 3 tons in 1952. Populations of these species and others plummeted after the invasion of the lake by sea lamprey.

The Great Lakes fishery is currently worth $7 billion, and is based primarily on the native species yellow perch, whitefish, walleye, and the introduced species smelt (*Osmerus*), alewife (*Alosa pseudoharengus*), splake, a hybrid between

A rare view of all of the Great Lakes with no cloud coverage in this Sea-Viewing Wide Field-of-View Sensor (SeaWiFS) NASA image from 2000. Together, Lakes Superior, Michigan, Ontario, Huron, and Erie form not only the largest group of freshwater lakes on the planet but the largest group of lakes anywhere on Earth. (NASA)

a lake trout and a brook trout (*Salvelinus fontinalis*), and Pacific salmon (*Salmo* and *Onchorynchus*). Commercial fishing of lake trout is allowed only in Lake Superior, where populations appear stable. In addition, researchers estimate that 25 to 40 percent of the salmon and trout populations in Lakes Michigan, Huron, and Ontario are now self-reproducing, citing improved habitat, water quality, and stream conditions. A number of species, including coho and chinook salmon, are raised in hatcheries and stocked in the lakes. The Great Lakes Fishery Commission was established in 1955 by the Canadian/U.S. Convention on Great Lakes Fisheries. The commission coordinates fisheries research, controls the invasive sea lamprey, and facilitates cooperative fishery management among the state, provincial, tribal, and federal management agencies.

Navigation, Shipping, and Canals

Although the Great Lakes were naturally connected to the Atlantic Ocean via the St. Lawrence River, this passage was not available to vessels because of rapids and waterfalls, especially Niagara Falls. In 1959, the St. Lawrence Seaway, a series of canals, locks, and dams, was completed. As part of this system, the 26-mile (42-kilometer) Welland Canal connects Lake Ontario to Lake Erie, allowing ships to bypass Niagara Falls. The locks are not wide enough to allow passage of the widest freighters; much of the shipping moves materials among Great Lakes ports. The primary commodities are iron ore (more than 6 million tons annually), coal (more than 4 million tons annually) and limestone (more than 3 million tons annually). Other cargoes include cement, salt, grain, and sand. The fleet of 300 ships from the last century has been reduced to about 140 in recent years. The largest boats are 1,000 feet (304.8 meters) long.

The Great Lakes are also connected to the Mississippi River via the Chicago Sanitary and Ship Canal. Completed in 1900, the canal is 28 miles (45 kilometers) long, 202 feet (62 meters) wide, and 24 feet (7.3 meters) deep. The canal was constructed to divert Chicago's treated sewage from entering Lake Michigan, the source of the city's drinking water. The canal reversed the flow of water in the Chicago River, which naturally flowed into Lake Michigan. Once the canal was opened, approximately 2,068 million gallons per day from Lake Michigan began to flow into it, pushing Chicago's treated sewage toward the Mississippi River and the Gulf of Mexico. The Great Lakes have a long history of shipwreck, groundings, and collisions. From the 1679 sinking of *Le Griffon* with its cargo of furs to the 1975 loss of the *Edmund Fitzgerald*, thousands of ships and thousands of lives have been lost; many of these losses involved vessels in the cargo trade. The Great Lakes Shipwreck Museum estimates that 6,000 ships and 30,000 lives have been lost.

Invasive Species

The Great Lakes are inhabited by 161 nonnative aquatic species, including zebra mussels, sea lamprey, and rainbow smelt, all of which are a major problem for Great Lakes ecosystems. Many nonnative species have entered the lakes via the St. Lawrence Seaway and the Chicago Sanitary and Shipping Canal. One of the most established invasive species is the sea lamprey (*Petromyzon marinus*), which is native to the Atlantic coast of North America, the Finger Lakes in New York, and Lake Champlain in New York and Vermont. It is thought to have invaded Lake Ontario via the Erie Canal. It spread to Lake Erie by way of the Welland Canal, and to Lakes Huron, Michigan, and Superior soon afterward. Sea lamprey attack other fish by attaching to them with a suction cup–like mouth and scraping away their flesh with sharp teeth and tongue. These fish usually die from blood loss and infection. Sea lamprey decimated populations of lake trout, lake whitefish, chub, and lake herring in the 1930s and 1940s, taking a huge toll on the commercial fishing of the Great Lakes region. Control efforts include applying electric currents, chemical lampricide, and physical barriers.

The alewife (*Alosa pseudoharengus*) is another invasive species whose population increased greatly when lampreys reduced populations of native fish. Alewives are native to the Atlantic coastal areas, but entered the Great Lakes through the Welland Canal in the 1940s. From time to time, large numbers of alewives die when they are exposed to rapid temperature changes, lack of food, or stress associated with living in freshwater or spawning. Zebra mussels (*Dreissena polymorpha*) are believed to have come from the Caspian Sea in Europe. They first entered the Great Lakes in the late 1980s in the ballast of ships. Zebra mussels can be up to 2 inches (50 millimeters) long, have striped shells, and can live four to five years. They form dense colonies in sediment and drainage pipes, blocking water flow. They also consume plankton by filter-feeding, competing with fish for food. Annual cost to manage zebra mussel populations exceeds $250 million.

Other invasive species of concern include Asiatic clam (*Corbicula fluminea*), bighead carp (*Aristichthys nobilis*), bloody-red shrimp (*Hemimysis anomala*), European rudd (*Scardinius erythrophthalmus*), European ruffe (*Gymnocephalus cernuus*), New Zealand mud snail (*Potamopyrgus antipodarum*), quagga mussel (*Dreissena rostriformis bugensis*), round goby (*Apollonia melanostomus*), rusty crayfish (*Orconectes rusticus*), silver (or Asian) carp (*Hypophthalmichthys molitrix*), white perch (*Morone americana*), and the aquatic plants European frog-bit (*Hydrocharis morsus-ranae*), hydrilla (*Hydrilla verticillata*), and Eurasian watermilfoil (*Myriophyllum spicatum*).

Water Quality, Pollution, and Water Diversions

Great Lakes water and sediments contain more than 800 toxic contaminants. The U.S. Environmental Protection Agency has identified 43 areas that have water impaired by pollutants such as mercury and polychlorinated biphenyls (PCBs). These invisible chemicals, many of which are no longer discharged, inhabit the sediment of the lake bottoms. Wave action and the bottom-feeding organisms bring them back into the water column and the food chain, including fish. Pollutants enter Great Lakes waters from three main sources: point sources, nonpoint sources, and the atmosphere. Point sources of contaminants, such as heavy metals and human waste, include industrial discharge drainpipes and sewage treatment plants. Nonpoint source pollution, including agricultural pesticides and fertilizers, salts and oils from roads, sediments from construction sites and eroding stream banks, and animal waste from farm pastures, comes from many different sources and is transported to the Great Lakes when rain and snowmelt runoff pick up pollutants and drain them into waterways. Airborne contaminants, such as mercury, phosphorus, and PCBs originate from the burning of fossil fuels and waste and enter the lakes via rain and snow.

In March and April 1993, an outbreak of the protozoan Cryptosporidium occurred in the drinking water system of Milwaukee, Wisconsin. Approximately 403,000 people became sick with stomach cramps, diarrhea, and dehydration, and 104 people died, making this the largest documented disease outbreak caused by a waterborne pathogen in U.S. history. The United States draws over 40 billion gallons of water from the Great Lakes every day, much of it for electricity generation. Some towns, such as Pleasant Prairie, Wisconsin, and Akron, Ohio, are allowed to divert Great Lakes water for public supply. Some large-scale schemes to divert Great Lakes water to, for example, Great Plains states to recharge the Ogallala aquifer, to Wyoming for mining, or to Asia, have been denied.

Legislation, Treaties, and Policies

Several treaties, agreements, and laws govern use of water from the Great Lakes. The 1909 Boundary Waters Treaty between the United States and Canada deals with all matters concerning flow and quality of waters passing over the international border. It also established the International Joint Commission to resolve disputes and review any projects that would affect the lakes' flow and uses, including boating and navigation. In the 1972 Great Lakes Water Quality Agreement, the United States and Canada agreed to control pollution, research and monitor lake health, and work toward restoration with an ecosystem approach. The 1985 Great Lakes Charter is a nonbinding agreement between all Great Lake states and provinces to cooperatively manage the lakes. Accordingly, before any new diversion or consumptive use of more than 5 million gallons per day of any Great Lakes water is begun, all parties must agree. The 1986 Water Resources Development Act gives veto power to each of the eight Great Lakes states' governors to block diversion or export of water from the Great Lakes outside the region. This act was reauthorized in 2000.

The 2001 Great Lakes Charter Annex between the eight Great Lakes governors and two premiers set the intention to develop new protections to prevent harmful water withdrawals and diversions of Great Lakes water. The most recent agreement, the Great Lakes–St. Lawrence River Basin Compact and Agreement of 2008, seeks to ban the diversion of waters, with some limited exceptions, and to set responsible standards for water use and conservation within the basin. Each of the eight Great Lakes states must develop policies for water conservation, according to the compact. These accords have improved the management of the Great Lakes. In 2009, a record $475 million was authorized by the U.S. government to improve the condition of the Great Lakes for future generations.

Susan Moegenburg

Ponds

Ponds are small lentic or still bodies of freshwater. There is much disagreement about how to define a pond, as well as the proper method of distinguishing ponds from lakes. Some defining methods involve size requirements. For example, some areas of the United States define a pond as having surface areas of less than 10 acres. Other methods specify a maximum depth, stating that they must be as shallow as 12 to 15 feet (3.66 to 4.57 meters), or shallow enough that light can penetrate to the bottom uniformly across the water body so that plants may root all the way through the pond. This shallow depth also means that ponds do not have temperature layers or stratification caused by heat convection as lakes do; however, they may have thermally driven microclimates. Another criterion is that ponds must be a quiet body of water, lacking wave action on the shoreline, though they may have wind-driven currents. Regardless of these criteria, many ponds have been given the misnomer of lake and vice versa. Ponds can form in any depression in the ground that collects and retains enough groundwater or precipitation to pool. These depressions can result from a wide range of natural geological and ecological events. For example, kettle ponds were formed when glaciers retreating across North America and Europe ground out deep potholes in the landscape, which are now filled from underground aquifers. Many pond-forming activities are now constrained by human activities, such as filling or irrigation.

Vernal Ponds

Some ponds are temporary, only holding water for part of the year, and are known as ephemeral or vernal pools. Most vernal ponds fill with winter rains and snowmelt, meaning that they are at their peak level in the spring, hence the term *vernal*, meaning spring. Because these ponds tend to dry up annually, they are generally fishless, which allows for the safe development of amphibians and insect larvae that would normally be preyed upon. However, some species of fish, called killifish, have adapted to this environment by laying

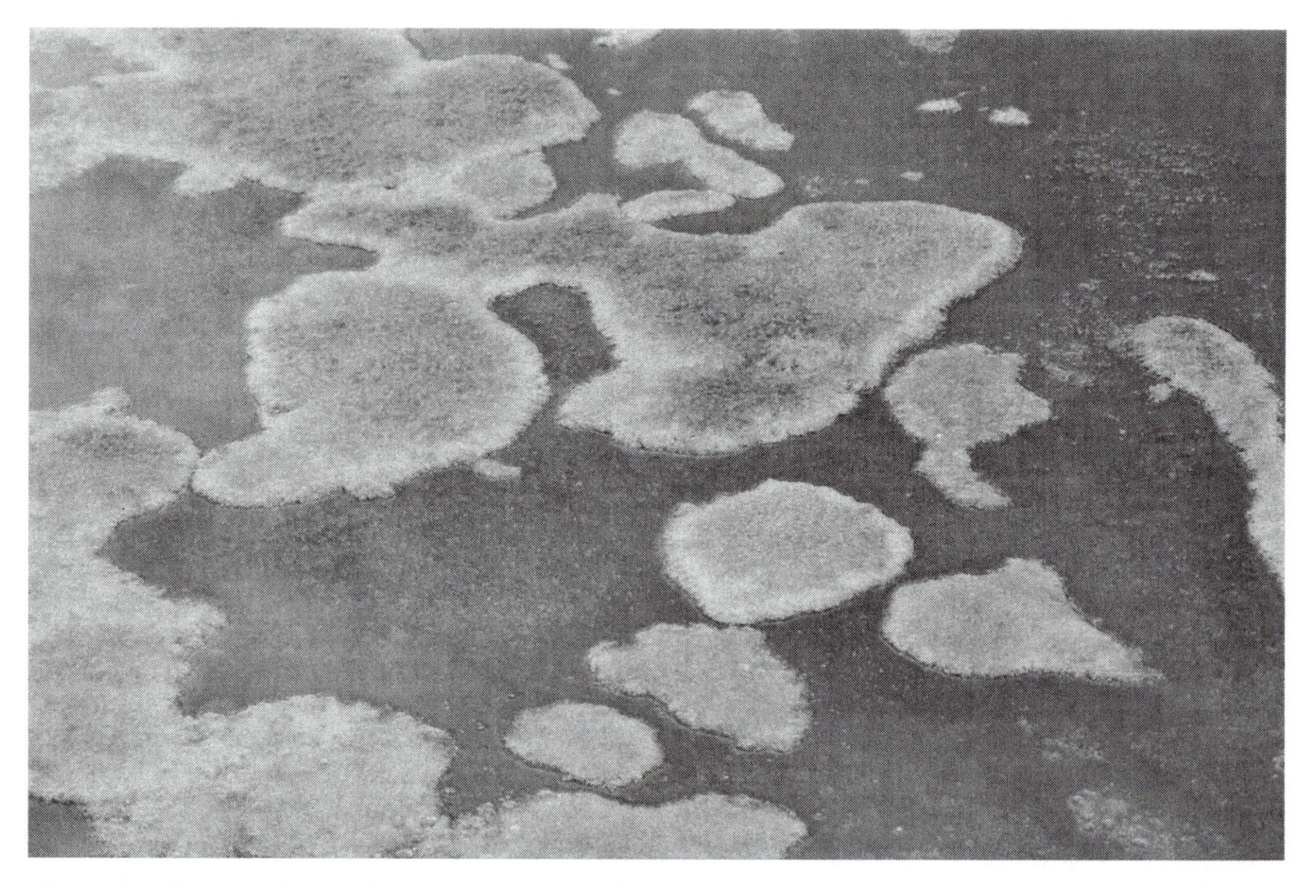

When a pond has no surface outflow, it becomes a closed, self-contained ecosystem. The most productive organisms in a pond system are phytoplankton, shown in the lighter areas above. As the microscopic algae float on the surface, they give the water a vivid green appearance. Phytoplankton helps produce oxygen for the pond bottom. (Thinkstock)

eggs that can survive partial dehydration. There are several species of organisms that rely upon these pools and are considered indicator species by scientists trying to identify pond types. Aquatic invertebrates and amphibians can serve as indicators of pond health and type because they have specific habitat requirements and different species respond differentially to changes in the biological, chemical, and physical makeup of a pond. Vernal pools are important breeding sites for frogs and toads, such as wood frogs (*Rana sylvatica*) and spadefoot toads (*Scaphiosus*), and mole salamanders like spotted salamanders (*Ambystoma maculatum*) and marbled salamanders (*Ambystoma opacum*). They are also habitats for daphnia and fairy shrimp, small invertebrates that survive the periodic dry periods by laying resting eggs in the dried mud of the pond bottom, which remain dormant until the pond refills with water in the spring. Some eggs can remain dormant for decades, ensuring the continuation of the species life cycle, even through years of drought.

Pond Seasonality

Most ponds retain water year round, and in temperate climates freeze partially or completely during the winter season. Aquatic organisms tend to tolerate a narrow temperature range, and have a variety of methods for dealing with seasonal changes. Cold-blooded animals, like reptiles and amphibians, can change their metabolism and reduce their activity levels to preserve energy stores. Turtles and frogs will often burrow in the mud of the pond or surrounding area and hibernate through the winter. Some frog species, such as the wood frog, can survive being partially frozen because of the presence of cryoprotectant chemicals that act like antifreeze in their blood and body tissues. Many insects also burrow and hibernate or lay winter eggs that will hatch in the spring. Fish and snails move to deeper waters and remain less active beneath the ice. Some aquatic plants form winter buds that drop to the bottom of the pond and germinate in the spring. Zooplankton, like rotifers or cladocerans, produce winter eggs, whereas protists and annelids protect themselves by forming cysts around their bodies.

Nutrient Cycling in a Closed Ecosystem

Many ponds have no surface outflow and are spring-fed, making them self-contained ecosystems. All ecosystems are made up of interactions among the biotic, or living, plants, animals, and microorganisms relating among themselves and to the abiotic, or nonliving, physical and chemical factors of the environment. In ponds, the abiotic factors include: water, carbon dioxide, oxygen, phosphoric salts, amino acids, and nitrogen. Like most closed ecosystems, ponds function through the interlocked relationship of two food chains cycling these nutrients through their environment. The first food chain consists of light-dependent primary producers or autotrophic organisms, and the organisms that eat them, called heterotrophs. Autotrophic organisms produce complex organic compounds from simple inorganic molecules. Though a few autotrophs can do this by using chemicals like hydrogen sulfide, most ecosystems, including ponds, derive their energy from the sun and are dominated by photosynthetic autotrophs. Photosynthesis involves the reduction of carbon dioxide molecules through the addition of oxygen, which creates chemical energy and sugars that are used by organisms to create biomass or growth.

The primary producers in ponds are phytoplankton, periphytic algae, submerged plants, floating plants, emergent plants, and shore plants. Phytoplankton, whose name means "wandering plants," are microscopic algae that float in open water, giving it a green appearance. Periphytic algae are microscopic algae that attach to substrates, making pond rocks a slimy, greenish brown. This algae helps to produce oxygen at the bottom of ponds, which is important for decomposers found in the depths. Submerged plants grow with their entire structure completely under the water. Floating plants are rooted at the bottom of ponds, but extend to float on the surface. Emergent plants are rooted in shallow water and have their stems and leaves above the water most of the time, whereas shore plants are completely above the water most of the time.

The energy stored in the algae and plant biomass is moved up the food chain by grazing organisms. These organisms are called consumers, or heterotrophs, which include animals, bacteria, and fungus. Ponds are filled with tiny zooplankton, microscopic animals that eat phytoplankton or other small zooplankton. They include single-celled animals like the foraminifera, tiny crustaceans such as copepods, or immature stages of larger animals. Algae and zooplankton are eaten by macroinvertebrates, larger animals that have no backbone, including thousands of species of pond dwelling insects.

Plants, zooplankton, and microinvertebrates are then eaten by larger vertebrates, animals having a backbone, which include fish, frogs, salamanders, turtles, snakes, water fowl, wading birds, and semi-aquatic mammals like muskrats, beaver, and otters, as well as raccoons. Consumers that eat only plants, herbivores, are considered primary consumers. Consumers that eat primary consumers are called secondary consumers, and can be strict carnivores, eating only animals; or omnivores, which eat both animals and plants. Tertiary consumers are considered apex or top predators, and can feed on both primary and secondary consumers. An example of a tertiary predator in a pond is the otter, which may eat both herbivores, such as some fish, as well as carnivores like frogs.

When the nutrients in a pond cycle up through the food chain as organisms are eaten by larger and larger animals, they are lost to the surrounding environment only when animals leave the pond through external predation or migration. The rest of the nutrients are cycled back down to the abiotic pool of nutrients by a second food chain. The second food chain is made up of non-light dependent organisms that feed on and break down the detritus from the first food chain. These detritivores are bacteria, fungi, and other organisms that feed on the waste products of the producers and consumers. In turn, the detritivores break the organic waste back into their inorganic components, which the primary producers, plants and algae, can then again use in the process of photosynthesis. In this way, the nutrients are cycled continuously through the pond ecosystem.

Pond Habitats

Ponds are made up of four distinct habitat zones: the shoreline, the surface film, the open water, and the bottom water. The surface film and the open water make up the pelagic or upper habitats of the pond. The surface film refers to the surface of the pond on which insects, such as water striders, and free-floating organisms range. It also includes the water just below the surface, as well as the underside of floating plants.

The open water is home to large, free-swimming organisms like fish, and drifters like plankton. The bottom water and shoreline make up the benthic or lower level habitats in the pond. The bottom water habitat varies depending on the pond depth. Quiet, standing water ponds often have muddy and silted bottoms into which crayfish, mayfly nymphs, and other macro- and microorganisms burrow. The shoreline is considered the area of the pond where water meets land, and is dominated by rooted plants and air-breathing organisms that vary depending on the substrate type. This is the richest part of the pond in terms of biodiversity because of the dense plant life, which is used for foraging and breeding by many species.

Though not strictly considered part of a pond, the area of land around a pond is essential to the health of the pond and the organisms within it. It is called a buffer because it often serves to protect or buffer the pond from exposure to harmful external forces such as pollution. Wide vegetated buffers around ponds provide essential nesting sites, winter habitat, and escape cover for wildlife. Species of amphibians like the spotted salamander are especially dependent on large and healthy buffer zones around ponds because, though they are dependent on the ponds to lay their eggs and for their larvae to develop, the adults only remain at the pond long enough to breed. Afterward, they return to the surrounding terrestrial habitats to live and feed for the rest of the year and over winter. Reptiles like snakes and turtles may depend on the pond for food and cover, but they require the surrounding area to lay their eggs buried on dry land.

Without a healthy buffer zone of at least 50 feet, many species cannot survive, despite the presence of a pond. A vegetated buffer also controls soil erosion and increases the amount of water filtration through the soil. It helps to improve water quality, filtering out runoff and pollutants like fertilizers. In addition, buffers can isolate ponds from intensive wading from livestock, such as cattle, which can disturb pond vegetation, increase water turbidity (the amount of suspended particles in the water), accelerate bank erosion, and increase nitrogen levels in the water.

Pond Eutrophication

Life in ponds depends on a pool of nutrients that includes nitrogen and phosphorous. The amounts of these nutrients present in a pond can limit the amount and type of organisms that can survive in a pond. The rate of release of these compounds into the pond from the surrounding environment can vary depending on the temperature cycle and seasonality of the pond, as well as the amount of sunlight it receives and the regional climate. Sometimes, ponds receive excess amounts of nitrogen and phosphorous through both natural and artificial means. Natural cases of increased nutrient intake occur when nutrients accumulate through naturally occurring processes, such as depositional environments wherein sediments are laid down or where ephemeral or seasonal nutrient flows occur. Artificial nutrient enrichment occurs from human activities such as untreated sewage, fertilizer runoff, or allowing livestock direct access to ponds and rivers, into which they expel fecal matter.

"Because vernal ponds tend to dry up annually, they are generally fishless, which allows for the safe development of amphibians and insect larvae that would normally be preyed upon."

Spikes in nitrogen and phosphorous levels can lead to eutrophication, which is the increase of plant biomass in a body of water. This process usually favors simple phytoplankton or algae over more complex plants, and often leads to algal blooms. The excess nutrients allow these photosynthetic species to convert even more sunlight into biomass, meaning that they can increase their populations rapidly. Algal blooms can coat water surfaces or make water cloudy, often turning ponds a shade of green, yellow, brown, or red depending on the species of algae. This overgrowth limits the amount of sunlight that can reach bottom-dwelling organisms, and can cause drastic changes in the amount of dissolved oxygen in the pond.

When photosynthesis occurs, oxygen is released from the breakdown of carbon dioxide and water. In a healthy pond, more oxygen is made during the day than can be consumed by respiring organisms and builds up in the pond. This oxygen reserve is then used by the pond organisms at night, without which they would otherwise suffocate. During an algal bloom, large excesses of oxygen are made during the day, but the unnaturally large populations of phytoplankton continue to undergo cellular respiration at night and can deplete oxygen stores within a pond. This oxygen depletion leads to hypoxic or oxygen-poor conditions in the pond, which can cause large die-offs of fish and other animal populations, especially immobile bottom dwellers. Through such die-offs, eutrophication can cause large changes in the species composition and richness of a pond. Eutrophication in a pond can also threaten species through toxic effects. When large amounts of certain species of algae die and are eaten, they can release both neuro- (affecting the brain) and hepato- (affecting the liver) toxins. These toxins can cause the death of many animals, including species of zooplankton. Livestock can also become sick from ingesting toxic algal blooms, and the toxins can even pose a threat to human health.

Laws have been successful in reducing some instances of eutrophication by regulating the discharge and treatment of sewage, thereby reducing a source of nutrient inflow to groundwater sources. However, there is need for regulation dealing with the use of agricultural fertilizer and animal waste. Fertilizer and pesticides that are sprayed on crops often end up washed by rains into nearby water bodies, especially when applied liberally. Reducing the amount, frequency, and timing of application of agricultural chemicals could help to reduce the occurrence of eutrophication. Vegetated buffers can again serve the health of a pond by helping to filter out chemicals that would otherwise wash into ponds, as well as keeping livestock from directly accessing water bodies.

Pond Pollution

Other types of pollution can also be harmful to ponds, changing the chemical makeup of the pond and affecting organisms. Foamy areas, fluorescence, or distinctive odors can signal pollution in a pond, but can also be misleading. Foam can be produced in ponds by harmless inorganic salts or some species of phytoplankton. Fluorescence and distinct odors can be caused by certain protozoan species. Carbon dioxide levels in water can be effected by pollution and are important to determining the pH of a pond, which can determine which types of life can survive in that environment. Agricultural pesticides and synthetic estrogens have been found contaminating ponds in agricultural and urbanized environments where frogs and fish are experiencing external and reproductive deformities, including the feminization of males. Ongoing research and policy development are needed to determine the effects of pollution on pond ecosystems and the best ways to protect water bodies like ponds from contamination.

Hannah Bement

Rivers

Rivers are defined as lotic biomes (from the Latin *lotus,* meaning "washed"); that is, inland watercourses that unidirectionally flow through topographic gradients in the landscape. The English word comes from the Vulgar Latin *riparia* ("riverbank, seashore, or river"). This term is frequently utilized to indicate relatively large water streams. The structural and functional characteristics of rivers are best understood at the watershed level. A watershed (or drainage basin, or catchment) is a topographic depression drained by a hydrogeographic network of tributary channels, which coalesce into a main river channel. Within the watershed, surface water flows from the area of maximum potential energy of the depression (the topographic divide; or below the ground level, the phreatic divide) to the point of minimum energy, which usually is another water body (e.g., a lake or sea).

Water influx entering the watershed as ice melt, snow, or rain (precipitation) is either intercepted by the vegetation, absorbed, transpired, or evaporated at the plants' surface; or by the substrate, where it can infiltrate the soil in a compartment where water moves by capillarity (the vadose zone), which is renewed by evapotranspiration. Below the vadose zone, groundwater is collected in a water-saturated compartment superiorly delimitated by a plane called the water table: the phreatic zone. Water exceeding the soil infiltration capacity (which in turn depends on the storage capacity of the vadose zone) can move in the gravitational field as overland flow; similarly, lateral subsurface flows can occur within the vadose zone, and base flows within the phreatic zone, which can also surface as springs. Hydrological connections are present among all these compartments. Therefore, the hydrology, chemistry, and biology of a river largely depend on the climate, geology, and vegetation coverage of the watershed and its groundwater connections.

Global Patterns

River and stream channels cover about 0.1 percent of land surfaces and contain only approximately .0001 percent of the total volume of water on Earth (about 1/100th of that contained in lakes), or approximately 288 cubic miles (approximately 1,200 cubic kilometers). Nonetheless, the total drainage area of the 30 major river basins on Earth (about 20 systems, mostly longer than 1,243 miles, or 2,000 kilometers) covers about 25 percent of the total land area of all continents, or 14.4 million square miles (37.2 million square kilometers), 1.9 million square miles (5 million square kilometers) by the Amazon River basin alone. Approximately 23,991 cubic miles (100,000 cubic kilometers) of water precipitates on continental masses each year: 65 percent of this amount returns to the atmosphere by evapotranspiration, and 35 percent reaches the oceans as stream and base flows as discharge. Rivers erode large amounts of dissolved and particulate materials collected by runoff within their watersheds, transporting them from the land to the sea. Such dynamic nature is exemplified by their water residence time, which is on average much smaller (throughflows, or renewal rates: 12 to 20 days) than in other aquatic systems, such as freshwater and saline lakes (1 to 1,000 years) or oceans (3,100 years). At the landscape level, hydrological regions where rivers originate and reach the sea are called exorheic; those where rivers originate and do not reach the sea are called endorheic; and those where rivers do not originate (e.g., desert regions) are arheic. Changes of hydrological regimes can occur during regional or global climatic changes.

"For thousands of years, rivers have been used for navigation (transportation, commerce, exploration, and tourism), bathing and other recreational activities, and waste disposal."

Geomorphology of River Channels

The river channel is a through-shaped geomorphic entity defined by soil or rock boundaries (banks and bottom). The river flow, whose velocity is determined by the balance of gravitational and frictional forces, erodes such solid boundaries, transporting materials on the bottom (bedload) or in the water column (suspended load), and dissipating energy. For this reason, the maximum flow velocity is in the middle of the channel, rapidly decreasing toward the water–air and water–substrate interfaces (boundary layers). This flow also interacts with the lateral and vertical control determined by the topography (valley), which determines whether the river is either relatively free to move laterally (unconstrained), thereby forming meanders and flood plains; or not free to move laterally (constrained), such as in incised or entrenched valleys, where scarce or absent riverine vegetation and no flood plains occur. The topographic gradient and other factors, such as the extent and type of sediment supply, then determine the channel morphology (e.g., straight, meandering, braided, or anostomosing).

The degree of water penetration in the interstitial space of the bottom and bank sediments of river channels is determined by the topographic gradient, sediment composition, and flow characteristics. Below the bottom, there is a soil layer where water actively infiltrates by advection in the hyporheic zone, an inch (few centimeters) to 3 feet (1 meter) deep; further below, there is a water-saturated layer where groundwater prevails (the interactive hyporheic zone). The boundary between these two zones changes spatially and temporally because of several factors, such as the penetration of animals and plants (bioturbation), and floods.

In general, base flow inputs into the river from the phreatic zone dominate upstream (e.g., in headwaters); while downstream, as the topographic gradient is reduced, subsurface flows from the river water into the streambed become more important (e.g., in flood plains and deltas); in these latter conditions, near to the minimum energy level, groundwater inputs to the stream flow are minimal, and hydrological boundaries between river and groundwater extend laterally. Geomorphological changes occur along the river as the kinetic energy decreases and the sedimentary budget becomes positive, such as when it is rapidly reduced by flowing into a larger body of water forming a delta; or after a sudden transition from a narrow valley to a plain, forming a fluvial fan. Flow patterns also undergo considerable temporal changes. Periodic or irregular temporal changes of the stream flow (e.g., floods) are related to climatic events (e.g., storms), drainage area, and type of interactions between overland and base flows.

Temperature

Apart from water velocity and turbulence, one of the most important physical factors affecting the composition of fluvial biotic communities and metabolism is temperature. Rates of temperature change in rivers are mainly determined by solar and atmospheric radiation; the back radiation released into the atmosphere at the water surface; and changes of volume. Hydrology, climate, and insulation thus interact and determine rivers' thermal regime. River ecological zonations are heavily affected by downstream changes in temperature along the elevational gradient, and biotic fluctuations are affected by the cumulative number of days over a year when the temperature is above 32 degrees F (0 degrees C). The energy provided by turbulent conditions makes ice formation rarely complete in running waters, especially at riffles; small ice crystals can form in turbulent and slightly supercooled conditions in the water column (frazil or slush ice) or on substrates (anchor ice), affecting both the sedimentary balance (erosion of bottom and banks) and the life of benthic organisms.

The Amazon River in South America carries more water than the Mississippi, Nile, and Yangtze Rivers combined. The 4,000-mile river begins in the Andes Mountains as a small stream called the Apurímac River. (Thinkstock)

Metabolism and Nutrients

The metabolism of fluvial ecosystems is mainly affected by the spatial and temporal budgets of a few chemical compounds (oxygen, inorganic carbon, nitrogen compounds, and phosphorous compounds), the abundance of which is determined by both biotic and abiotic processes. In streams of lower order, with colder and more turbulent waters, oxygen is often approximately at saturation, declining during periods of higher temperature (e.g., summer in temperate regions), as equilibrium solubility decreases. However, oxygen levels can also markedly fluctuate because of chemical and biological processes, especially in larger, well-illuminated, and relatively sluggish rivers, where bacterial respiration and photosynthetic activity by plants and phytoplankton are more intense. Such changes are mainly associated with spatial and temporal changes of inputs and outputs of organic matter, which rapidly reacts with oxygen in chemical or biochemical processes. Examples are anoxic groundwater inputs, periodic floods, or inputs of leached materials from leaf fall from adjacent terrestrial systems. Hypoxic or anoxic conditions are often found in microenvironments within microbial communities attached to solid substrates, and in the interstitial water of saturated sediments.

Respiration and decomposition dominate over primary production within rivers; furthermore, surface and groundwater inputs from runoff are often both anoxic and hypercarbic. These effects are more evident in smaller rivers of lower stream order. As a result, variations in carbon dioxide (CO_2) concentration and pH are common. Lower pH and lower alkalinity levels can occur during periods of reduced flow or in areas with slower flows, and in correspondence with loadings of dissolved and particulate organic matter. In particular, softwater rivers with low alkalinity (e.g., flowing on granitic or sandstone sediments) can be rapidly acidified by modest inputs of acidic water from atmospheric sources or runoff. On the other hand, acidic inputs in hardwater rivers (e.g., flowing on carbonatic rocks) may increase alkalinity and equilibrium pH because of increased rates of carbonate weathering in the watershed.

The concentration and net downstream transport of dissolved nutrients depend on their rates of uptake and release between the streambed and the water column along the river, in a process called nutrient spiraling. The retentiveness of nutrients depends on the extent by which their downstream transport is delayed by physical, chemical, and biological processes (mainly from microbial action), relative to the transport of water. Nutrient limitations of primary production and of heterotrophic microbes' production are associated with faster cycling rates and higher biological demands of a particular nutrient (e.g., nitrates or dissolved phosphorous).

Lower stream order rivers are often oligotrophic, both because of the smaller size of their drainage basins and the lower anthropogenic impact, which limit nutrient loadings. Nonetheless, nutrient limitations are less common in these latter systems, since most biota are attached to the substrate, and recycling within the community is more efficient than in larger and higher stream order rivers, where nutrient retention is reduced.

Dissolved inorganic and organic phosphorous is more controlled by biological uptake than by discharge, while the opposite is true for particulate phosphorous, which is more slowly metabolized. On the other hand, accumulated particulate phosphorous can be released downstream in large pulses, during precipitation-mediated events of higher discharge.

System Ecology and Productivity

At the ecosystem level, terrestrial and aquatic systems are connected within watersheds by water movements, and downstream fluvial sections are unidirectionally influenced by upstream sections. Especially in streams of lower order, the majority of the organic matter is allochthonous, deriving from adjacent floodplains, wetlands, and terrestrial systems. Spatial and temporal changes of flow patterns are accompanied by changes of the associated biotic communities, affected by the different types of habitats that flowing patterns create; in particular, stream biota are generally adapted to flowing waters. For the rest, spatial and temporal variations characterize flow, chemical and biological conditions and generalizations are rather difficult. Nonetheless, general models were proposed (e.g., the river continuum concept, or RCC) to describe how spatial and temporal changes of the functional composition of fluvial communities, organic matter supply, and resource partitioning can be related to stream order, current velocity, and intensity of biological activity.

Within the water column in the river channel, photosynthetic activity (e.g., phytoplankton) is inhibited by turbulence and turbidity induced by water flow and shading from the riparian vegetation. The productivity of submersed macrophytes is generally only slightly higher than that of phytoplankton, with higher contributions from attached algae and biofilms in shallow and less turbulent lotic habitats. Also, the biomass of zooplankton is generally negatively correlated with flow velocity, and planktonic animals compensate for the effect of flow and downstream transport with reduced life cycles, benthic lifestyles, and colonization of littoral backwater habitats. The bacterioplankton heavily relies on allochtonous organic matter, and its abundance in rivers is generally higher than in lakes, with increasing productivity in higher stream order rivers. The metabolism and trophic level of the water column generally changes along the river as it is affected by illumination, turbulence, turbidity, and the relative abundance of suspended coarse and fine particulate organic matter (CPOM, FPOM, respectively). For example, river metabolism can be heterotrophic both in headwaters (production/respiration, or P/R ratios < 1), which are often characterized by higher turbulence, shaded banks, and higher inputs of CPOM in the riparian zone; and in lower fluvial sections, which are often characterized by higher turbidity levels. Autotrophic metabolism ($P/R > 1$) can occur in intermediate sections.

In fact, most of the fluvial metabolism generally takes place in the hyporheic zone, which is dominated by microbial benthic activity, mainly through anaerobic processes and heterotrophic consumption of allochthonous dissolved and particulate organic matter. Most of this primary production derives from the terrestrial, emergent, and floating vegetation found in the riparian zone, floodplains, and associated wetlands. Aquatic insects often dominate the invertebrate benthic macrofauna of rivers, and are primary food sources for fish and other vertebrates. In general, production rates of fish in rivers is higher than in lakes, since rivers are characterized by proportionally wider riparian zones and associated transitional systems, where most of the highly diverse and productive benthic habitats are found.

Human Uses

The connection between human civilization and large rivers is witnessed by the numerous human settlements, both extant and archaeological, distributed along their banks or in closely associated systems, such as fertile reclaimed floodplains and wetlands. In general, increasingly larger human settlements developed downstream, as the topography becomes more flattened, and near coastal areas, where estuaries offer preferential access to the sea. For thousands of years, rivers have been used for navigation (transportation, commerce, exploration, and tourism), bathing and other recreational activities, and waste disposal. Rivers have also been extensively utilized as political boundaries and defensive barriers, such as the Mississippi in North America, and the northern border of the Roman Empire along the Danube, respectively, easily controlled by crossing points (bridges). Riverbanks provided caves of gravel and sand for building purposes, and river channels provided sources of food and water, both for drinking (for humans and livestock) and irrigation, from at least 5,000 years ago. Since the 1900s, and most markedly after the 1960s, the green revolution increased human dependency on irrigation for agriculture, which is now responsible for 87 percent of the total water used globally.

Since the ancient Romans, and with a rapid acceleration since the late 1800s to early 1900s, dams and reservoirs have been built as water management tools (hydromodification) for agricultural, industrial, and domestic consumption; for production of hydropower (e.g., watermills, blacksmiths' forges, mining, quarrying, and hydroelectric power facilities); and protection from floods. Channelization is a river engineering procedure that substitutes straight cuts for the meandering river course. Straightening and deepening the river channel increases drainage in lowland agricultural systems or make streams suitable for navigation, especially for larger vessels.

Anthropogenic Impacts

The biodiversity, stability (e.g., flood predictability and sensitivity to headwater disturbing events), and productivity of large river ecosystems largely depend on floodplains and associated habitats, which greatly increase the spatial and temporal heterogeneity of these systems, efficiently store water and nutrients, and provide most of the organic matter that derives from plant and microbial production. Dams and reservoirs drastically disconnect rivers from flood plains and wetlands, thus reducing floodplain habitats and disrupting the natural cycle of inundations. This makes rivers' metabolism entirely dependent on downstream transport from headwaters. Channelization has several negative effects, mainly deriving from the reduction in stream length, increase of flow velocity, and reduction of ecological connectivity with terrestrial ecosystems. Many habitats are lost (riffles, pools, riparian vegetation, and wetlands), and biodiversity is drastically reduced. Other effects include lowering of the water table, increased runoff and erosion, nutrient losses, greater temperature and flow fluctuations, and increased turbidity. The reduction of water retention capacity can also result in increases in frequency and destructivity of downstream flooding events.

Human impacts on transitional flood plain and wetland systems result in overall declines of many fluvial ecosystem services, such as destruction of fisheries, species extinction, reduced nutrient uptake, lower water retention capacity, and higher sensitivity to pollution events. In the attempt of increasing the accessible runoff for human use, by 1996, about two-thirds of fluvial waters were regulated by 800,000 dams, 40,000 of which were higher than 50 feet (15 meters). Higher dams are currently built at a rate of about 500 per year, while hundreds of thousands of rivers are regulated by levees, embankments, or dikes. Global climate warming will likely exacerbate these effects, with several models predicting decreased precipitation frequency and increased intensity of precipitation events, which would determine higher fluctuations of downstream flows and major declines of habitat quality of the still functional wetlands and flood plains.

Degradation of fluvial waters mainly occurs as eutrophication, siltation, acidification, and introduction of toxic pollutants. Nutrient loadings (e.g., phosphorous and nitrogen) mainly result from organic pollution (e.g., sewage discharges, industrial and agricultural runoff, and nutrient losses in the watershed from harvesting and fire), which cause nutrient enrichment and massive growth of algae and macrophytes. Soil erosion and inputs of suspended sediments result from deforestation and agricultural activities in the watershed, reducing the volume of aquatic habitats and increasing water turbidity. Loadings of hydrogen ions from strong acids result from gases produced by combustion of fossil fuels, which enter the system as fallout or precipitation (rainfall and snowfall), affecting water pH and alkalinity. Direct inputs of toxic substances (e.g., heavy metals, pesticides, chlorinated hydrocarbons, and radioactive materials) from industrial and urban wastes variously affect the biotic components and the metabolism of fluvial ecosystems. All these forms of degradation ultimately affect water quality for human use and consumption.

Management and Restoration

Restoration and/or management of riparian areas and flood plains play a major role for an efficient management of large rivers. In fact, marginal habitats make rivers naturally more resistant and resilient to natural and human-induced disturbances. For this reason, river restoration activities often imply modifications of the channel that encourage development of wetlands and riparian vegetation on buffer strips along the banks; reduce the slope gradient at the edges; and promote channel meandering. Even limited restoration or rehabilitation of transitional systems can yield considerable improvements of fluvial ecosystem services. For instance, it has been estimated that watersheds including 5–10 percent of wetlands would experience a 50 percent reduction of peak floods, compared to watersheds without wetlands. Long-term reduction of the inputs of pollutants implies complex social, political, administrative, and economic problems, especially when reduction of the sources is attempted, and international boundaries are crossed. Treatment of polluted waters is often highly expensive, and most remedial actions aim at diluting and dispersing pollutants.

Gianluca Polgar
Laura Ribero

Saline Lakes

Saline lakes narrowly fit into the biome concept because of their heterogeneous origin and fluctuating water regime; instead, they are azonal ecosystems, meaning that their ecological features are not predictable from the general climatic zone in which they occur. Despite their stepping-stone geographic distribution, the global volume of saline lake water (24,950 cubic miles, or 104,000 cubic kilometers) is almost as great as that of the world's freshwater (29,989 cubic miles, or 125,000 cubic kilometers). Saline lakes are common landscape features on every continent, except in warm deserts, as this biome is in general too dry to allow their presence.

However, as river catchments become somewhat less arid, there are depressions containing saltwater for at least some time after episodic rain events. Asian steppes and neighboring semiarid regions located southward are spattered with many of the world's saline lakes, livening up the ancient Mongols' home turf. Some are large and permanent lakes, such as the Caspian and Aral Seas, and Lakes Balkhash, Issyk-Kul, Chany, Alakul, and Tengiz. Others are small and temporary, like some Central Anatolian *gölü* (Turkey) and many *nuur* of Mongolian steppes.

Out of Asia, the wealth of saline lakes increases with the many potholes of the Canadian prairies, the *playas* of the United States and Mexico, the high-Andean *salares*, the *lacke* of southeast Austria, the Hungarian *sodic szék*, north African and Middle East *chotts* and *sabkhas*, the Ethiopian Rift Valley salt lakes, the South African *pans*, the lagunas of the Mancha Húmeda Biosphere Reserve and Andalusian *campiñas*, and the *saladas* of the Monegros region (Spain). Saline lakes are common even in some regions of Antarctica, like the Vestfold, Bunger, and Larsemann Hills, the McMurdo Dry Valley, and west of the Antarctic Peninsula.

How Saline Lakes Form and Their Function

Saline waters have a brackish taste. Their salinity is more than 10 percent that of seawater, which has a fairly constant sodium chloride composition. However, inland saline waters show a variable ionic composition because their dissolved salt content does not come from either dilution or concentration of seawater. Many saline lakes are like bathtubs without a drain, from which water can only get out one way, through evaporation. This usually occurs when the lake basin is the bottom of a large catchment, like Great Salt Lake (United States) or Lake Eyre (Australia). On the other hand, saline lakes may be fed with salts from either saline groundwater discharges (e.g., the Davsnii Lake in Mongolia), or the washing out of other salt lakes, evaporite rocks like gypsum, or some volcanic rocks (e.g., Mono Lake in the United States). All of these sources of salts combine unevenly in all saline lakes. Thus, terminal depressions at the end of large catchments used to be tectonic, where associated fractures act as pipes through which groundwater discharges to the lake bed, like the Sidi El Hani Sebkha in Tunisia.

Nonetheless, salinity of groundwater discharging into a lake may be the result of geochemical evolution of water in its flow along regional paths within large aquifers; strikingly, this might be composed of rocks with a low salt content, like quartz sandstone; this is the case of Laguna de las Torres in Spain. The diversity in the origin of salt lake basins and in their water inputs is a forcing function of the sharp patterns in spatial heterogeneity and fluctuations through time, controlling saline lake ecosystems. Thus, the water regime of endorheic saline lakes with large catchments are most sensitive to one or more global circulation phenomena, such as the El Niño Southern Oscillation (ENSO), the North Atlantic Oscillation, or monsoons; the higher the number of global phenomena involved, the sharper the fluctuations. Accumulated effects of positive and/or negative anomalies (wet and dry periods) provide complex fluctuation scenarios at different timescales.

This is the case, for example, of Mar Chiquita (province of Córdoba, Argentina), where three different climates affect its huge catchment, which are controlled by ENSO, the South Atlantic anticyclone, and the Amazonian tropical cell. Its flooded surface shifts from about 772 square miles (2,000 square kilometers) in dry periods to more than 3,861 square miles (10,000 square kilometers) during wet periods, making it the fifth-largest salt lake in the world. Moreover, fluctuations of local scope, rather than global, are more likely to affect saline lakes with small catchments, such as those of Mediterranean zones. These are narrow strips of land in the west coasts of the continents and around the Mediterranean Sea, with short-river, densely drained watersheds.

Summer drought and late-summer to early-autumn heavy storms feature in the Mediterranean climate, but their occurrence is very irregular through time and space. For example, the Laguna de Salicor (Mancha Húmeda Biosphere Reserve, Spain) suffered a severe thunderstorm (super cell) between May 20 and 26, 2007, that poured 9.4 inches (240 millimeters) of rain in 24 hours; as a result, the water column depth shifted from less than 6.3 inches to 7.9 feet (16 to 240 centimeters) between April and June 2007, and water salinity decreased from 172 to 2.38 $g\, l^{-1}$.

Life on Mars

Saline lakes are like different planets for Earth biota. Oceans contain plenty of life, despite their salinity. Nonetheless, environmental gradients are extreme in saline lakes, where marine organisms would feel like a fish out of water. Saline water biota must avoid dehydration because of the difference in osmotic pressure between the cells and the environment. Many microbials do this by accumulating "friendly" solutes in their cells. However, osmotic concentration rockets during desiccation in temporary salt lakes, and water salinity increases beyond that of seawater (hypersaline waters). Whether saline lakes are always or eventually hypersaline, extremophiles are then the dominant characters; they do not just tolerate hypersaline waters, but actually require them for growth. Ecosystems of permanently flooded, relatively deeper saline lakes are also commonly controlled by a strong, yet vertical gradient, called chemocline. This forms when a water layer lies over a denser one in a lake water column because of their differences in chemistry, that is, salinity (meromixis). Lakes Lyons and Mahoney (Canada), Big Soda (United States), Chaunaca and El Molino palaeo-lakes (Bolivia), the Salada de Chiprana (Spain), Lakes Van (Turkey), Shala (Ethiopia), Qinghai Hu (China), Panggong Tso (Indian Tibet), Gnotuk (Australia), and Bonney (Antarctica) are classic examples of meromictic salt lakes.

Like thermal stratification, meromixis avoids deepwater deficient in oxygen (O_2) mixing with upper, aerated water layers. In low, dark water layers, sulfate reduction is an alternative to photosynthesis and O_2-consuming decomposition to obtain energy. Sulfate-reducing bacteria can do that by "breathing" the large amounts of sulfur present in salt lakes as either organic matter or dissolved sulfate (SO_4^{2-}). This form of anaerobic respiration gives off hydrogen sulfide (H_2S) as a waste; it is toxic for O_2 respiring organisms, and

The Dead Sea, also known as the Salt Sea, is actually a saline lake that is bordered by Jordan, the West Bank, and Israel. Here, with salt formations in the foreground, a woman floats on the Dead Sea because the lake's high salt content makes people naturally buoyant. The lake's high mineral and salt content make it uninhabitable for fish or plants. (Thinkstock)

its rotten-egg odor is a marker of this process in saline lakes. Most H_2S reacts with metal ions in the water to produce metal sulfides, such as ferrous sulfide (FeS); these are dark colored and not soluble in water, leading to the typical black or brown mud of salt lakes.

Still, some H_2S escapes upward and supports surprisingly dense life crowds at the chemocline interface. Phototrophic green and purple sulfur bacteria bunch there, taking advantage of both the sunlight from above and the H_2S below, as an electron donor in photosynthesis, instead of H_2O. Accumulations of phototrophic bacteria are usually accompanied by similar layers of other organisms, either slightly above or below them. This microbial loop replaces or just couples to the conventional big-fish-eats-small-fish food chains in saline lakes; it provides the means for cycling significant amounts of energy and matter under such harsh conditions for life. The resulting layered structure is analogous to that of microbial mats, which are typical of some saline aquatic habitats where sunlight reaches the sediment–water interface, but spanning centimeters instead of millimeters, respectively.

"Many saline lakes are like bathtubs without a drain, from which water can only get out one way, through evaporation."

Knowledge about the microbial component led to scientists disregarding former misconceptions of extremely saline lakes as sterile, like the Dead Sea. Actually, bacterial photosynthesis often exceeds algal production during the winter months; for example, up to 85.5 percent of total production in Deadmoose Lake (Canada), or 17.1 percent of the total annual planktonic primary production. Regarding maximum daily rates, gross primary productivity may vary between 7.67 and 7.34 g C m^{-2} d^{-1} in macrophyte- and microbial mats-dominated saline lakes of the Mancha Húmeda Biosphere Reserve (Spain), and 10.15 g C m^{-2} d^{-1} in phytoplankton-dominated salt ponds in Israel. These figures indicate that maximum productivity rates are similar, whatever the functional group of primary producer and the lake salinity. These are comparable to or higher than productivity rates of other types of lakes and wetlands that are highly productive ecosystems. On the other hand, respiration rates may also be highest in saline lakes because of increased needs for energy to cope with extreme salinities. As a result, the range of net primary productivity rates in saline lakes is very wide, even in the same site across time.

The subsequent pulses in primary production of saline lakes may result in such contrasted trophic patterns along time that a single site may behave as two distinct ecosystems in different moments. This occurs in both permanently and temporarily flooded saline lakes. For example, in the Laguna de Mar Chiquita (Argentina), food webs may involve combinations of either microbial-planktonic-waterfowl communities that are usual in saline lakes, which are dominated by the Andean flamingo, or planktonic-fish communities dominated by the Argentinian silverside. Similarly, primary production in the Laguna de Salicor (Mancha Húmeda Biosphere Reserve, Spain) regularly supports a waterfowl population that is hardly higher than 600 individuals per month; however, in the breeding seasons 2007–09, a few thousand birds were recorded during several months; species richness also increased from 7 to 21 species, respectively.

Although biodiversity patterns are rarely documented under such contrasted trophic patterns, it may be logically induced that saline lake fluctuations increase their biodiversity, whether the changes involve primary productivity, hydrology, and/or hydrochemistry. Nevertheless, the negative correlation observed between species richness and salinity, over a broad range of salinities, underlay once upon a time the general assumption that salinity was an important determinant of saline lakes biota. Salinity is negatively correlated with species richness and community composition when salinity is lower than 10 g l^{-1}, but not at intermediate or high ranges of salinity (i.e., 10–30 and 100–200 g l^{-1}, respectively), according to more recent and precise studies. In other words, salinity is not the only, or perhaps not the most important factor, which determines the occurrence of a particular taxon in a saline lake.

Saline lakes are select attributes of certain landscapes in all continents; they are usually found in groups of saline water islands surrounded by land. At the scale of the continents, these groups are linked by biogeographical relationships. For example, saline lakes of the Mediterranean basin, east-central Europe, the Middle East, and central Asia were refuges for biota during glaciations; genetic exchange between them was then very active, but it decreased during subsequent interglacial periods, such as in the present time. As a result, new species appeared, and endemisms and vicariant distributions became increasingly common. It is the case of the submerged herbs of the genus *Althenia*; although its complete distribution is not known, it extends from the Atlantic coast of France and Morocco through the Mediterranean to Asia, with isolated records from Turkey, Iran, and south-central Siberia.

Similarly, the presence of large branchiopod crustaceans up to 0.86 inch (22 millimeters) long in the Mancha Húmeda Biosphere Reserve was thought to be unique among Spanish salt lakes, and reveals unsuspected affinities with those of central Asia. These communities appear in temporarily flooded habitats, unlike their relative (the brine shrimp *Artemia*), which is common in permanent hypersaline waters. Branchiopods are ancient crustaceans; they are too large in size to escape predators; as a result, their distribution is restricted to extreme environments that are saline, temporarily flooded, turbid, and/or cold. The three former attributes feature many saline lakes, like those of the Mancha Húmeda.

Ecologist Ramón Margalef did field observations of these communities; he found out how the islands of water that saline lakes remain in are presently connected. One of the most typical shoreline birds of saline lakes is called *Charadrius alexandrinus* (Kentish plover). Birds and flying insects exploit saline lake resources in an alternating way; they move around looking for available resting, refuge, food, and/or breeding sites. Usually, these resources are not available in all sites of a saline lake district at the same time, which increases overall biodiversity. For this reason, the biota of saline lakes usually follow a metapopulation pattern, that is, a great population (saline lake district) consisting of small populations (sites). Trip distance is extreme in the case of intercontinental seasonal migrations, which are reserved to waterfowl.

Insect migrations are not limited to the exchange between locations within a saline lake district. Soil beetles of the carabid group, for example, are able to fly and disperse locally. However, their populations also move both horizontally and vertically within a single lake, depending on the temporal variations in soil salinity and water content. Actually, carabids are major cyclers of matter and energy in the shores of saline lakes, thanks to their diverse feeding strategies, including omnivores, predators, carrion feeders, and granivores. The outermost belt of saline lake biota is featured by the presence of plant communities that support a varying degree of occasional flooding and soil salinity. These include many glassworts of the *Salicornia* genus and *Limonium* sea lavenders, and *Micronecmum coralloides*, another example of east–west Mediterranean disjunct species. Some are considered halonitrophilous because they indicate relative soil enrichment in nitrogen after decomposition of the high amounts of organic matter accumulated in their shores.

Habitats on a Human Scale

Saline lakes are not marginal habitats for humans. One of the most common species found in saline lakes is an introduced one, cattle skeletons. The present inability to adapt to saline lake environments, in which large-scale economy is poor, plays a crucial role in the evolution of human attitudes toward saline lakes. As a result, saline lakes are frequently used for spilling solid urban waste and sewage water. Adaptation to marginal habitats may be limited by consequences of demographic characteristics of marginal populations. Salt lakes are not found in the most densely populated parts of the planet, but even so, 500 million people live in regions where saline lakes are a common feature of the landscape. Genetic, developmental, and functional constraints also limit the adaptation to marginal habitats. However, functions, values, services, and products of saline lakes are abundant and valuable. Many have market value. Saline lakes are important sources of minerals such as uranium, lithium, zeolites, and sodium chloride. The brine shrimp *Artemia* and its cysts are a base for the aquaculture industry, the common green algae *Dunaliella* is a very important source of carotene and glycerol, and the cyanobacteria *Spirulina* is used for the food industry.

Salt lakes and their muds also have a great interest for salt and mud therapy. Important spa resorts were born in saline lakes, like in Mar Chiquita (Argentina) and Laguna de la Hijosa (Mancha Húmeda Biosphere Reserve, Spain), and some of them remain active as locally strategic economic alternatives, like the Playas de La Mancha (Spain). Occasionally flooded salt flats are aesthetic icons that are frequently used by the publicity and film industries, or even for land speed records. This is the case of the Bonneville Salt Flats (United States), where portions of the movies *Pirates of the Caribbean: At World's End*, *Independence Day*, and many others have been filmed. Desiccation polygons of salt flats throughout the world are extensively used as symbols of drought.

However, the cultural values of salt lakes are not just a matter of Hollywood and the mass media. Salt caravans are common in all continents, like that going from Salinas Grandes to Buenos Aires (Argentina) between the 17th and 18th centuries, or the 2,000-year *azalai* of Tuaregs between Bilma and Agadez (Niger). Cities, transport ways, territorial networks, and cultures were born at both sides of salt caravans. If populations of saline lakes were marginal, they would be sparse, fragmented, prone to isolation and civilization collapse, or they would be demographic sinks subject to immigration from highly populated areas. These and many more functions, values, services, and products of saline lakes are only possible thanks to their particular traits, including the flatness, desiccation, salinity, physiological adaptations to osmotic potential, protection against excessive sunlight, accumulation of proteins, extremophilous fauna, landscape uniqueness, and chromatic richness.

Saline lakes have been conserved in a relatively good condition until recent times. Irrigation agriculture was not feasible in their surroundings or with their water. However, technology has allowed intensive exploitation of groundwater and great diversions of surface waters, which were the main inputs to some saline lakes. As a result, the level of the Aral Sea, formerly the fourth-largest saline lake in the world, has fallen some 66 feet (20 meters) since 1960, increasing salinity and affecting its characteristic processes and communities. Similar impacts occur elsewhere in the world. On the contrary, the excess of water also affects many saline lakes.

Usually, it is because of the input of wastewater, after having passed (or not) through a water treatment plant. In this case, the impact is double, involving a change in hydrology and a decrease in water quality. Even worse is the deliberate filling of a salt lake from these water sources, allegedly for "restoration ecology" purposes; these projects aim at getting huge populations of waterfowl in a single site, whatever its hydrological and trophic functioning was. To make things even worse, the excess of organic matter, lack of recycling, and pollution frequently result in large waterfowl kills because of avian botulism outbreaks.

Máximo Florín

Streams

A stream is defined as a body of water with a current (a lotic environment), confined within a bed and stream banks. But a stream is much more than that. A stream is a dynamic, living entity that links ecological processes across space and time, creating habitat for a diverse assemblage of organisms. Streams are important as conduits in the water cycle, as instruments in groundwater recharge, and as corridors for fish and wildlife migration. Lotic waters are diverse in their form, ranging from springs only a few centimeters wide to major rivers kilometers in width.

Although streams and rivers have been critical to the development of civilization, they represent but a small fraction of the world's freshwater supply. Nevertheless, streams play an important role in connecting fragmented habitats and supporting biodiversity. Like many ecological systems, streams are highly imperiled by natural and anthropogenic stressors.

A stream is governed by its physical setting. Local geology determines its substrate type and sediment load. Rocks and sediments are transported by water in solution (dissolved load), as particles in suspension (suspended load), and/or as particles moving along the streambed (bed load). The size of particles determines how they are transported and where they are ultimately deposited. Stream channels vary in curvature and cross-sectional profile, alternating between pools, riffles, and runs. The degree of habitat variability is often used as a measure of stream health. Current is the defining physical variable of rivers and streams, shaping the streambed habitat and interacting with many other biotic and abiotic variables. Discharge is determined by climate and altitudinal gradient, and varies horizontally and vertically within the stream. Obstructions such as logs and boulders create variation in current that form different types of habitat. Even in smooth channels, water near the sides and bottom travels at lower velocities because of friction. Flow transports oxygen, carbon dioxide, and nutrients, and conveys food to waiting organisms.

To shelter from the current, organisms must either develop attachment devices or perform energetically costly work to hold position. Stream dwellers exhibit a wide variety of adaptations to the current, including a flattened profile, silk and other sticky secretions, hooks, claws, and suckers. In addition, many organisms adapt behaviorally, "holding station" in slow-moving water adjacent to faster-moving water, where they can dart out to capture passing prey. Others shelter near the streambed, where virtually no movement of water molecules occurs.

Other important physical factors affecting streams include sunlight, chemical constituents, and temperature. Water temperature is first affected by air temperature, then by the relative mix of surface water and groundwater contributing to the stream, the degree of shading, and the overall stream size. Small streams warm more by day and cool more by night because of solar heating of the streambed and their relatively small mass of water. Since the majority of stream organisms are cold-blooded, temperature strongly influences where organisms can live and their metabolic rates. While most streams range from 32 to 77 degrees F (0 to 25 degrees C), subtropical and tropical streams can reach 86 degrees F (30 degrees C), and some desert streams can reach 104 degrees F (40 degrees C).

Stream Vegetation

Energy flow through stream food webs is complex. In some streams, aquatic plants are the dominant energy sources; in others, terrestrial plant production dominates. Allochthonous sources of energy include leaves, fruits, and other plant materials that fall or are blown into the stream. This nonliving organic matter (detritus) is also called particulate organic matter (POM). Terrestrial energy sources also include detritus that decomposes on the forest floor, enters soil water, and eventually enters the stream as dissolved organic matter (DOM). Stream ecologists refer to energy produced within the stream channel as autochthonous production, and organic matter produced outside the stream as allochthonous production. Autochthonous energy sources include microscopic algae (though large colonies can be seen with the naked eye), cyanobacteria, and macrophytes (mosses, liverworts, and the true vascular plants, or angiosperms).

Studies suggest that in woodland streams, allochthonous inputs may be even more important than instream production. Terrestrial organic matter such as leaves, fruit, twigs, and logs produce coarse particulate organic matter (CPOM, greater than 1 millimeter in size) that is in many streams a major source of energy input. Through a combination of wetting, physical abrasion, microbial colonization, and shredding, CPOM is converted to fine particulate organic matter (FPOM). FPOM is readily available for bacterial colonization, serving as a food source for animals that collect it from the substrate or filter it from the water column. Found both within and outside the stream, dissolved organic matter (DOM) is the largest carbon pool in stream food webs. DOM comes dissolved in water and reaches the stream via canopy drip during rain, as surface flows, and via subsurface pathways.

Diatoms, algae, and cyanobacteria provide important direct sources of energy to stream heterotrophs (organisms that cannot produce their food) since most macrophytes are unpalatable (though they do contribute to instream production of DOM and FPOM). Unicellular diatoms create a brown, slippery coating comprised of millions of cells on stones or other instream objects exposed to sunlight. Diverse and abundant, they are the most important autochthonous energy source to stream food webs. Algae proliferate in streams with sufficient nutrients, a stable substrate, and adequate sunlight. They are common, occurring as single cells, colonies, or long filaments, and are generally palatable.

Cyanobacteria have the ability to "fix" atmospheric nitrogen that can be utilized by other autotrophs (organisms that can produce their food), but are probably less important as an energy source to stream food webs than diatoms and algae. Both autotrophs and heterotrophs benefit from their close association in biofilms, which are complex microecosystems and important regions of energy production. A self-contained ecosystem, biofilms—surface slime composed of bacteria, algae, and fungi bound within a polysaccharide (complex sugar) matrix—are in turn inhabited by protozoans and micrometazoans that consume the matrix. Within the matrix, dead algal cells and organic exudates nourish heterotrophic bacteria that, in turn, convert organic compounds into inorganic compounds used by algae for continued photosynthesis.

Mosses (commonly *Fontinalis* and *Fisidens*) are usually found growing on rocks and logs in and around the colder, well-shaded areas of streams. They can tolerate low light and low temperatures, and have a rapid rate of nutrient uptake and high resistance to being dislodged by high flow events. Poorly studied, mosses may rival algae for productivity, but because of their limited distribution, are probably of minor importance as an overall energy source. Macrophytes gain a toehold as streams increase in size and falling current velocities cause silt to settle out. Common genera include *Potamogeton, Elodea, Ranunculus,* and *Nuphar,* as well as watercress (*Nasturtium*) in smaller springs and brooks. Macrophytes play various roles in the ecology of streams: as an energy source both before and after death, as a substrate for attachment of other organisms, and as cover from predators. Their greatest role as an energy source is probably in the production of fine particulate organic matter (FPOM, less than 1 millimeter in size) produced after they die and their remains are broken down.

Stream Insects

Aquatic insects are by far the most abundant organisms in streams, and have the greatest diversity of feeding methods. Stream ecologists classify these macroinvertebrates into functional feeding groups that encompass both their feeding roles and how their food is acquired. Functional feeding groups include shredders, grazers and scrapers, collectors (both gatherers and filterers), and predators. In addition, aquatic insects are excellent environmental indicators because they display a range of tolerances for environmental conditions.

Shredders are found in well-oxygenated waters with large accumulations of CPOM. Shredders obtain energy from the leaf and from the microbes, primarily fungi, which colonize it. Using their mouthparts, they shred and tear CPOM particles into smaller FPOM particles and also contribute FPOM via their fecal pellets. Common shredders include some stoneflies (e.g., the famous salmon fly *Pteronarcys californica*), many dipteran larvae (e.g., crane flies of the family Tipulidae), and most of the caddisflies that construct cases of organic material (Limnephilidae).

Grazers and scrapers are typically found where light reaches the stream bottom, promoting algal growth, their main food source. Grazers and scrapers have mouthparts adapted for scraping the film of algae growing on the surfaces of rocks and other large surfaces (periphyton) and produce copious amounts of FPOM through the production of fecal pellets and dislodgement of

A stream runs through moss-covered rocks in the Great Smoky Mountains on the Tennessee–North Carolina border. Mosses thrive in low light and temperatures, so they are most often found in shaded areas of streams. Streams in the Great Smoky Mountains (also called the Smokies) are part of the Tennessee River watershed. (Thinkstock)

algal cells during feeding. Grazers (e.g., mayflies) browse the overstory of loosely attached algae, while scrapers close-crop the diatoms adhered to the substrate. One common grazer is *Glossosoma*, a small caddis that carries its small, tortoise shell-shaped case of coarse cemented sand grains around on its back while feeding. It is a general rule that caddis larvae with cases made from inorganic matter are scrapers, and those made from organic matter are shredders.

The largest and perhaps most interesting functional group, the collectors, can be further divided into filtering-collectors and gathering-collectors. Filtering-collectors filter FPOM from the water using nets (mainly caddisflies) or specially adapted body parts. Like fishermen, net-spinners construct a net across their nonportable case or fixed retreat and wait for current-borne FPOM to collect in the fine mesh. Then, they either feed on the trapped particles, or consume the net and its contents altogether. This ubiquitous group is highly variable with respect to stream order, temperature, and FPOM size, though body and net size generally decrease with distance downstream as CPOM inputs decrease and FPOM increases. Other filtering-collectors have specialized body parts to enable them to filter FPOM from the current. The caddis *Brachycentrus* sits inside its small, four-sided case with its hairy-fringed forelegs extended, ready to intercept FPOM from the current. Once the fringes are full, it combs the captured material from its forelegs with its mouthparts. Blackfly larvae anchor their abdomens to rocks and throw themselves into the current, filtering FPOM from the water with their fan-like mouthparts that are fitted with fine hairs and sticky mucous.

The mayfly *Isonychia* crouches in the current, holding its fringed front legs in front of its body like a basket. When the basket is full, it raises its front legs and consumes the contents with its mouthparts. Unlike filtering-collectors, gathering-collectors (such as mayflies of the genus *Baetis)* are typically generalists that scurry around the stream bottom, picking up particles wherever they find them or ingesting their way through the sediments like earthworms.

Predatory insects occur throughout the stream community and have many different adaptations to enable them to pursue and capture prey. Common predators include stoneflies (Plecoptera), dragonflies and damselflies (Odonata), hellgrammites (Megaloptera), and one family of free-roaming caddisfly (Rhyacophylidae) that does not construct cases. Other types of instream insects include miners that feed on detritus buried in fine sediments, piercers with sucking mouthparts that feed on plant fluids, and gougers that burrow into large woody debris while feeding on their associated fungal and bacterial colonies.

Animals in Streams

Fish also assume functional roles. Fish can be classified into feeding guilds—top-feeders, midwater-feeders, and bottom-feeders. Different species occur along a stream's length. Overall, species richness and variation in body size increases with distance from headwaters to river mouth. Species that prefer clean, swift gravel runs and feed on the aquatic insects found in such habitat are generally absent from low-gradient, soft-bottomed lower river reaches. As with macroinvertebrates, indices of stream health have been built on assessments of fish diversity and abundance.

Other instream organisms include mollusks, crustaceans, and meiofauna, which include protozoans, rotiferans, annelids, and the microcrustaceans (harpactacoid copepods and ostracods). Some minor groups include the macroscopic freshwater sponges (Porifera), freshwater hydroids and jellyfish (Coelenterata), flatworms (Turbellaria), horsehair worms (Nematomorpha), and mosslike animals (Bryozoa), and the microscopic gastrotrichs (Gastrotricha) and tardigrades (Tardigrada).

Reptiles, and especially amphibians, utilize streams where in certain cases, such as with the Pacific giant salamander (*Dicamptodon ensatus*) and tailed frog (*Ascaphus truei*), they assume a dominant role in terms of biomass. The tailed frog is the only New World amphibian that exhibits internal fertilization, presumably as an adaptation to rapidly flowing mountain streams where currents would wash eggs and sperm away.

Most mammals and birds visit streams at one time or another. Some, like beavers, river otters, and mink, and diving and wading birds, are intimately associated with them. The dipper or water ouzel (*Cinclus mexicanus*), a small songbird, is entirely dependent on stream ecosystems. A dipper actually walks underwater to pick aquatic invertebrates off submerged substrates, maintaining its position on the streambed by adjusting its wings against the force of the current. Other songbirds and bats consume emergent aquatic insects that are a large part of their food supply.

River Continuum Concept

The River Continuum Concept (RCC) describes how a hypothetical stream might change physically and ecologically along its length as a result of variable energy inputs and their resultant changes to the community of organisms. This model uses stream order, which describes the different links in a stream network, and can give an indication of their discharge. For example, a first-order stream lies at the very headwaters of a stream network where the stream begins to flow year-round. Further downstream, first-order streams merge into larger, second-order streams, and so forth. The RCC generates many useful predictions about patterns that can be seen in any geographical region or biome. For example, a stream's headwaters (orders 1–3) might flow through a heavily shaded forest reach, then through a midreach of more open country (orders 4–7), and finally through a lowland reach of relatively flat, open country as a large, deep, heavily silted river (orders 8 and above).

Upper reaches are generally narrow and well-shaded by streamside trees. The stream substrate may be rocky or sandy, depending upon the geology of the region. Here, mosses are the dominant aquatic primary producers because of insufficient light for instream algal growth, and the current is usually too swift, nutrients too low, and the substrate too unstable to support the growth of macrophytes.

Considerable CPOM enters the stream from leaves, twigs, and branches. This matter is important in meeting the energy demands of instream organisms, shredders who break down the coarse organic matter into FPOM, and collectors who gather this suspended material from the water column. Fishes in upper reaches include minnows, trout, sculpins, and other species tolerant of high-current velocities and seasonal and daily cold temperature regimes. Upper reaches export a considerable amount of FPOM to middle reaches where the stream widens, warms, and is well-lit by direct sunlight. These conditions lead to a proliferation of algae, filamentous greens, and/or diatoms. The diatom layer makes middle-reach rocks slippery. Rooted macrophytes find purchase in protected places where the current slows and sediments accumulate. Biomass of collectors is generally similar to upper reaches (though the individual species may differ), but shredders are much reduced. However, grazer biomass is much increased to take advantage of the proliferation of algae on the stream bottom.

"Streams are important as conduits in the water cycle, as instruments in groundwater recharge, and as corridors for fish and wildlife migration."

Unlike other reaches, organisms in middle reaches produce more food than they consume, exporting large amounts of FPOM to lower reaches. The majority of the world's trout streams are located in these highly productive middle reaches. Lower reaches of streams generally fall into the river category. The RCC is well-supported in most streams around the world. Exceptions occur in streams with different geographical or geological conditions, and/or where changes occur because of human alterations to the pristine pattern. Even in relatively undisturbed streams, reset mechanisms interrupt the predictable patterns of the river continuum. These can occur through natural means—such as when tributaries join larger streams, resetting the conditions in the receiving stream to those higher in the continuum—or through human activities, such as damming or deforestation.

Examples of streams around the world include cold-water trout streams that support a wide variety of organisms and provide world-class recreational opportunities for anglers; midwestern United States warm-water streams that are, in their pristine state, the prototype of the RCC; southeastern United States blackwater streams that, while entirely heterotrophic, provide habitat for a wide variety of threatened and endemic species; mineral-laden thermal springs and streams and their resulting unique biota; ephemeral hot-desert streams driven almost entirely by seasonal precipitation; cold-desert spring streams that store excess primary productivity until released by high-flow flushing events; alpine streams that, despite a short growing season, support flourishing communities of algae, macroinvertebrates, and fishes; cave streams in which energy flows are based almost entirely on detritus or chemosynthesis; and tropical streams that differ from temperate streams in their seasonality regimes, high interannual variation, frequent storm events, and differences in species composition. For example, many tropical streams are dominated by decapod crustaceans (mainly atyid shrimp), rather than insects.

Human activities have a profound influence on streams. Dams alter flow, temperature, and sediment regimes. Dams also fragment river systems, block fish passage, and isolate populations. Nonpoint source pollution is the dominant contributor to nutrient loading in streams and rivers. Acid rain has lowered pH in many streams in western Europe, Scandinavia, the northeastern United States, and some regions of the Rocky Mountains, killing much of the life in these systems. Mining also produces acid and toxic metal runoff that has poisoned many streams in Appalachia, the Rockies, and elsewhere, virtually extinguishing life. Introduced species displace native species. And climate change is almost certain to alter thermal and hydrologic regimes in streams, with associated changes to the ecology of organisms in these systems.

Melanie L. Truan

Wetlands

Wetlands are productive ecosystems that are formed when water is present for a sufficient period of time to create low-oxygen conditions in the soil, resulting in the creation of unique soil properties and a suite of wetland plants capable of living and reproducing in these environments. Most wetland definitions exclude permanently flooded areas such as lakes, rivers, and oceans or temporary puddles, such as those that occur on driveways, because of a lack of wetland plants.

Wetlands may be associated with streams and lakes, or they may be isolated from surface water. Wetlands include both fresh and saltwater systems and have a variety of local names such as marsh (dominated by emergent herbaceous vegetation), salt marsh (emergent herbaceous vegetation in saltwater), swamp (trees and shrubs), peatland (accumulating dead plant material), bog (peat accumulating wetlands with no significant water inflow or outflow), fen (peat wetland with some water flow), seep (sloped wetlands), wet meadow (herbaceous vegetation with waterlogged soils), oxbow (abandoned river channel), vernal pools (seasonally flooded pools in forested environments), and playa (arid or semi-arid wetlands). Wetlands may be completely natural systems, or highly altered and managed systems, such as rice (*Oryza*) fields and cranberry (*Vaccinium*) bogs. Although diverse in name and appearance, all wetlands generally have water, wetland vegetation, and unique soils.

Water or hydrology is central to the functioning of a wetland. The cumulative effects of water level, flow, seasonal pattern, and frequency of inundation dictates the plant communities and species composition, and the rate of chemical processing, which influences animal communities and other attributes within the wetland. Under most circumstances, wetlands must have standing water or be saturated within 11.8 inches (30 centimeters) of the soil surface for two weeks or more during the growing season, at least every other year. However, there is a large gradient in the hydroperiod (water regimes) of wetlands, from permanently to intermittently flooded. Permanently flooded wetlands are flooded throughout the year in all years, whereas in intermittently flooded wetlands, the surface is usually exposed with surface water present for variable periods, without detectable seasonal patterns. The overall hydroperiod or water budget results from the balance between inflows and outflows of water, which is dictated by the surface contour of the landscape, subsurface soil, geology, and groundwater conditions. Sources of water to a wetland include precipitation from rain and snowfall, surface inflows from overland or from streams and groundwater, and tidal inflow in coastal wetlands. Water is lost from a wetland from surface outflows, groundwater outflows, evaporation, plant transpiration, and tidal outflow.

Wetland Plants and Animals

Wetland plants, technically called hydrophytes or hydrophytic vegetation, are important components of wetland systems. Hydrophytic vegetation are plants that grow in water or in soil that at least periodically has low oxygen levels as a result of excessive water content. Hydrophytes have unique adaptations to survive in wetland environments. Aerenchyma, porous tissues in roots and stems that are filled with large gas-filled spaces, allow efficient storage, exchange, and movement of oxygen, carbon dioxide, and ethylene. Adventitious roots, roots that arise from tissues other than root tissues, spread onto the surface layer of the soil or grow above the soil surface to increase access to the air. Bald cypress (*Taxodium distichum*) has modified erect roots that grow from the roots above the water surface to aid in efficient carbon dioxide and oxygen exchange. Many wetland tree species have shallow root systems as a mechanism to grow in waterlogged soil, and take advantage of occasional dry periods to obtain atmospheric oxygen. However, shallow roots make trees susceptible to wind, so large trees must maintain a balance between remaining upright and rooted, and efficient respiration.

Many hydrophytes provide the basis for the complex wetland food web (the interwoven pathways by which the plant materials are consumed by other trophic levels of organisms) and also can contribute substantially to the food webs in terrestrial and aquatic systems. Many waterfowl species, as well as muskrats (*Ondatra zibethicus*) and beaver (*Castor canadensis*), consume the seeds or tubers of wetlands plants. Wetland managers manipulate water regimes to produce specific species of seeds, such as millets (*Echinochloa*) and smartweeds (*Persicaria*), to attract waterfowl. Much of the plant material, particularly from leaves and stems, enters the food chain as detritus as the plants break down and decompose at the end of the growing season. Detritus are small plant particles resulting from the breakdown and decomposition of the plants, and are subsequently consumed by various organisms such as invertebrates, which are then consumed by fish and wildlife. A variety of organisms also use plants as cover or habitat. Wetland plants enhance water quality by removing nutrients and some toxins from the water and storing them. Moreover, hydrophytes promote wetland function by reducing peak flood events and stabilizing soils.

Hydrophytes are classified into primary categories based on their growth form and appearance: emergent, floating, floating-leaved, scrub-shrub, trees, and submerged. Emergent wetland plants, including cattails (*Typha*), common reed (*Phragmites australis*), and common rush (*Juncus effusus*), are herbaceous species rooted in soil with basal portions that typically grow beneath the surface of the water, but with aerial leaves, stems (photosynthetic parts), and reproductive organs. Floating plants are not rooted in wetland soil, and instead float on the surface of the wetland. Roots may or may not be present, and there is no connection to the bottom substrate.

Water lettuce (*Pistia stratiotes*), duckweeds (*Lemna*), and water hyacinths (*Eichhornia*) are highly productive examples of the floating growth form. Water lilies (*Nymphaea*) and other floating-leaved plants have leaves that float on the water's surface and roots that are anchored in the substrate. Stems connect the leaves, which are circular or oval, and have a tough leathery texture designed to repel water to the bottom. Scrub-shrub plants are rooted in the soil, like emergent plants, but have a woody stem, growing up to six meters tall. Scrub-shrub plants may be true shrubs or small trees. Trees are large woody plants greater than 19 feet (6 meters) tall. Submerged plants, often referred to as submerged aquatic vegetation, spend their entire life cycle beneath the surface of the water. Nearly all are rooted in the substrate. Submerged plants take up dissolved oxygen and carbon dioxide from the water column and are well known for their water clearing abilities.

Hydric soils are unconsolidated organic and mineral matter on the Earth's surface where all pore space is filled with water, the soil surface is temporarily covered with flowing water, or water stands in a depression for a long enough period of time during the growing season to promote the absence of molecular oxygen in the soil. As a soil becomes saturated with water, it progresses through a series of chemical and physical steps involving inundation with water, promotion of anaerobic conditions, sequencing of a set series of biochemical reductions, slowing of organic matter decomposition, accumulation of distinct mottles or spots of different colors within the dominant soil color, and accumulation of organic matter. Many wetland soils are organic, but some are mineral. Soil color, texture, structure, and even smell all provide clues to the type of hydric soil present in a wetland. For example, some wetlands have a rotten-egg smell under anaerobic conditions because of the presence of sulfate-sulfur.

Location of Wetlands

Wetlands occur throughout the world. An estimated 5 to 8 percent of the land surface is covered by wetlands, with greater amounts in tropical (31 percent), subtropical (25 percent), and boreal (30 percent) regions, and lesser amounts in polar (2 percent) and subboreal (11 percent) regions. About 95 percent of wetlands are freshwater and only 5 percent are saltwater. The Pantanal in Brazil, Paraguay, and Bolivia, covering about 88,803 square miles (230,000 square kilometers), is often considered the largest wetland in the world. Wetlands are often viewed unfavorably as a hindrance to development and progress, or as disease reservoirs for mosquitoes carrying and spreading malaria, yellow fever, and other maladies. Wetlands have been drained and filled to facilitate agriculture, development of residential and industrial complexes, road building, mineral extraction, and other human infrastructure.

More than 50 percent of wetlands around the world have been destroyed because of human activities, and many more are affected by water pollution and groundwater withdrawal. The conterminous (lower 48) United States has lost 53 percent of wetlands, which is probably the most accurate estimate available for any region. New Zealand has lost 90 percent, Europe and China 60 percent, and Australia 50 percent of their wetlands. Wetland loss is most rapid during periods of growth and development in countries. Coastal areas and agricultural regions have suffered the most rapid and extensive wetland losses because of high human populations, high land use conversion rates, and a propensity of wetlands. In developed countries, wetland loss has slowed because of contemporary laws designed to protect wetlands and because wetlands easy to drain and fill are now gone. However, the United States still lost 5,590 hectares annually between 2004 and 2009. From 1998 to 2004, the United States actually gained 89,140 hectares of wetlands because of national policies. Wetland losses and gains only tell part of the story. Open-water wetlands often increased, whereas more ecologically valuable vegetated wetland types continued to be lost. Wetland loss rates in developing countries will likely continue to remain high as technological advances and human population growth continues.

Benefits of Wetlands

Wetlands produce disproportionately high benefits, compared to the size of their presence on the landscape. Wetlands provide numerous ecological services to people, including providing a source of clean water, reducing flood damage, enhancing biodiversity, providing recreational opportunities, and producing food and fiber. Wetlands are highly effective at purifying and storing water. Because of their ability to remove pollutants from the environment, they are often referred to as the kidneys of the landscape. Wetlands are effective at removing pollutants and sediment because of a reduction in water velocity based on their position in the landscape and vegetation, which slows water velocity, in addition to the numerous chemical processes occurring in wetlands during alternating wetting and drying cycles. Because of their effectiveness, wetlands are specifically constructed to treat industrial and household wastewater and mine land runoff. Generally, the slower the water passage and the higher the complexity in vegetation and substrate, the better the system will function. Wetlands substantially reduce the impacts of floods by intercepting storm runoff and storing it. The ability of wetlands to store flood water reduces the sharpness of peak water flows (peak flows cause the most damage to buildings and other infrastructure) and allows a slower discharge of water over a longer period of time. This results in less severe flooding and less flood damage. Wetlands associated with rivers generally provide the most benefit in regard to flood control.

Wetlands are the most productive systems in the world, with net primary productivity reaching up to 1,200 grams per square meter. This high productivity results in complex food webs and abundant plant and animal life. Waterfowl, which includes ducks, geese, swans, and other waterbirds including herons, egrets, and shorebirds, are the most widespread inhabitants of wetlands. The Prairie Pothole Region of the Northern Great Plains of North America is among the most productive waterfowl-producing areas in the world. It is known as the North American Duck Factory because of the large numbers of migratory dabbling and diving ducks that the prairie potholes (glacial wetlands) produce. Wetlands with a balanced ratio of open water to emergent vegetation are most productive for both breeding and wintering birds. Wetlands also provide habitat for numerous amphibian (frogs, toads, and salamanders), reptile (turtles, snakes, and crocodilians), mammals, fish, and invertebrate species. Wetlands provide a unique resource for recreational use, including hunting, birding, angling, hiking, exploring, and botanizing. An estimated $10 billion is spent annually by 50 million people observing and photographing wetland-dependent wildlife. In the United States, waterfowl alone attract over 15.4 million observers. There are over 1.3 million waterfowl hunters who spend over 13 million days hunting and $780 million in expenditures annually in the United States. Everglades National Park, a 6,500-square-kilometer complex of wetlands in Florida, receives around 1 million visitors annually.

Wetlands provide numerous economically valuable products, including peat, pelts, hides, fish, shellfish, timber, fiber, fruits, and grains. Peat is harvested extensively for use in the horticultural industry, and to a lesser degree as a fuel source. Fur-bearing mammals that depend principally on wetlands include beaver, mink (*Neovision vision*), muskrat, otter (*Lutra*), and nutria (*Myocaster coypus*). American alligator (*Alligator mississippiensis*), saltwater crocodile (*Crocodylus porosus*), and spectacled caiman (*Caiman crocodilus*) are harvested for their valuable hides, which are used to make boots, belts, and other leather items. An estimated 95 percent of commercially harvested fish and shellfish, including pink shrimp (*Farfantepenaeus duorarum*) and a variety of crabs, oysters, crayfish, salmon, catfish, and menhaden are wetland-dependent species. Frogs are also commercially grown and harvested for their legs in some regions. Bald cypress trees produce commercially valuable timber that is used in construction, and also are important sources of mulch. Numerous other trees, such as pecans (*Carya illinoinensis*), swamp white oak (*Quercus bicolor*), and sweetgum (*Liquidambar*) are also important for timber, and pecan trees also produce valuable nut crops. The productivity of many herbaceous wetland plants, such as cattails, common reed, and paper reed (*Cyperus papyrus*), produce material useful for making baskets, mats, and even roofing material. Rice production in specialized, manipulated

American alligators laze in the sun in Everglades National Park in Florida. According to the U.S. National Park Service, alligators are an important part of the Everglades ecosystem and are considered a key species for the park. The nesting activity of female alligators plays an important role in the creation of peat, and the retained water in old or abandoned alligator holes can be used as a nesting spot for several turtle species or a refuge for a variety of wildlife. (Thinkstock)

wetland systems called rice paddies provide 20 percent of the human diet worldwide. Wild rice is also harvested for consumption, primarily by indigenous cultures, and is also often packaged and sold for premium prices. Cranberries are commonly grown in northern wetland systems or upland systems converted into wetlands specifically for growing cranberries.

Protection of Wetlands

Because of the valuable ecosystem services and functions that many wetlands provide, political entities, including the United States, have passed laws and regulations to protect wetlands. The primary purpose of these laws is to avoid impacting wetlands. However, avoidance is not always possible. In many cases, permits are needed from local, state, and federal officials if a wetland is to be filled. In the United States, the federal Clean Water Act is the primary mechanism for protecting wetlands. Wetlands are protected under Sections 401 and 404 of the Clean Water Act, which regulates the discharge of dredged or fill material into the waters of the United States. If an area meets the technical definition of a wetland, it is illegal to fill it in, unless a federal and state permit is obtained. The U.S. Environmental Protection Agency has the technical lead for issues dealing with wetlands. However, the U.S. Army Corp of Engineers is the primary agency responsible for enforcing wetland regulations. The U.S. Department of Agriculture's Natural Resources Conservation Service (NRCS) deals with most wetland issues on farmlands, but it is not a regulatory agency.

"Wetlands may be completely natural systems, or highly altered and managed systems, such as rice fields and cranberry bogs. Although diverse in name and appearance, all wetlands generally have water, wetland vegetation, and unique soils."

Mitigation in the form of creating a new wetland is often required as a condition to receive the permit. This requirement has a spawned an industry built around the delineation, creation, and restoration of wetlands. Wetland delineators determine the boundary between wetland and uplands to determine the size of a wetland and to determine where development should and should not occur. Most created wetlands are built near the site of impact as a means of replacing the lost wetlands function within the same watershed. However, function is not readily assessed or easily measured. Most mitigation in the past has been performed on a size basis, but functional assessment mitigation is increasing in popularity. A size-based mitigated strategy results in the loss of more complex vegetated wetland types, requiring larger wetlands to be built. Hence, forested wetlands require a larger ratio of created to destroyed land than emergent wetlands, and both require more than open-water wetlands. Recently, more emphasis has been placed on mitigation banking. Mitigation banks have been designated as the preferred means of compensatory mitigation for unavoidable wetland losses. Wetland banks are generally built by companies willing to make long-term investments for future potential profit by creating wetland prior to destruction. Once a wetland is impacted, a company or individual then buys credits from the bank. Large public agencies also may produce wetland banks for their use. Mitigation banking is more cost effective than traditional compensatory mitigation, larger wetland sites provide more ecological value, mitigation occurs before impacts so functional success is already known, and project review time is reduced, which reduces permit processing and enhances regulatory agency effectiveness.

In an April 2001 court case (*Solid Waste Agency in Northern Cook County v. U.S. Army Corps of Engineers*) the Supreme Court ruled that the U.S. Army Corp of Engineers does not have jurisdictional authority to enforce wetland regulations on wetlands that are not associated with navigable waters. Currently, there is no protection for these isolated wetlands, except for disincentive programs associated with wetlands on farmlands or individualized state laws. The "swampbuster" provision of the 1985 Farm Bill withholds farm program benefits if a farmer drains, dredges, or otherwise manipulates a wetland to make production possible. Wetlands manipulated before 1985 are termed prior converted wetlands and are exempt from the swampbuster provision. Internationally, some wetlands are protected and receive recognition through the Convention on Wetlands, which was signed in Ramsar, Iran, in 1971. The Ramsar Convention was signed as an international treaty for member nations to maintain the ecological character of wetlands of international importance. The Ramsar definition is broader than most wetland definitions, but is an effective means to help protect wetlands on international boundaries, and also provides additional recognition to wetlands within partner countries. There are 160 member nations to the convention, with 1,960 sites encompassing 190 million hectares of internationally significant wetlands.

The interest in wetlands has continued to grow over the past 40 years. Formal academic courses in wetland science started in the early 1970s and continued steady growth through the 1980s and 1990s. After 2000, the number of wetland courses increased substantially, which is indicative of the popularity and acceptance of wetland science as a discipline. The Society of Wetland Scientists, an international organization of professional scientists and managers dedicated to promoting wetland science, education, and management, was formed in 1980. There are also more specialized groups interested in particular aspects of wetlands, such as the International Peat Society, which formed in 1968; its vision is to be the authoritative international organization on all peat and peatland issues. Since 1981, the Society of Wetland Scientists has published a scientific peer-reviewed journal called *Wetlands*. In 1989, the journal *Wetlands Ecology and Management* was first published by Springer. The journal *Mangroves and Salt Marshes* was published from 1996 to 1999, and in 2000 was merged into *Wetlands Ecology and Management*. Moreover, numerous ecological and environmental management journals publish articles on wetlands. The popularity of wetland science is expected to continue to grow over time because of their ecological importance.

Wetlands will continue to be lost in the future because of development pressures. Climate change has the potential to exacerbate wetland loss problems, causing additional losses and leading to degradation in others. Increased temperature results in higher evaporation and plant transpiration rates, reducing water levels and changing the hydroperiods of wetlands. Rainfall and snowfall amounts and intensity are also predicted to vary. These changes have important implications for species of plants and animals that depend on specific water regimes and hydroperiod lengths for completing their life cycles. Higher temperatures also impact species with specific temperature tolerances such as fish and amphibians. These species either have to adapt, move, be moved by humans, or become extirpated.

Climate alteration also will impact the suite of pathogen and insect stressors affecting particular wetland communities. Some of the greatest potential for wetland loss is in coastal areas. Increased melting rate of glaciers leads to sea-level rise and the subsequent loss of coastal salt marshes. There is also much concern in northern peat bogs about drying wetlands releasing carbon back to the atmosphere and increased methane emissions. Wetland managers are working on solutions to address climate change issues, many of which involve comprehensive land planning, increased human assistance, and adopting an adaptive management strategy.

James T. Anderson

Desert Biomes

Coastal Deserts

Coastal deserts are sometimes called mild deserts because the presence of the nearby ocean is a moderating influence on temperature and narrows the distance between the experienced extremes. They are most frequently found along the western coasts near the tropics. Major coastal deserts include the Atacama Desert, the Namib Desert, the western part of the Sahara, and the Baja California portion of the Sonora Desert. In biological terms, deserts are areas with sparse vegetation and wildlife as a result of prevailing aridity, and in which native life has adapted to survive in arid conditions.

Those arid conditions may result from temperatures hot enough to evaporate precipitation too quickly for it to accumulate, or from naturally low levels of rainfall: In either case, evaporation exceeds precipitation. In some cases, cold ocean currents can dessicate coastal terrestrial ecosystems because they contribute little atmospheric moisture through evaporation, and therefore limit coastal precipitation. This is the case with the Namib Desert (along southern Africa's Atlantic coast) and the Atacama Desert (along South America's Pacific coast). Many of the world's deserts are caused by the subtropical high pressure belts at roughly 30 degrees north and south. There may be multiple causes for a desert's aridity—the Atacama, in addition to being on a cold coast with a polar current and falling within one of those subtropical belts, is in the rain shadow of the Andes Mountains, which block the flow of humid air from the Amazon River.

Variations in Aridity

Deserts are generally arid or hyperarid, and usually semi-arid or dry semi-humid areas with similar characteristics are called semi-desert or desert fringe (and they usually appear in the transitions between deserts and less-arid ecosystems). There are exceptions: Most of the Kalahari Desert is semi-arid, for instance, and coastal deserts may be semi-arid or dry semi-humid for certain parts of the year, or in certain geographic portions. Climatically, coastal deserts are one of the three fundamental divisions of deserts. Though cold deserts (e.g., those of central Asia) and hot deserts (e.g., those of Australia) are so categorized because of their prevailing temperatures, coastal deserts are those which, because of their location on or proximity to a coastline, experience frequent fogs. These fogs, which deliver moisture to an otherwise arid environment,

A dense morning fog rolls through the sand dunes of the Namib Desert along southern Africa's Atlantic coast. In coastal desert biomes, these fogs may be the only source of moisture available in an otherwise arid climate. Rainfall, when it does occur, can be spotty and may strike heavily in a relatively small area. (Thinkstock)

are as characteristic a feature of the coastal desert ecosystem as the summer highs and winter lows of hot and cold deserts. In some years, many coastal deserts experience more foggy days than rainy days, and this fog may be the main source of moisture. Coastal deserts also experience fewer temperature extremes because of the moderating effect of the ocean. While hot and cold deserts often experience a diurnal range—the difference between cold nighttime and hot daytime temperatures—of 95 degrees F (35 degrees C), coastal deserts typically have a diurnal range of only 50 degrees F (10 degrees C), comparable to less-arid regions.

Though it is easier to describe deserts in terms of their vegetation and soils, because of the scarcity of climate data as a result of the deficiency of permanent human settlements in deserts, soil and vegetation are in turn impacted by, if not outright determined by, climate. Desert soils are predominantly mineral soils with low organic content. They're generally fine-textured and porous, with good drainage. Coastal desert plants tend toward extensive root systems to take advantage of rainstorms, fleshy leaves and stems for water storage, and in some cases longitudinal ridges and grooves, which allow stems to swell when full of water. Salt bush, buckwheat bush, rice grass, and black sage are all common coastal desert flora. Rainfall is variable in deserts, enough to make annual average statistics less than elucidating. A single heavy storm in the Namib Desert in 2006 brought six times more rainfall than its annual average. Desert rainfall also tends to be "spotty" because desert storms are often convective; rainfall may strike heavily, but affect a very small space, perhaps only a mile or two across.

The role of moisture in the desert is complex because there is so little of it, and what little there is plays an extraordinarily large role in characterizing the ecosystem and the landscape. For instance, outside of a few subareas—such as along a riverbank winding through the desert—deserts are devoid of plant life, which is dependent on regular water access. Even the life in coastal desert fog zones is specially adapted to that fog, and differs from the plant life that depends on periodic rainfall. Because that plant life is not present, there are fewer grasses and root systems to hold soil in place, and less microbial activity and decomposition of organic material impacting the character of that soil. When rain occurs, therefore, it has a much greater erosive effect because the landscape has fewer protections against erosion; the 12 inches (30 centimeters) of annual rain in a desert may have much greater effect than four or five times that much rain in another environment, and a rainstorm can completely change the shape of the land and wipe out fragile ecosystems.

Similarly, the presence of even trace moisture increases the efficiency of temperature extremes, contributing to the fragmentation and weathering of rocks; and in coastal deserts where such extremes are less common, the more saline precipitation leaves salt crystal deposits in the soil, which can pry rocks apart as they grow. Even in noncoastal deserts, salt is more common in the soil than in nondesert environments because there is so little rainfall to wash it away. Coastal desert soil salts can include common or "table" salt (sodium chloride); gypsum [$CaSO_4 \cdot 2(H_2O)$], which is more likely to appear in fog zones or semi-arid areas; and calcrete, a duricrust rich in calcium carbonate. A famous form of desert salt is the sand rose, an assemblage of gypsum crystals that are flat and blade-shaped, resembling rose petals. In the soil of the coastal Atacama Desert, not only common salt and gypsum are found, but also deposits of perchlorates, iodates, and nitrate deposits so concentrated that since the 1830s they have been mined for use in fertilizer.

Inhabitants of Coastal Deserts

There is more life in a desert than is at first apparent. A fundamental part of the coastal desert ecosystem is the biofilm, the thin layer of organisms that forms on the surfaces of rocks, in the cracks that form in them, and in their pore spaces. These organisms include cryptogams—organisms that reproduce by spores, most commonly lichen, algae, and mosses in deserts—as well as funguses and bacteria. The biofilm not only makes a living space out of the pores and cracks in rocks, but in fact contributes to their formation through the acids it excretes; biofilm and weathering are major contributors to the desertification of arid spaces and the breakdown of rock into the characteristic desert soils.

Microbes also play a role in the formation of "desert varnish," which appears in many environments, but is vividly noticeable in deserts, where a paper-thin coating of mineral clay, manganese, and iron paints the surface of exposed rocks, leaving them brown or black. Microbes also thrive in coastal deserts' fog zones, where the dominant vegetation is cryptogams, which can make better use of fog moisture than vascular plants. Desert fog water has been studied for its chemical composition and ion concentrations in order to learn more about the types of life it can and does support; studies have found that although the acidity, salinity, and levels of chemicals such as manganese, calcium, and nitrous oxide can be higher than the norm, they are all within the levels allowed by the World Health Organization for water collection for human consumption. They are safe for most animal life, in other words; the differences come in part from the way the moisture is collected (plants do best when they can take it from the air or from condensation, rather than relying on a root system underground), and the way that trace elements accrete over time, such as the accumulation of salt in soil.

Much of the fauna in coastal desert fog zones obtains its water through what it eats; while desert cultures have always had erroneous beliefs about various large mammals (such as gazelles) never needing to drink water, it holds true for many small mammals who subsist on insects, larvae, and plant matter. There are also animals that have adapted to obtain their water directly from the fog. The Namib Desert has 13 different species of *Onymacris* (darkling beetles, or tenebrionid beetles).

They have longer legs than other beetles, resembling spiders but for the number of limbs, which allows them to escape the hot boundary layer of air clinging to the desert sand. They obtain their moisture through "fog basking."

When the fogs appear in Namibia's coastal desert, two species in particular—*Onymacris unguicularis* and *O. bicolor*—have been observed climbing to the tops of dunes in great numbers in order to allow the fog to condense on their abdomens (cooler than the sand temperature because of those long legs) and drip into their mouths. Fog basking has been shown to have evolved independently on two separate occasions, and to constitute a true adaptation to desert conditions. There has been no conclusive proof that this is also true of *stilting* (the term entonomologists use for the tenebrionids' long legs).

The fauna present in a coastal desert may depend in some way other than the fog on the nearby presence of the ocean. The black widow spider (*Latrodectus indistinctus*) thrives in the Namib Desert, and although the principal effect of this within the desert is to reduce the herbivore population on the dunes (which protects dune vegetation, as well as the microbial ecosystem surrounding it), these high spider populations are maintained not by prey in the desert, but by the detrital-algae-feeding flies the black widow consumes, which have flown into the environment from the adjacent marine biome. In Baja, California, coastal spiders are six times more abundant than inland, and they subsist on a principally marine diet.

"Coastal deserts experience fewer temperature extremes because of the moderating effect of the ocean."

The Baja Desert system is also home to 3–24 times more insects, scorpions, lizards, rodents, and coyotes on the coast than inland, and studies of coastal desert mammal populations suggest a great dependence on marine life, which constitute more than half the diet of the coastal coyote. Marine-subsidized coyotes also depress coastal rodent populations; on desert islands off the coast of Baja, coyote-less islands have far more rodents than the coyote-rich islands. The proximity of the ocean also leads to changes in the types of algae present in the ecosystem. In the Atacama Desert, the marine green alga *Ulva* is foraged by many of the invertebrates, who in turn are preyed upon by scorpions (*Brachistosternus ehrenbergii*), solifuges (*Chinchippus peruvianus*), and geckos (*Phyllodactus angustidigitus*). Coastal deserts play a key role in nutrient transfer from marine biomes to terrestrial biomes, because some of these desert fauna will in turn be preyed upon by predators in the fringe, and so on as those nutrients move further and further inland as they pass through the food chain. It has even been demonstrated that the productivity of Amazon rainforests depends on fertilization from phosphorous-rich dust blown in from the coastal western Sahara Desert, 3,107 miles (5,000 kilometers) away.

Desert Adaptations

The Atacama Desert is thought to be the driest in the world, receiving no rainfall from 1570 to 1971, over 400 years. While other coastal deserts often receive periodic floods, they are not always able to make use of that suddenly available moisture. Some deserts develop temporary lakes as a result of flooding, and if this happens regularly enough, an ecosystem can develop around those lakes. However, when the floodwaters are the result of overflowing rivers or other bodies of water outside the desert environment, they will bring marine life with them, and in some cases may introduce new microbial life or other fauna, or attract new migratory birds or other fauna. Because there is so little life in the desert, the introduction of a new species can have drastic consequences; in Australia, a common problem is the threat to grasses holding desert dunes together because of overgrazing by newly introduced rabbits.

Desert centers of endemism—species peculiar to a given ecosystem, in other words, deserts with a great number of unique species—are always near coasts (which includes coastal deserts), which is believed to be because of Pleistocene expansions of the central deserts. Endemic species are very often well-adapted for a particular ecosystem, but unlike species found in a wide variety of environments, their life cycle may depend heavily on the balance of that ecosystem; they may be ill-equipped to deal with a new disease, predator, or competitor for resources, where a more commonplace species might be more robust, having weathered those challenges many times in its evolutionary past.

The pulse-reserve paradigm of describing rain events in the desert considers deserts as pulse-driven ecosystems. A rain event triggers a pulse of biological activity; some of it is lost to consumption or mortality, and the remainder is committed to a reserve, such as seeds or the water storage of vegetative life like geophytes or succulents. Life in the desert has adapted to dealing with rainfall that arrives in pulses, rather than continuously, and which is sometimes too much and usually too little. For example in the Arctic, the life cycle of various microflora and microfauna may all but pause during the winter, when the sea is covered in ice and neither new nutrients nor sunlight are available, only to pulse back to life during the summer thaw. In the desert, many species conserve their energy during the dry periods in order to act on the sudden availability of rainwater to replenish their reserves.

The manner of adaptation varies considerably; even in the same desert, a kangaroo is adapted to move long distances from food source to food source in the most energy-efficient way possible, whereas smaller rodents are adapted to move very slowly in order to consume less of their energy with unnecessary movements. Some amphibians in coastal deserts have accelerated larval stages, which allow them to reach maturity faster, improving their chances of survival; on the other end of the spectrum, some burrow-dwelling toads essentially hibernate for the dry season, sealing their burrows off with gelatinous slime that is washed away by the next rainy season. While coastal deserts receive some moisture from fog, this is true only for a portion of the desert, and the pulse-reserve paradigm still holds true for the desert's response to rainfall. Pulses replenish soil water better than fog can, leading to plant growth, and perhaps triggering the mating of certain fauna.

The ecosystems that make the best use of floodwaters include shrubs with grasses at their bases instead of bare soils, because the root matrix of the grasses and shrubs creates macropores in the soil, encouraging greater infiltration of water. Otherwise, most of the water will simply run right off the dusty, hard-packed surface of the desert, compacted by weathering and wind, glazed by desert varnish, and will eventually evaporate or continue to flow into another environment at a lower elevation.

In the Namib Desert, plants with different types of C_4 photosynthesis, all of which initially fix carbon dioxide (CO_2) in the mesophyll cells to form oxaloacetate, have different responses to rainfall. Some convert the CO_2 mainly into aspartate, with an inner bundle between metaxylem elements and the Kranz sheath. Others primarily convert the CO_2 into malate, with a single chlorenchymatous or Kranz sheath and centrifugal chloroplasts formed around vascular bundles. The malate converters lack the well-developed grana and higher mitochondrial frequency of the aspartate converters, and increase in abundance as rainfall increases, while the aspartate species decrease as rainfall increases.

Bill Kte'pi

Mid-Latitude Deserts

Mid-latitude deserts are characterized by their dry environments, with distinct topographic and physical features. A suite of different organisms have adapted to living in this harsh environment, and some classic examples range from plants that have adapted to different forms of photosynthesis in order to reduce water loss during the hottest part of the day, to animals that have adapted behavioral patterns to thermoregulate. Humans have also historically inhabited these deserts, and have left behind a rich cultural legacy, even in areas no longer populated. Major threats to this ecosystem include: urbanization, desertification, erosion, and resource overexploitation. Mid-latitude deserts are deceiving at first, because they appear desiccated and devoid of life. On a geologic time-scale, when compared with temperate forest ecosystems, deserts do not jump to the top of the list for primary productivity, but are instead weathered remnants that are severely resource constrained. However, mid-latitude deserts are rich centers of life, often with organisms that have adapted to become especially suited to these harsh environments. Mid-latitude deserts are also home to a diverse range of human inhabitants, who have similarly adapted to the desert environment. While deserts can seem unforgiving, it is worthwhile to traverse the dry landscape. Some great examples of the diversity of desert dwellers can be best exhibited through environmental adaptations and defense mechanisms, ranging from the spines commonly found on cacti to the unique defense strategies employed by horned lizards. Deserts are visually stunning and extreme environments, well worth the effort of exploring firsthand.

Desert Formation

Mid-latitude deserts are globally distributed. They occur approximately 25–35 degrees north or south of the equator (often referred to as the horse latitudes). They are perhaps the most common type of deserts and account for some of the most famous deserts in the world, like the Sahara in Africa, the Great Sandy Desert in Australia, and the Sonoran Desert in North America. Mid-latitude deserts are formed when dry air circulates down to the ground after having emptied its moisture in the tropical latitudes of the world. As the sun hits the Earth, the tropical latitudes receive the most solar radiation; this causes warm wet air to rise to the atmosphere.

As the air rises, the wet tropics receive nearly all of the precipitation in it, leaving dry cool air to circulate down from the atmosphere at about 30 degrees north or south of the equator. This dry air accounts for the high aridity and precipitation patterns observed in most mid-latitude deserts. Most deserts can be characterized by the ratio of precipitation to evaporation. In most mid-latitude deserts, evaporation will frequently be greater than the total precipitation that hits the ground. Since "mid-latitude desert" is a very broad classification, and they are widely distributed across the world, there is quite a bit of variation in total precipitation that these deserts receive. Most of these deserts are extremely arid, and tend to have less than 9.84 inches (25 centimeters) of annual precipitation. The low precipitation also results in low cloud cover in most of these deserts. Low cloud cover leads to high daily temperature variation. In some of the most extreme cases, temperatures can reach 104 degrees F (40 degrees C) during the day and close to 32 degrees F (0 degrees C) at night. The cool, dry air that flows into mid-latitude deserts often can become very violent and triggers massive sandstorms, with walls of dust reaching .93 mile (1.5 kilometers) into the air, and with winds reaching 37–62 miles (60–100 kilometers) per hour.

Mid-latitude deserts have distinctive geophysical features that are indicative of this type of habitat. Often, mid-latitude deserts have vast stretches of sand dunes; some of the largest dunes in the world occur in the western Sahara Desert, and some of the tallest reach nearly 0.62 mile (1,000 meters) in height. Much of mid-latitude deserts are covered in dry, rocky outcrops, or even vast salt flats. The largest of these salt flats is Salar de Uyuni in the northern extent of the Atacama Desert, and is approximately 4,054 square miles (10,500 square kilometers) in area. Rivers are rare in mid-latitude deserts, but some, like the Nile, flow through vast portions of the Sahara. Usually, rivers that run through mid-latitude deserts tend to originate in high-elevation regions that receive more precipitation. Oases are often another source of freshwater in mid-latitude deserts, and are typically formed by aquifers or underground river systems. Both rivers and oases are areas where most of the diversity of organisms in mid-latitude deserts can be found.

Life in Mid-Latitude Deserts

Most mid-latitude deserts are less than 20,000 years old, making them relatively new ecosystems. Mid-latitude deserts grew in size after the glaciers that covered the northernmost and southernmost latitudes of the world during the last glacial maximum began to retreat, and historically temperate regions underwent desertification. However, even while mid-latitude deserts are relatively young in geologic and evolutionary timescales, they are still home to many endemic taxa with very interesting evolutionary histories.

Mid-latitude desert organisms have evolved traits that allow them to survive the harsh conditions of limited water availability, high temperatures, scarce food resources, high competition, and high predation. The evolution of life in the desert makes mid-latitude deserts interesting places for evolutionary biologists to consider evolutionary phenomenon like convergent evolution. Convergence is when similar traits evolve in organisms that do not share a common ancestor, mostly because of shared selection pressures in different regions of the world. Examples of organisms that have evolved convergent traits in mid-latitude deserts are the Kangaroo rats (*Dipodomys)* of North American mid-latitude deserts, and the Australian mid-latitude desert Hopping mice (*Notomys*). To prevent herbivory, succulent mid-latitude desert plants have also evolved convergent traits, like the thick spines in North American cacti and African *Euphorbia* species.

Life in the mid-latitude deserts is heavily dependent on water availability and the organisms' physiological and behavioral adaptations to deal with high temperature fluctuations. Water in mid-latitude deserts is restricted to a few main sources like aquifers, oases, and rivers. It is along these riparian areas where most of the biodiversity in mid-latitude deserts can be found. However, unique organisms have adapted to extreme aridity by drawing water from fog or water condensation from morning dew, as is the case with *Euphorbia* in the Atacama Desert (the driest mid-latitude desert in the world) in western South America. Behavioral adaptations also allow organisms, especially vertebrates, to cope with the high temperature fluctuations in mid-latitude deserts. Many desert organisms are primarily nocturnal, and simply avoid activity during the warmest periods of the day. Adaptation and evolution have resulted in deserts rich in endemic biodiversity. The Sonoran Desert is North America's most biodiverse mid-latitude desert.

Biodiversity in mid-latitude deserts is threatened by habitat disturbance, depletion of water resources, and urbanization. As people continue to colonize the mid-latitude deserts of the world, water resources are depleted, activities like overgrazing change the structure of the vegetation and animal communities of the desert, and cities expand into areas where human life may not be sustainable. All three of the anthropogenic disturbances devastate the endemic and rare biodiversity of desert organisms, and further research is required to fully understand how these disturbances impact the native species of the mid-latitude deserts of the world.

The zebra-tailed lizard in Arizona's biodiverse Sonoran Desert is a heat-tolerant lizard that can remain active in midday when high temperatures force other types of lizards to seek shelter. (Thinkstock.

Natural Resources of Mid-Latitude Deserts

Mid-latitude deserts are homes to the largest terrestrial oil deposits in the world. The oil reserves in the Middle East and in northern African mid-latitude deserts are among the largest in the world. A total of six of the 10 largest oil-producing nations are in the deserts of these regions, including: Iraq, Kuwait, the United Arab Emirates, Iran, Saudi Arabia, and Libya. In a world so heavily dependent on oil and other fossil fuels for transportation and energy production, mid-latitude deserts are important global regions for production of these fossil fuels. Drilling for oil and improper management of oil production has impacted the deserts in Iraq and Kuwait, especially from recent wars and political conflicts. During the Gulf War in the early 1990s, oil fields in Kuwait were burned and millions of barrels of oil were released into the sands of the desert, devastating wildlife and vegetation communities. Restoration efforts are still ongoing. Oil production and regulation in the mid-latitude deserts of the world should be carefully monitored and regulated to protect endemic and threatened organisms that occupy these parts of the world.

"Biodiversity in mid-latitude deserts is threatened by habitat disturbance, depletion of water resources, and urbanization."

Mineral resources are abundant in mid-latitude deserts. Because of high rates of evaporation, minerals leached from rainwater and groundwater are frequently accumulated in the most arid regions of mid-latitude deserts. Salts are frequently accumulated near the surface in many mid-latitude deserts. One of the most salt-producing mid-latitude deserts is the Atacama Desert in South America. Other evaporites (mineral sediments that accumulate after evaporation brings the elements out of solution) commonly found in mid-latitude deserts are gypsum and boron. Gypsum also accumulates belowground, and is responsible for the giant crystals in the vast cave systems in the deserts of the southwestern United States (e.g., Lechuguilla Cave). Important metals are also often accumulated in the soils of mid-latitude deserts. In North America, copper production in the Sonoran and Chihuahuan Deserts is a major industry. Other metals of importance that can be found in mid-latitude desert are zinc, iron, and lead. Economically valuable metals like gold, silver, and uranium are also frequently mined from many of the mid-latitude desert regions of the world.

Management, extraction, and processing of these mineral resources frequently take a toll on the delicate desert ecosystems. In the United States, open-pit copper mining, especially in deserts of Arizona and New Mexico, modifies vast areas of the landscape, which will likely never recover from such a major disturbance. The processing of these resources in refineries also takes a toll on desert ecosystems, and often results in heavy-metal contamination of the soils and bioaccumulation of metals like lead, copper, and zinc in the plants and animals of the desert. Often, the heavy metal contamination and bioaccumulation can be detected up to 40 miles from a refinery.

Mid-latitude deserts provide natural resources that can be harvested with minimal ecological impact and multiple global benefits. Historically, the major rivers that dissect many mid-latitude deserts have been regulated to produce hydroelectric power. Even though damming rivers to produce hydroelectric power is damaging to the desert ecosystems, the sustainability and overall ecological consequences of energy production from these sources may be better and more sustainable than the burning of fossil fuels. Globally, about 2 percent of energy production comes from hydroelectricity, and only about 0.1 percent comes from dams in mid-latitude deserts.

However, mid-latitude deserts are ideal for the development of solar and wind energy production. These sources of energy have the potential to meet approximately 33 percent of the global energy demand, and mid-latitude deserts have the potential to become major producers of sustainable energy production across the globe. Major solar energy production ventures are already underway in the Mohave Desert in the United States (e.g., by solar energy generating systems). The mid-latitude deserts have the highest solar energy production potential in the world. Places like the Sahara, the Arabian Peninsula, and the Australian deserts all are great regions for collecting solar energy. The geomorphology of the mid-latitude deserts makes these areas suitable for development of wind-energy production. Desert areas in Texas, California, and New Mexico in the United States are already being developed for major wind-energy production. Currently, 2.5 percent of U.S. energy comes from wind production; with increased development, up to 20 percent of the energy production in the United States can come from sustainable wind energy production by 2030.

People in Mid-Latitude Deserts

Mid-latitude deserts have a rich history of civilizations that have developed in these arid global regions. Among the most notable are the civilizations of ancient Mesopotamia that developed along the major fertile river valleys in the mid-latitude desert regions of the Middle East. The heart of ancient Mesopotamia revolved around the riparian habitat provided by the Tigris, Euphrates, and Nile, but the people of these civilizations developed their cultures and societies in the harsh conditions of the mid-latitude deserts.

In North America, the mid-latitude deserts of the southwestern United States and northern Mexico are homes to ancient Native American tribes, like the Anasazi (which are ancestors to the more recent Pueblo people), the Mogollon, and the Hohokam. Until the early 1900s, native aboriginal cultures (Wangkangurru people) thrived in the Simpson Desert (an Australian mid-latitude desert) by moving between reliable well and groundwater resources. In the Sahara, the Berber people, and most notoriously the Tuareg people, have also historically lived nomadic herding and trading lifestyles in this vast desert region. Currently, mid-latitude deserts are occupied by a wide variety of people. In the United States, Native American tribes have been drastically restricted to reservations and now account for less than 5 percent of the total population in the desert southwest.

However, some mid-latitude deserts have retained many of the people's ancestral traditions. For example, in the Gobi Desert, most people live a nomadic herder lifestyle. The descendants of the Atacama and Aymara native people of the Atacama Desert still herd llamas and alpacas through this mid-latitude desert. Urbanization has dramatically changed the populations of the mid-latitude desert. Some of these regions of the world are experiencing rapid population growth as cities grow and infrastructure connects previously isolated regions. This urbanization movement is particularly pronounced in areas where mineral and other natural resource exports stimulate the regional economy. One example is the growth of major population centers in the major desert oil-producing nations. Major economic growth has stimulated the population and economic growth in major desert cities like Dubai in the United Arab Emirates. In the United States, mid-latitude deserts' major economic and urbanized areas include cities like Phoenix and Tucson, Arizona. Additionally, there are major border cities that are important for North American trade, including the cities of El Paso, Texas, and Ciudad Juarez, Chihuahua, Mexico. These cities all have populations ranging from 450,000 to 1,500,00 people. This growth has major impacts on the terrestrial ecosystems of mid-latitude deserts.

Threats

There are four major categories of threats to mid-latitude desert ecosystems, including: urbanization, desertification, erosion, and resource exploitation. Additionally, there are miscellaneous threats, usually tied to anthropogenic pollution. These include nuclear waste and solid waste disposal, increased irrigation pressures, increased grazing, and irresponsible recreational vehicle use. While all of these threats are imminent and pose great danger to the maintenance of healthy deserts, if steps are taken now, scientists can mitigate damage to these unique ecosystems and preserve desert ecosystems for future generations to enjoy.

Urbanization is one of the key threats to desert ecosystems. Development and construction practices can be highly detrimental to fragile biotic crusts that cover desert soils. Along with development and construction come the associated needs for water and land resources, both of which stress already limited resources (water is very limited in desert ecosystems, and land well suited for development may also be limited). Urbanization pressures are also inherently tied to sprawl, which increases the footprint of urban centers on surrounding landscapes—a good example of this is the sprawl associated with cities such as Las Vegas, which required enormous resource infrastructure to survive in environments that are not naturally suited to support large populations. Anthropogenic pollution is also likely to increase with urbanization, and additional public works networks (to supply people with water, or provide waste processing) will also need to be established as new areas become urbanized.

With global warming there are increasing rates of desertification, which is often associated with increased likelihood of drought, loss of biodiversity, and decreased soil moisture (depending on specific geographic positioning). Desertification is the process of land changing from productive or arable land to less productive deserts because of climate change and human-induced land changes such as deforestation or clearing for agriculture. This is a counterintuitive threat, since desertification effectively creates deserts; however, desertification usually leads to nutrient-depleted deserts that do not support native healthy desert animal and plant communities, and are usually located in regions where deserts would not naturally occur.

Correlated with urbanization and desertification comes land degradation and increased erosion pressures across desert landscapes. Biotic crusts, which cover many desert landscapes, can support a great diversity of microorganisms, but are very sensitive to environmental changes. With landscape disturbances, soil crusts may fragment, decrease functionality within desert ecosystems, and lead to a negative feedback loop, leading to less productive desert ecosystems. Deserts, while not necessarily noted for their abundance of vegetation, are highly susceptible to erosion because of low vegetation cover, and with increased disturbances they can erode even faster. This means that the desert does not only become drier in light of climate change, but as a result could have significant changes in vegetation, decreased soil moisture, and increased erosion. Over time, land degradation and erosion can completely destroy productive desert ecosystems.

Resource exploitation can rapidly diminish healthy ecosystems. Deserts are no exception; even if on the surface they look dry and covered with spiky cacti, they offer a wealth of below-ground resources. Human energy needs are only increasing as the world's population surpasses 7 billion people. Oil extraction and associated extraction technologies are not only greatly disruptive to these ecosystems, but also pose hazardous environmental threats in the form of oil spills and other chemical contamination. Deserts can also be filled with other valuable resources that are useful for energy, such as petroleum oil, coal, and natural gas extraction, and precious metals, such as gold and diamonds.

Israel Del Toro

Paleodeserts

Data on ancient sand seas (vast regions of sand dunes), changing lake basins, archaeology, and vegetation analyses indicate that climatic conditions have changed considerably over vast areas of the Earth in the recent geologic past. During the last 12,500 years, for example, parts of the deserts were more arid than they are today. About 10 percent of the land between 30 degrees north and 30 degrees south is now covered by sand seas. Nearly 18,000 years ago, sand seas in two vast belts occupied almost 50 percent of this land area. As is the case today, tropical rainforests and savannas were between the two belts.

Fossil desert sediments that are as much as 500 million years old have been found in many parts of the world. Sand dune–like patterns have been recognized in presently nonarid environments. Many such relict dunes now receive from 3.15 to 5.9 inches (80 to 150 millimeters) of rain each year. Some ancient dunes are in areas now occupied by tropical rainforests. One of these is the Nebraska Sand Hills, which is an inactive 22,008-square-mile (57,000-square-kilometer) dune field in central Nebraska. The largest sand sea in the Western Hemisphere, it is now stabilized by vegetation and receives about 19.69 inches (500 millimeters) of rain each year. Dunes in the Sand Hills are up to 394 feet (120 meters) high.

Colorado River Basin

The Grand Canyon is a huge rift in the Colorado Plateau that exposes uplifted Proterozoic and Paleozoic strata, and is also one of the 19 distinct physiographic sections of the Colorado Plateau province. It is not the deepest canyon in the world (Kali Gandaki Gorge in Nepal is far deeper), nor the widest (Capertee Valley in Australia is about 0.6 mile, or 1 kilometer wider, and longer than Grand Canyon); however, the Grand Canyon is known for its visually overwhelming size and its intricate and colorful landscape. Geologically, it is significant because of the thick sequence of ancient rocks that are beautifully preserved and exposed in the walls of the canyon. These rock layers record much of the early geologic history of the North American continent.

"Some ancient dunes are in areas now occupied by tropical rainforests. One of these is the Nebraska Sand Hills, which is an inactive 22,008-square-mile dune field in central Nebraska."

Uplift associated with mountain formation later moved these sediments thousands of feet upward and created the Colorado Plateau. The higher elevation has also resulted in greater precipitation in the Colorado River drainage area, but not enough to change the Grand Canyon area from semi-arid. The uplift of the Colorado Plateau is uneven, and the Kaibab Plateau that the Grand Canyon bisects is over 1,000 feet (300 meters) higher at the North Rim than at the South Rim. Temperatures on the North Rim are generally lower than the South Rim because of the greater elevation (averaging 8,000 feet, or 2,438 meters, above sea level). Heavy rains are common on both rims during the summer months.

The Colorado River basin (of which the Grand Canyon is a part) has developed in the past 40 million years. A recent study places the origins of the canyon at some 17 million years ago. Previous estimates had placed the age of the canyon at 5 to 6 million years. The result of all this erosion taking place through the years is one of the most complete geologic columns on the planet. The major geologic exposures in the Grand Canyon range in age from the 2-billion-year-old Vishnu Schist at the bottom of the Inner Gorge to the 230-million-year-old Kaibab Limestone on the rim. There is a gap of about 1 billion years between the stratum that is about 500 million years old and the lower level, which is about 1.5 billion years old. This large unconformity indicates a period of erosion between two periods of deposition.

Many of the formations were deposited in warm, shallow seas, nearshore environments (such as beaches), and swamps as the seashore repeatedly advanced and retreated over the edge of a proto–North America. Major exceptions include the Permian Coconino Sandstone, which contains abundant geological evidence of aeolian sand dune deposition. Several parts of the Supai Group were also deposited in nonmarine environments. The great depth of the Grand Canyon, and especially the height of its strata (most of which formed below sea level) can be attributed to 5,000 to 10,000 feet (1,500 to 3,000 meters) of uplift of the Colorado Plateau, starting about 65 million years ago (during the Laramide Orogeny). This uplift has steepened the stream gradient of the Colorado River and its tributaries, which in turn has increased their speed, and thus their ability to cut through rock. Weather conditions during the ice ages also increased the amount of water in the Colorado River drainage system. The ancestral Colorado River responded by cutting its channel faster and deeper.

The base level and course of the Colorado River (or its ancestral equivalent) changed 5.3 million years ago when the Gulf of California opened and lowered the river's base level (its lowest point). This increased the rate of erosion and cut nearly all of the Grand Canyon's current depth by 1.2 million years ago. The terraced walls of the canyon were created by differential erosion. Between 3 million and 100,000 years ago, volcanic activity deposited ash and lava over the area, which at times completely obstructed the river. These volcanic rocks are the youngest in the canyon. The ancient Pueblo people were a Native American culture centered on the present-day Four Corners area of the United States. The ancient Pueblos were the first people to live in the Grand Canyon area. There are approximately 1,737 known species of vascular plants, 167 species of fungi, 64 species of moss, and 195 species of lichen found in.

Grand Canyon boasts a dozen endemic plants (known only within the park's boundaries), whereas only 10 percent of the park's flora is exotic. A total of 63 plants found here have been given special status by the U.S. Fish and Wildlife Service. Of the 34 mammal species found along the Colorado River corridor, 15 are rodents and eight are bats.

Sahara Desert

Some 12,000 years ago, the only place to live along the eastern Sahara Desert was the Nile Valley. It was so crowded that prime real estate in the Nile Valley was difficult to find. Disputes over land were often settled with the fist, as evidenced by the cemetery of Jebel Sahaba, where many of the buried individuals had died a violent death. But around 10,500 years ago, a sudden burst of monsoon rains over the vast desert transformed the region into habitable land. This opened the door for humans to move into the area, as evidenced by 500 radiocarbon dates of human and animal remains from more than 150 excavation sites. "The climate change at [10,500 years ago] which turned most of the [3.8-million-square-mile] large Sahara into a savanna-type environment happened within a few hundred years only, certainly within less than 500

The Colorado River passing through the Grand Canyon in Arizona, which has been in continuous human use and occupation since the Paleo-Indian period. The rock layers at the park range from the lowest layer of Zoroaster Granite (from the Precambrian-archeozoic era), which is almost 2 billion years old, to the top layer of Kaibab Limestone, which is the remains of a shallow warm sea occuring at the end of the Paleozoic Period (65 million years ago). (U.S. Geological Survey)

years," said study team member Stefan Kroepelin of the University of Cologne in Germany.

In the Egyptian Sahara, semi-arid conditions allowed for grasses and shrubs to grow, with some trees sprouting in valleys and near groundwater sources. The vegetation and small, episodic rain pools also enticed animals well adapted to dry conditions, such as giraffes, to enter the area. Humans also frolicked in the rain pools, as depicted in rock art from southwest Egypt. In the more southern Sudanese Sahara, lush vegetation, hearty trees, and permanent freshwater lakes persisted over millennia. There were even large rivers, such as the Wadi Howar, once the largest tributary to the Nile from the Sahara. Wildlife included species such as elephants, rhinos, hippos, crocodiles, and more than 30 species of fish up to 6.5 feet (2 meters) long. A brief history of the Sahara Desert follows:

- 22,000 to 10,500 years ago: The Sahara was devoid of any human occupation outside the Nile Valley, and extended 250 miles further south than it does today.
- 10,500 to 9,000 years ago: Monsoon rains begin sweeping into the Sahara, transforming the region into a habitable area that was swiftly settled by Nile Valley dwellers.
- 9,000 to 7,300 years ago: Continued rains, vegetation growth, and animal migrations lead to well-established human settlements, including the introduction of domesticated livestock such as sheep and goats.
- 7,300 to 5,500 years ago: Retreating monsoonal rains initiate desiccation in the Egyptian Sahara, prompting humans to move to remaining habitable niches in Sudanese Sahara. The end of the rains and return of desert conditions throughout the Sahara after 5,500 years ago coincides with population return to the Nile Valley and the beginning of pharaonic society.

At the end of the last Ice Age, the Sahara Desert was just as dry and uninviting as it is today. However, sandwiched between two periods of extreme dryness were a few millennia of plentiful rainfall and lush vegetation. During these few thousand years, prehistoric humans left the congested Nile Valley and established settlements around rain pools, green valleys, and rivers.

GHAZALA NASIM

Polar Deserts

One-third of the Earth's surface is desert, with polar deserts covering approximately 1.93 million square miles (5 million square kilometers). Polar deserts are areas that receive less than 10 inches (250 millimeters) of precipitation per year, and have a mean temperature of less than 50 degrees F (10 degrees C) in the warmest months of the year.

Frequently, the temperature fluctuates, and it often crosses the freezing point. Polar deserts can be found around the world, including the continent of Antarctica and the Arctic, including portions of Alaska, Canada, Greenland, Iceland, Norway, Sweden, Finland, and Russia. The Antarctic is the largest desert in the world, covering approximately 5.5 million square miles (14.25 million square kilometers). Because they are among the coldest places on Earth, these areas are also among the driest. Only the heartiest of plant life can survive, and animal life is centered in coastal areas where food is abundant and available from the surrounding waters.

The Arctic is a polar desert, and is the second-largest desert in the world, encompassing approximately 5.3 million square miles (13.73 million square kilometers). The Arctic region near the North Pole is primarily ocean beneath its thick layers of ice, whereas the Antarctic region includes the South Pole and the continent of Antarctica. The Arctic Ocean is warmed by the waters of the Atlantic Ocean, thus areas of the Arctic support human, plant, and animal life. Antarctica is cut off by a cold ocean current that flows around the continent, making the region even colder. Antarctica has no native human population. The Arctic Ocean has an average winter temperature of 15 degrees F (minus 26 degrees C) and an average summer temperature of 26 degrees F (minus 3 degrees C). On top of the Antarctic ice cap, the average winter temperature falls to minus 76 degrees F (minus 60 degrees C) with an average summer temperature of minus 26 degrees F (minus 32 degrees C). Central Greenland is covered with ice that is approximately 5,000–8,000 feet (1,524–2,438 meters) thick, making for extreme cold temperatures and an extremely dry environment.

Only about 3 inches (8 centimeters) of snow falls here each year, and as a result of the extreme temperatures, it does not melt. Small amounts may be lost through *sublimation,* where the precipitation vaporizes before actually reaching the ground and accumulating. Even this small amount of yearly precipitation is thickening the ice by approximately 0.8 inch (2 centimeters) each year. The Greenland ice sheet of the polar desert region encompasses an area of 708,069 square miles (1.8 million square kilometers), roughly the size of Texas, New Mexico, Arizona, California, and Mississippi combined.

Antarctica encompasses a larger area, approximately 4.8 million square miles (12.4 million square kilometers), equivalent to an area roughly the size of the United States and Mexico combined. Antarctica is the highest, windiest, coldest, and driest of all the continents, covered by a sheet of ice that ranges from 1 to 3 miles (1.6 to 4.8 kilometers) thick in places. The Antarctic Peninsula and coastal regions receive approximately 8 inches (20 centimeters) of precipitation each year. Most of the continental interior lands receive as little as 2 inches (5 centimeters) of precipitation each year. Like Greenland, Antarctica is dry because of the extreme cold temperatures. Coastal areas may reach 32 degrees F (0 degrees C) for a short period during the summer months, but never in the interior land. The warmest month is December, and the coldest is August. The North or South Poles never face the sun directly, and the sunlight that does reach the area is spread over a large surface area, providing little warmth.

Polar deserts are dry because of the cold temperatures. The amount of water vapor the air can hold is in direct correlation to the air temperature. Cold air simply cannot hold moisture, which results in a dry desert condition. In addition, the high altitude of the regions of polar deserts contributes to its dryness. Solar radiation and heat are reflected off, rather than being absorbed by, the ground. Cold, dense air produces permanently high atmospheric pressure, which prevents moister air from entering the regions. As air rises, it cools, turning moisture in the air to rain or snow. Sinking air has lost its moisture through the precipitation process. In the polar desert regions, dry air is continually sinking, resulting in little or no precipitation, blue skies, and a dry environment. Winds are strong and continually blow across the polar desert biome.

Polar Regions and the Midnight Sun

There is technically one day and one night each year at (or very near) the North and South Poles, with each day lasting six months. The sun cycles through the sky every 24 hours, remaining above the horizon for a six-month period. Once the sun does set below the horizon, the region is plunged into a six-month period of darkness and bitter cold. At the North Pole, the dark winter lasts from September 21 through March 21, whereas at the South Pole it will be the opposite with summer lasting from September 21 through March 21.

This strange pattern of day and night occurs because the North and South Poles are located close to Earth's axis. While the Earth spins, turning our day into night over a 24-hour period, the North and South Poles stay virtually in the same place. The six-month day and night is a result of the slight tilt in the Earth's axis, allowing for a slight movement where day turns into night. Six months of daylight seem as if they would warm the areas and melt the ice and snow, but it is not so. The sun hangs low in the sky, where the ice and snow reflect away much of the light and the sun's heat.

McMurdo Dry Valleys of Antarctica

Found in Antarctica near the Transantarctic Mountains, this rocky polar desert area has been free of ice for millions of years. The Dry Valleys receive no precipitation or moisture of any kind. Included in these ice-free valleys are the Taylor, Wright, McKelvey, Balham, Victoria, Barwick, Miers, Marshall, Garwood, and Salmon Stream Valleys. Several lakes can also be found in this region. Although frozen most of the year, these lakes are over 98 feet (30 meters) deep and contain ice that is several meters thick.

Lake Vanda, found in the central region of the Wright Valley, is approximately 3 miles (4.8 kilometers) long and up to 250 feet (76 meters) deep. This is a hypersaline lake, which allows for the water to reach a temperature of 80 degrees F (about 27 degrees C) near or at the bottom of the lake. A *thermal inversion* occurs as a result of the saline content, where the water is heated through the ice.

With the warmer temperature, a variety of algae, bacteria, and microorganisms live within its waters on and in the ice. Lake Vida, one of the largest lakes found in the Victoria Valley, is presumed to be frozen solid. Lake Vostox, buried under 2 miles (3.2 kilometers) of ice, is around the size of Lake Ontario, but is twice as deep. Its makeup has yet to be explored.

"Only the heartiest of plant life can survive [in a polar desert], and animal life is centered in coastal areas where food is abundant and available from the surrounding waters."

In the Dry Valley region, Antarctica's only river can be found. The Onyx flows only in the summer months. Lake Vanda is fed by the Onyx, which carries glacial runoff. Lake Vanda is considered a closed-basin lake because water may enter, but none will exit. Numerous shallow ponds can also be found in the McMurdo Valleys. Don Juan Pond, located in the Wright Valley, remains liquid year-round because of the hypersaline content of the water, even with surrounding temperatures reaching as low as minus 40 degrees F (minus 40 degrees C). Taylor Valley is home to Blood Falls. Iron oxide has turned the hypersaline water source blood red in some areas. Water sporadically emerges from small fissures in the ice, forming red cascades; thus, the Blood Falls name.

The rocks in the Dry Valley are home to bacteria called cyanobacteria, which can lay dormant for many years. Cyanobacteria have been found in the most barren area of Antarctica and in the high Arctic. When the rocks are damp, the cyanobacteria spring to life, producing food from the sun's energy. Cyanobacteria appear in the springtime, and with their green photosynthetic material, they turn the semi-permanent snowfields and glaciers blue-green in color. In the 1970s, the McMurdo Dry Valleys area was thought to resemble the surface of the planet Mars, with its dry desert climate and landforms. NASA utilized this Dry Valley region to test its space vehicles before launching them into space.

Polar Plant Life

Numbing cold, icy winds, darkness, and lack of shelter all contribute to the plant life of the polar desert region. When the polar desert areas surrounding the North and South Poles are dark, the plant life will either become inactive under the snow and ice or die after setting seed. These seeds will then become active the following summer, and the inactive plant life will become active as the temperature rises. Plants need sunlight, water, and soil to root. Skeletal soil composition and stony ground are common in a polar desert region. Freeze-thaw cycles result in cracking and the formation of a patterned texture to the ground. Physical weathering is the prominent erosion process. Woody plants are absent. There is not enough water in a polar desert region to sustain their growth.

In addition, the constant freezing wind without shelter stunts the growth of taller plants. Vegetation covers less than 5 percent of the ground surface, and most growth is less than 4 inches (10 centimeters) high. As a result of the severe wind and cold, plants within a polar desert have developed a vertical plant structure, growing sideways close to the ground, rather than upward. Cold desert plants are scattered throughout the region, where areas with significant sunlight will have roughly 10 percent of the ground covered with plants. All of

A polar desert is an ocean shore environment biome, meaning that animals living here use the ocean as their main food source. The arctic fox has adapted to conserve heat in this environment with round, compact bodies, and short muzzles, ears, and legs. (Thinkstock)

the plants growing in a polar desert are deciduous, and may contain spiny leaves. Permafrost is permanently frozen ground found in a polar desert region, which makes it extremely difficult if not impossible for root systems of plants to take hold, thus anchorage and absorption of water and nutrients from the ground is impossible in much, if not all, of the area.

Grasses, lichens, algae, and mosses are the main component of a polar desert plant community. Algae are plantlike organisms that live by photosynthesis. They have no leaves, stems, or roots, and by definition are not true plants. On the rocks and ice of Antarctica, brilliant red, green, and yellow patches of algae can be seen. Mosses and lichens lack a significant root system, making them *bryophytes,* or nonvascular plants. This also makes them dependent upon their immediate environment for water because they are unable to absorb any through a nonexistent root system. Lichens are small and slow growing plants, *poikilohydric* in nature, capable of surviving extremely low levels of water. They thrive in places where higher plants have difficulty growing such as bare rock, sterile soil or sand, and structures such as walls, roofs, and monuments.

Like mosses, they do not have a root system, and like mosses, lichens have the ability to enter a metabolic state called *cryptobiosis,* where they are able to dehydrate themselves to suspend biochemical activity. In this state, lichens are able to survive extreme weather conditions in areas that are often inhabitable. Lichens are made of fungus and algae living together. They are a slightly more sophisticated life form than algae, and they are among some of the hardiest living creatures on Earth. They can grow on any surface, turning them brilliant shades of orange and gold. Lichens are capable of growing on bare rocks, and they especially like darker rocks as they warm up in the sunlight quicker than pale rocks. Lichens have survived temperatures of minus 460 degrees F (minus 273 degrees C) in a laboratory setting. They also provide the starting point for the polar desert mosses. Lichens coat rocks, causing a rough surface where the mosses can then grow.

Mosses are soft, small, nonvascular, herbaceous or nonwoody plants, which tend to grow close together in clumps or mats in damp or shady areas. They have thin, wiry stems covered by leaves, and are without flowers, fruits, cones, or seeds. Rhizoids, thin rootlike filaments, are used for anchorage and absorption. Mosses reproduce with spores. Food is produced through photosynthesis, and mosses have the ability to dehydrate and rehydrate themselves as necessary for survival, sometimes for months when enough water and sunlight become available to restart the photosynthesis process. Both mosses and lichens trap windblown dirt, creating a small layer of soil.

Antarctica's only two flowering plant species grow as a result of this process, where they can find nutrients and a foothold to survive within this small amount of soil. Antarctic hairgrass and pearlwort (colobanthus) can be found on the South Orkney Islands, the South Shetland Islands, and along the western Antarctic Peninsula, where they will grow within moss-lined crevices of rock. Here, the climate is milder than within the interior of the continent. All living things in the polar desert have adapted to their environment. The biggest challenges are staying warm, finding water, locating shelter from the extreme conditions, and producing food. Polar plants have adapted to the environment, where the stems and leaves of plants are covered with tiny furry hairs. These hairs are used to trap warmer air around them. Plant life must also contend with gale-force winds and freezing blizzards. Utilizing a vertical growth pattern, arctic plants grow closer to the ground, where the wind speed is slower and the temperatures are slightly warmer.

Polar Animal Life

The biome for a polar desert is an ocean shore environment. All the animals found here feed from the ocean. Some animals may go onto land to rest or reproduce, but their food comes from the icy ocean waters. Primary producers include tiny diatoms, crustaceans, and protozoa. These microscopic unicellular creatures find their nutrients from the bottom of the ocean.

Primary consumers include small fish and squid, which in turn eat the primary producers. The most important are krill, which are small squid-like animals found in great numbers. Second consumers include blue and humpback whales, seals, and fish that feed off the krill and smaller fish. Penguins and seabirds also feed on krill, making them a very important part of the food chain of the polar region. Top predators include polar bears and killer whales. They will consume whatever they can find, often feeding on penguins and seals. In the polar desert of Antarctica, birds live along the coastal regions, feeding from the ocean. Two penguin species nest in Antarctica: the Adelie and Emperor Penguins. Large colonies can be found roosting on the rocky coasts and islands. Snow petrels and South Pole skuras, predatory birds, hunt and threaten the penguin breeding grounds. Snow petrels and South Pole skuras fly inward and nest on the rocky outcrops and in the dry desert valleys. Ross seals live deep within the sea ice and Weddell seals live on permanently frozen ice shelves.

Adaptation to the cold temperatures is essential for life. The ptarmigan is one of the few arctic bird species that does not fly south for the winter months. They change color from brown to white as the wintery weather arrives. They have developed fluffy feathers on their feet and dig tunnels in the snow to keep warm. Seals and walruses utilize blubber, along with an extra insulating feature. They have the ability to change their blood flow, allowing them to keep their blubber layer cool and their deep body core warmer. On land, walruses risk overheating because of this feature, but they rest a great deal on land as it uses less energy to stay warm than swimming in the icy waters. Arctic foxes have thick coats of fur that are able to hold a layer of warm air around their body to help maintain a normal core body temperature. Arctic fox have developed small ears and a small nose, which aids in heat preservation. Their thick fur coats even cover the bottom of their feet. Adaptation for arctic animals includes white fur. Polar bears, Dall's sheep, Peary caribou, and some gyrfalcons are naturally white in color. Some turn white in the winter months, such as the arctic hare, ermines, lemmings, arctic fox, and ptarmigans. This helps the animals blend in with their environment, making it easier to hide from dangerous predators. White fur also keeps the animal warmer because it reflects body heat back toward the body.

People of the Arctic

Eskimos, Inuit, Inuk, Inupiat, Chukchi, and Nenet are all native peoples of the Arctic. Transportation, shelter, hunting, and fishing have all changed with the advancement of technology. Snowmobiles and trucks are now used, instead of foot and dogsled traffic. *Umiaks,* which are large boats used by the people of the Arctic, are still used today, although most likely they are powered by a gas outboard motor. Hunters can be seen using both harpoons and telescopic rifles. Modern grocery stores are also available, a modern addition. Many people think of an igloo as the primary housing for this region, but people today live in wooden houses with central heating. Igloos are still used as shelters for hunting trips. Established in 1960, McGill Arctic Research Station (MARS) is located at Expedition Fjord, Nunavut, in the Canadian High Arctic. MARS is one of the longest-operating field research facilities in the High Arctic. Vital to the understanding of the Earth's polar biomes, this facility's goal is the understanding of physical and biological processes in cold polar desert and glacierized environments. Current research includes glaciology, climate change, permafrost hydrology, geology, geomorphology, limnology, planetary analogues, and microbiology.

Global Warming

Global warming is a threat to the polar deserts. The accumulation of dangerous greenhouse gases trap heat and light within the Earth's atmosphere. As a result, the temperature is rising on land and sea. The pollution causing this process has actually eroded portions of our ozone layer. This layer protects the Earth from the harmful rays of the sun. The first hole in the ozone layer was noted in 1985 over Antarctica, and another was found later over the Arctic. As a result, scientists have noted that the Arctic is shrinking, permafrost is melting, glaciers are receding, and sea ice is disappearing. The ice on Greenland is melting, and the meltwater is slowly raising sea levels. The polar ice cap is said to be melting at the rate of 9 percent per decade, and Arctic ice thickness has decreased 40 percent since the 1960s. While Antarctic ice is growing due to increased snowfall from warmer air that holds more water vapor, this net increase is less than the net loss in Arctic ice; the global balance remains a net loss of ice. In turn, low-lying coastal areas are disappearing underwater. These changes are expected to impact ecosystems around the world.

With a warmer climate, forests are predicted to expand northward into the arctic tundra, and the arctic tundra will expand into the polar desert regions. Warmer climates will also change the growth patterns of plants, as the climate will support the growth of taller and fuller varieties. The expectation is a smaller tundra, reducing breeding areas for birds and grazing animals of the region. Changes in bird, fish, and butterfly species have already been noted in some areas. The Arctic presently supports more varieties of mosses and lichens than anywhere else in the world. With an increased temperature, this number will likely decrease. Insects, diseases, and weeds are likely to increase throughout the Arctic as the temperatures gradually warm. The shifting of both plants and animals species northward is expected to increase, thus changing the current ecosystems.

Sandy Costanza

Rain Shadow Deserts

Formally defined by Stephen Marshak in 2001, a desert is a region that is so arid or dry that it contains no permanent streams, except for rivers that bring water in from temperate regions elsewhere, and supports vegetation on no more than 15 percent of its surface. In general, a desert is a landscape or region that receives an extremely low amount of precipitation, less than enough to support growth of most plants. Most deserts have an average annual precipitation of less than 10 inches (25 centimeters) per year. A common definition distinguishes between true deserts, which receive less than 10 inches (250 millimeters) of average annual precipitation, and semi-deserts or steppes, which receive between 10 inches (250 millimeters) and 16 to 20 inches (400 to 500 millimeters). Deserts can also be described as areas where loss of water by evapotranspiration is more than gained as precipitation. Measurement of rainfall alone cannot provide an accurate definition of what a desert is because being arid also depends on evaporation, which depends in part on temperature.

Deserts are sometimes classified as hot and cold deserts. Cold deserts can be covered in snow or ice—frozen water unavailable to plant life. These are more commonly referred to as tundra if a short season of above-freezing temperatures is experienced, or as an ice cap if the temperature remains below freezing year-round, rendering the land almost completely lifeless. In some parts of the world, deserts are created by a rain shadow effect in which air masses lose much of their moisture as they move over a mountain range; other areas are arid by virtue of being very far from the nearest available sources of moisture. Deserts are also classified by their geographical location and dominant weather pattern as trade wind, mid-latitude, rain shadow, coastal, monsoon, or polar deserts. Former desert areas presently in nonarid environments are paleodeserts.

Deserts take up about one-fifth (20 percent) of the Earth's land surface. Hot deserts usually have a large diurnal and seasonal temperature range, with high daytime temperatures and low nighttime temperatures (due to extremely low humidity). In hot deserts the temperature in the daytime can reach 113 degrees F (45 degrees C) or higher in the summer, and dip to 32 degrees F (0 degrees C) or lower at nighttime in the winter.

Rain shadow deserts form when tall mountain ranges block clouds from reaching areas in the direction the wind is going. As the air moves over the mountains, it cools and moisture condenses, causing precipitation on the windward side. When that air reaches the leeward side, it is dry because it has lost the majority of its moisture, resulting in a desert. The air then warms, expands, and blows across the desert. The warm, desiccated air takes with it any remaining moisture in the desert.

What Is a Rain Shadow?

A rain shadow is an area of dry land that lies on the leeward (or downwind) side of a mountain. Winds carry air masses up and over the mountain range and as the air is driven upward over the mountain, falling temperatures cause the air to lose much of its moisture as precipitation. Upon reaching the leeward side of the mountain, the dry air descends and picks up any available moisture from the landscape below. The resulting profile of precipitation across the mountain is such that rainfall and moist

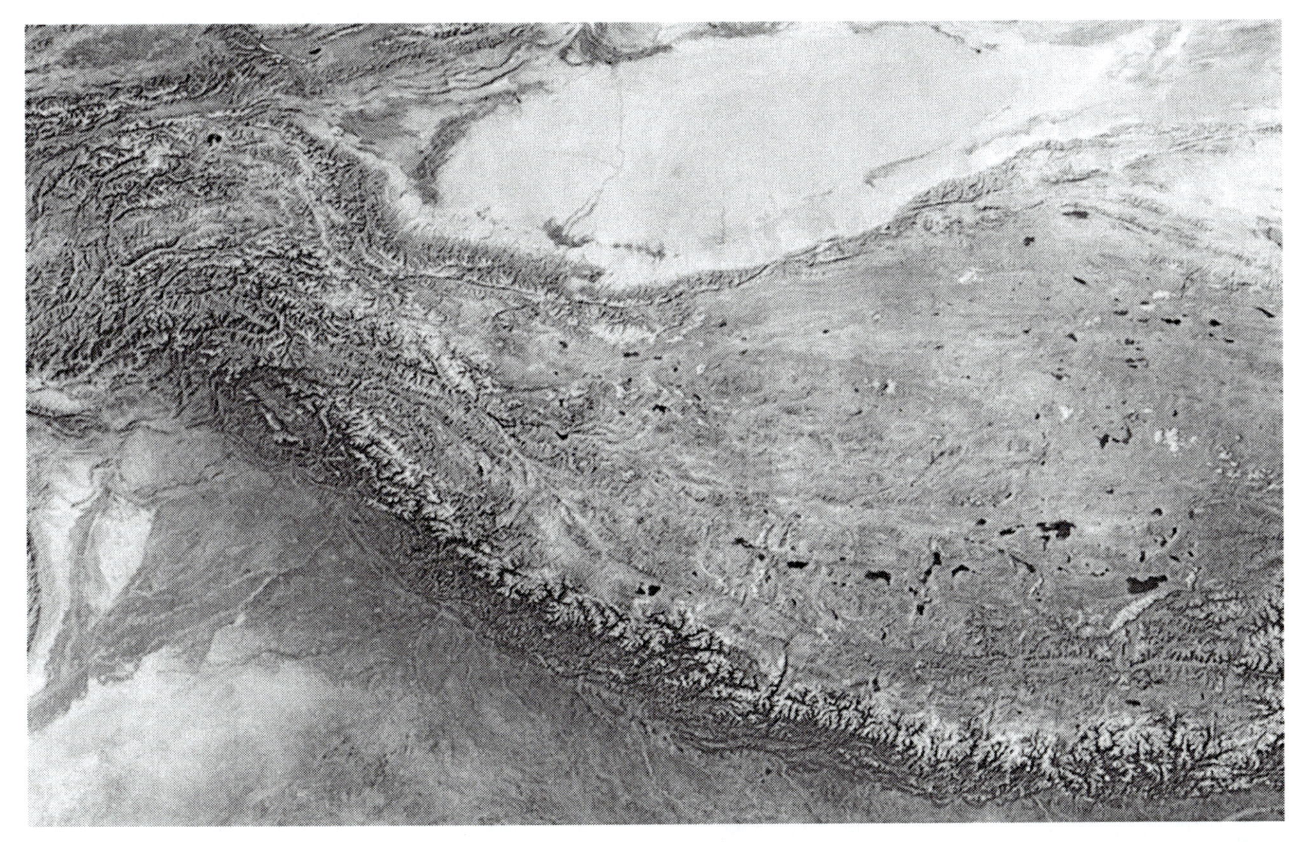

One of the best examples of a rain shadow desert occurs at the Tibetan Plateau, which is located between the Himalayan and the Kunlun mountain ranges. Rain does not occur past the Himalayas (mountain ridge shown in the darker areas in the lower left), resulting in the dry, desert conditions on the downwind side of the mountain range. (NASA)

air prevail on the windward side of a mountain range, while arid, moisture-poor air prevails on the leeward side of the mountain range. Mountain ranges act as barriers to the flow of air across the surface of the Earth. They act to squeeze moisture out of the air.

When a parcel of warm air reaches a mountain range, it is lifted up the mountain slope, cooling as it rises. This process is known as orographic lifting and the cooling of the air often results in large clouds, precipitation, and even thunderstorms. Orographic lifting is a fascinating process that keeps the windward sides of mountain ranges moist and filled with vegetation but the leeward sides dry and deserted.

The following locations are examples of dry, rain shadow regions and the mountain ranges that shield them:

- The Gobi Desert lies in the rain shadow of the Himalayas.
- The Atacama Desert lies in the rain shadow of the Andes.
- The Patagonia region lies in the rain shadow of the Andes.
- Death Valley lies in the rain shadow of the Pacific Coast Ranges of California and the Sierra Nevada.
- The city of Spokane in the state of Washington lies in the rain shadow of the Cascade Mountain Range (Spokane receives little rainfall). Seattle, Washington, lies on the windward side of the Cascades (it receives generous amounts of rainfall).

The rain shadow condition exists because warm moist air rises through orographic lifting to the top of a mountain range or large mountain. Because atmospheric pressure decreases as altitude increases, the air has expanded and adiabatically cooled to the point that the air reaches its adiabatic dew point. At the adiabatic dew point, moisture condenses onto the mountain and it precipitates on the top and windward sides of the mountain. The air descends on the leeward side, but because of the process of precipitation, it has lost much of its initial moisture. Typically, descending air also gets warmer because of adiabatic compression down the leeward side of the mountain, creating an arid region.

There are regular patterns of prevailing winds found in bands around the Earth's equatorial region. The zone designated the trade winds is the zone between about 30 degrees north and 30 degrees south, blowing predominantly from the northeast in the Northern Hemisphere and from the southeast in the Southern Hemisphere. The westerlies are the prevailing winds in the middle latitudes between 30 and 60 degrees latitude, blowing predominantly from the southwest in the Northern Hemisphere and from the northwest in the Southern Hemisphere. The strongest westerly winds in the middle latitudes can come in the Roaring Forties between 30 and 50 degrees latitude.

Asia

Examples of rain shadowing in Asia include the Tirunelveli (India), cut off from the monsoons by the Agasthiyamalai hills, creating a rain shadow region. The peaks of the Caucasus Mountains to the west, the Alborz Mountains to the south, and the ranges tied to the Himalayas to the east rain shadow the Karakum and Kyzyl Kum Deserts east of the Caspian Sea, as well as the semi-arid Kazakh Steppe. The Judean Desert and the Dead Sea are rain shadowed by the Judean Hills. The Himalaya and connecting ranges also contribute to arid conditions in central Asia, including Mongolia's Gobi Desert, as well as the semi-arid steppes of Mongolia and north-central to northwestern China. The Great Indian Desert, or Thar Desert, is bounded and rain shadowed by the Aravalli ranges to the southeast, the Himalaya to the northeast, and the Kirthar and Sulaiman ranges to the west.

South America

The Atacama Desert in Chile is the driest non-Antarctic desert on Earth because it is blocked from moisture on both sides (by the Andes Mountains to the east and high pressure over the Pacific at a latitude that keeps moisture from coming in from the west). The Argentinian wine region of

Mendoza is almost completely dependent on irrigation, using water drawn from the many rivers that drain glacial ice from the Andes. The nearby Chilean wine region of Valle Central, on the other hand, is situated on the Chilean side of the Andes and experiences a maritime climate. Patagonia is rain shadowed from the prevailing westerly winds by the Andes range and is arid (e.g., in Santa Cruz few spots are capable of cultivation, the pastures being poor, water insufficient, and salt lagoons fairly numerous). The Guajira Peninsula in northern Colombia is in the rain shadow of the Sierra Nevada de Santa Marta and despite its tropical latitude is almost arid, receiving almost no rainfall for seven to eight months of the year and being incapable of cultivation without irrigation.

North America

Most rain shadows in North America are because of mountain ranges, notably the Sierra Nevada and Cascades, that intercept rain and snowfall that would otherwise reach a valley in the lee of the mid-latitude prevailing westerlies. The deserts of the Basin and Range Province in the United States and Mexico, which includes the dry areas east of the Cascade Mountains of Oregon and Washington, and the Great Basin, which covers almost all of Nevada and parts of Utah, are rain shadowed. The dry precipitation regime of the Great Plains of western Canada and the central United States can be attributed in large part to the rain shadow of the North American Cordillera. The Mojave, Black Rock, Sonoran, and Chihuahuan Deserts are all rain shadowed. The aptly named Death Valley in the United States, behind both the Pacific Coast Ranges of California and the Sierra Nevada range, is one of the driest places on the planet.

"A rain shadow is an area of dry land that lies on the leeward (or downwind) side of a mountain."

The Colorado Front Range is limited to the rainfall that makes it over the continental divide. The Rocky Mountain Front Range in Montana shows areas of limited precipitation as long as the systems passing over the mountain range come from the west. The clouds dry out considerably by the time they reach the peaks of the mountain range. The east slopes of the Coast Ranges in central and southern California also cut off the southern San Joaquin Valley from enough precipitation to ensure desertlike conditions in areas around Bakersfield. The areas to the east of the Coast Mountains and the Cascade Range in the province of British Columbia, Canada, and the American states of Washington, Oregon, and Idaho are located in a rain shadow. The Dungeness Valley around Sequim, Washington, lies in the rain shadow of the Olympic Mountains. The area averages 10–15 inches of rain per year, less than half the amount received in nearby Port Angeles and approximately 10 percent of that which falls in Forks on the western side of the mountains.

The rain shadow effect even occurs in the eastern United States. Although much more humid than any obvious deserts or steppes, the Shenandoah Valley, mostly in western Virginia, lying between the Blue Ridge and the Appalachian Mountains, is drier than areas to the east and west because the modest mountains reduce rainfall within the valley.

Europe

The Pennines of northern England, the Welsh Mountains, and the Highlands of Scotland create a large rain shadow that covers almost the entirety of the eastern United Kingdom, with Glasgow and Manchester, for example, receiving around double the rainfall of Edinburgh and York, respectively. The contrast is even stronger farther north, where Aberdeen gets around a third the rainfall of Fort William or Skye. The Fens of East Anglia receive rainfall amounts similar to Seville. The Cantabrian Mountains make a sharp divide between "Green Spain" to the north and the dry central plateau. The northern-facing slopes receive heavy rainfall from the Bay of Biscay, but the southern slopes are in rain shadow. The most evident effect on the Iberian Peninsula occurs in the Almería, Murcia, and Alicante areas, each with an average rainfall of less than 12 inches (300 millimeters) and the driest spot in Europe, mostly a result of the mountainous range running through their western side, which blocks the westerlies.

Some valleys in the inner Alps are also strongly rain shadowed by the high surrounding mountains. The Plains of Limagne and Forez in the northern Massif Central, France, are also relatively rain shadowed (mostly the plain of Limagne, shadowed by the Chaîne des Puys, with up to 78 inches, or 2,000 millimeters, of rain annually on the summits and below 24 inches, or 600 millimeters, on Clermont-Ferrand, which is one of the driest places in the country). The Piedmont wine region of northern Italy is rain shadowed by the mountains that surround it on nearly every side; Asti receives only 20–21 inches (527 millimeters) of precipitation each year. Athens is shielded strongly by mountains from the strong moisture-bearing winds of the Adriatic Sea and receives only a quarter the rainfall of most of Albania.

The Scandinavian Mountains create a rain shadow for lowland areas east of the mountain chain and prevent the oceanic climate from penetrating farther east; thus Bergen west of the mountains receives almost 89 inches (2,250 millimeters) precipitation annually while Oslo receives only about 30 inches (760 millimeters), and Skjåk, a municipality situated in a deep valley, receives only 11 inches (280 millimeters).

Africa

The windward side of the island of Madagascar, which sees easterly onshore winds, is wet and tropical, while the western and southern sides of the island lie in the rain shadow of the central highlands and are home to thorn forests and deserts. The same is true for the island of Réunion. The formation of the Atlas Mountains has been deemed at least partially responsible for the climatic change that eventually created the Sahara. There is a strong rain shadow effect to the south side of the mountains.

Oceania

New Caledonia lies astride the Tropic of Capricorn, between 19 degrees and 23 degrees south latitude. The climate of the islands is tropical, and rainfall is brought by trade winds from the east. The western side of the Grande Terre lies in the rain shadow of the central mountains.

Hawaii also has rain areas of the islands classified as desert. Orographic lifting produces the world's second-highest annual precipitation record, 500 inches (12.7 meters), on the island of Kauai; the leeward side is understandably rain shadowed. The entire island of Kahoolawe lies in the rain shadow of Maui's East Maui Volcano.

In New Zealand is to be found one of the most remarkable rain shadows anywhere on Earth. On the South Island, the Southern Alps intercept moisture coming off the Tasman Sea. The mountain range is home to significant glaciers and about 250 inches (6,300 millimeters) to 350 inches (8,900 millimeters) liquid water equivalent per year. To the east and down slope of the Southern Alps, scarcely 30 miles (50 kilometers) from the snowy peaks, yearly rainfall drops to less than 30 inches (760 millimeters) and in some areas to less than 15 inches (380 millimeters).

In Tasmania, one of the states of Australia, the central Midlands region is in a strong rain shadow and receives only about a fifth as much rainfall as the highlands to the west. In New South Wales and Victoria (also states of Australia), the Monaro is shielded by both the Snowy Mountains to the northwest and coastal ranges to the southeast. In Victoria, the area around Port Phillip Bay is in the rain shadow of the Otway Ranges. The area between Geelong and Werribee is the driest part of southern Victoria: whereas the crest of Otway Ranges receives 79 inches (2,000 millimeters) of rain per year, the area around Little River receives as little as 17 inches (420 millimeters) annually, which is as little as Nhill or Longreach.

Western Australia's Wheatbelt and Great Southern regions are shielded by the Darling Range to the west: Mandurah, near the coast, receives about 28 inches (700 millimeters) annually. Dwellingup, about 24 miles (40 kilometers) inland and in the heart of the ranges, receives over 39 inches (1,000 millimeters) a year, while Narrogin, 80 miles (130 kilometers) farther east, receives less than 20 inches (500 millimeters) a year.

Ghazala Nasim

Trade Wind Deserts

The desert ecosystems of the world are often considered barren places where conditions change slowly and that support little life. In fact, these ecosystems, which cover about 20 percent of the Earth's surface and are broadly characterized by low annual precipitation rates and a high proportion of bare soil, support an incredible array of highly specialized plants and animals. They are also home to many diverse cultures and livelihoods and play a significant role in the global environment and economy.

However, desert ecosystems are coming under increasing human and environmental pressures that present serious challenges for both the ecosystems and the human populations they support. This article provides a description of biophysical and biodiversity features as well as the challenges and opportunities presented by deserts, particularly as they apply to a specific type: the trade wind deserts of the world. Deserts occur on virtually every continent, including Antarctica. They are classified into several types using different criteria, such as aridity or bioecological productivity.

The U.S. Geological Survey's (USGS) classification of deserts, based on geographical location and predominant weather patterns, includes trade wind, mid-latitude, rain shadow, coastal, monsoon, and polar deserts, as well as paleodeserts—former deserts that are now located in nonarid environments. The most important defining feature of deserts is their aridity, or the ratio between mean annual precipitation and mean annual evapotranspiration, and thus water is the main limiting factor in biological processes in deserts. Desert soils have little or no organic matter, are rich in nutrients, and yet they are unproductive in the absence of water and only become productive after rainfall events.

The trade wind deserts form in two fairly distinct belts on either side of the equator in the region known as the horse latitudes, where the northeasterly and southeasterly trade winds typically form. These winds are the result of warm air rising at the equator and then moving toward the poles as part of the Hadley cell circulation. As this warm air rises, it cools, compresses, and descends near 30 degrees latitude on both sides of the equator, creating a band of high atmospheric pressures.

Winds blowing from the northeast toward the equator in the Northern Hemisphere are deflected to the right due to the Coriolis force and are known as the trade winds. The trade winds blowing from the horse latitudes around 30 degrees north and south meet at the Intertropical Convergence Zone (ITCZ) in an area of calm winds known as the doldrums. The trade winds are dry winds that dissipate cloud cover, allowing more sunlight to heat the land and, in combination with the predominant high pressures caused by the descending air at the horse latitudes, have led to the formation of most of the major deserts of the world.

The world's largest desert, the Sahara of north Africa, also known as the Great Desert, is a typical trade wind desert even though parts of it to the north fall outside the direct influence of the trade winds. Other deserts that fall within this category include parts of the Namib and Kalahari Deserts in southern Africa, the Atacama Desert in South America, as well as large swaths of the Arabian and Middle Eastern deserts. Given the wide range of climatic conditions among these different regions, the temperature regimes and humidity characteristics of trade wind deserts in relatively close proximity to oceans tend to differ greatly from those located in continental interiors.

Diversity of Deserts

Based on climatic conditions, approximately one-third of the surface of the planet is classified as desert. As a result, the diversity of desert landscapes from a topographic and geomorphological point of view is quite astounding, ranging from variously shaped dunes, plains, and mountains to lakes, rivers, oases, and deltas. Even though typically one thinks of dunes and seas of sand when thinking of deserts, the actual proportion of the different trade wind deserts covered by eolian sand dunes varies greatly.

It is as high as 30 percent in the Sahara, southern Africa, and Arabian Peninsula and as low as less than 1 percent in South America. The landscape of most trade wind deserts is the result of

Tuareg nomads leading tourists on a camel caravan among the wind-swept Erg Chebbi dunes of the Saharan desert region of Morocco. Nomads living in trade wind deserts survive in these harsh conditions by making a living mainly by camel herding, caravan trading, crop growing, and hunting and gathering. (Thinkstock)

millennia of geologic and geomorphologic evolution and sculpting by winds, water, and weathering processes and one of the features under protection as world heritage sites around the world.

Their landscape is thus unique as it continues to be actively shaped and modeled primarily by wind processes that are responsible for creating a wide variety of desert dunes, yardangs, pans, deflation surfaces, and the material for dust storms. Episodic rainfall events can have a rather major landscape-sculpting influence, especially during episodes of high intensity, and lead to the formation of a variety of features such as badlands, ephemeral stream channels locally known as wadis, alluvial fans and debris flows, pediments, and natural arches. Third, different forms of physical and chemical breakdown of rocks and sediments (known as weathering) also create a variety of physical landforms such as desert varnishes and karst, caverns and crusts that in turn support different plant and animal forms and create environmental niches for unique adaptations and endemic characteristics.

Most of the present-day trade wind deserts became established as deserts during the mid-Holocene geologic time period, when changes in incoming solar radiation associated with shifts in Earth's orbit led to changes in global atmospheric circulation and shifts in precipitation regimes. Trade wind desert ecosystems are defined by their overall low levels and high spatial and temporal variability of water inputs that are thus intermittent.

Because water inputs pulsate throughout the year (rainfall is sporadic and mostly unpredictable and mainly occurs in "flashes"), one of the defining features of these ecosystems for both plant and animal life is their ability to create storages and reserves of water for later use. The larger these reserves, the more water that can be accessed in the soils and possibly the more abundant the life forms supported. If there is increased water availability, then there exists increased potential for more stable and potentially less sensitive characteristics of a system, especially in the absence of human-induced disturbances. Another defining feature of trade wind deserts is the spatial stratification in response to soil water availability. For instance, if water is only available in the soil at a depth less than 2 inches (5 centimeters), algae are the most likely life forms to colonize and/or survive.

Plant Life

Ephemeral plants have deeper roots that can access water in the soil at depths between 2 and 20 inches (5 and 50 centimeters), while any water deeper than 20 inches (50 centimeters) can potentially only be accessed by shrub species that develop deep rooting systems.Despite the harsh environmental conditions, deserts are home to complex ecosystems with unique, diverse, and fragile plant and animal associations. As a result of the vertical soil water and nutrient stratification and availability and harsh climatic and environmental conditions (wind, heat, and temperature), most plant species of trade wind deserts have developed a series of adaptations to aridity, climate variability, unpredictable rainfall pulses, and scant summer and winter patterns of precipitation that allow them to survive and sometimes even thrive. Plants escape, retreat short term, or learn how to tolerate desert conditions, or they may only tolerate them to very limited degrees by colonizing oases or wetland locations with perennial water sources.

The plant tolerance adaptations to arid conditions are of a morphological or physiological nature. Usually they are represented by exploitation of favorable microclimates and ephemeral life cycles accompanied by the capacity to leave behind drought-resistant forms of propagation and reproduction (bulbs, seeds, or dormant shrubs). Morphologically, plant adaptations to desert conditions are exemplified by cacti. These plants develop thick cuticles, store and harvest water from dew and fog, and present varied coloration or different shapes and sizes. Finally, among the physiological adaptations of plants, the most common are tissue tolerance to high temperatures, tolerance to cold, nighttime photosynthesizing to avoid water loss, columnar growth to maximize exposure to sunlight in the morning and late day and avoid the excessive heat during midday, or the loss of leaves during the extent dry seasons for woody species. Plants are generally represented by perennial grasses, some succulent species (cacti), and dwarf species of shrubs, as well as many species of colorful lichens depending on the geographic location and the degree of local endemism. As deserts transition toward less arid ecosystems at their margins, for instance, or are characterized by less arid conditions, annual species of grasses begin to establish themselves. They tend to lie dormant in the form of seeds during periods of extended droughts and, with any rainfall above about 0.8 inch (20 millimeters), they sprout rapidly, transforming the landscape into a sea of grass.

Animal Adaptation

Animal species have also learned to adapt in various ways to life in the trade wind deserts. Some of the most common adaptations, as is the case with plants, are centered around increased tolerance and specialization, and animals add behavioral adaptations to the morphological and physiological adaptations displayed by plants. For example, many vertebrates go into prolonged dormancy to survive long periods without water, while others migrate long distances to take advantage of resources available at different times in different locations. Other animals take refuge in burrows beneath the sand, while others sidewind to traverse hot surfaces.

Morphologically, ears, tails, feathers, and differently colored pelages are used to cool or warm animals at different times of day, and other adaptations include water uptake from food, tolerating dehydration, optional hypothermia, or dormancy. Overall, the trade wind deserts are inhabited by relatively few large mammals, mainly because they lack the ability to access and store sufficient amounts of water and withstand the heat and huge temperature fluctuations between day and night. Thus, trade wind deserts are mainly populated by nonmammalian animals such as reptiles, some with high rates of endemism such as desert lizards, several types of sand lizards, and the day gecko and barking gecko. Second are rodent species, such as the gerbil, bats, or an eyeless mole that can swim through the loose, dry sands. Few ungulates such as the gemsbok or springbok in the Namib Desert can be present, and relatively small predators such as cheetahs, foxes, jackals, or hyenas are also present depending on location. Finally, biotic species interactions are also an important component of adapting to and living with the harsh environmental conditions afforded by the world's trade wind deserts.

"Trade wind deserts encompass arid and hyper-arid regions of the world and expand in the areas of the globe crossed by the trade winds."

Most of these interactions are of a positive nature; for example, some plants may serve as "nurse plants" to other germinating seedlings or may provide shelter or shade for animal species. Animals, on the other hand, can assist flora species with seed dispersal or function as effective pollinators. As with the flora and fauna, the peoples of the world's trade wind deserts have not only inhabited these restrictive environments since times immemorial, but they have also uniquely adapted their livelihood, lifestyles, culture, technology, and behavior and have innovated superbly through time in order to call them home.

Human Desert Survival

Unlike plants and animals, people have been able to thrive in deserts, despite meager physiological adaptations, by actively modifying their microenvironments. Examples include the use of cooling shelters and livelihood strategies that fully maximize the available resources, such as hunting and gathering, nomadic pastoralism, and, on conditions, irrigated agriculture. Humans have also adapted their clothing, diets, and housing and have been involved in traditional ways of resource use and management that allow for continued sustainability in otherwise adverse conditions. The hunters and gatherers of the world's largest deserts have an impressive repertoire of plant and animal use and knowledge, even though in recent years their livelihoods are increasingly becoming more mixed.

The San tribes of the Kalahari and Topnaar people of the Namib Desert in southern Africa are the emblem representatives of this ancient lifestyle in today's world, and their traditional way of living is surviving in increasingly isolated spatial pockets. The nomads of the Sahara such as the Tuareg, Tibbu, and Moor tribes are among the best-known pastoralists making a livelihood in these ecosystems, and they often combine hunting and gathering with domesticated-animal grazing and crop growing.

Humans have also been the most active factor shaping and transforming these desert ecosystems, and their impact is becoming more important as population numbers grow and natural resources, especially relatively large animal species in some regions, are dwindling. The trade wind desert landscape covers relatively vast areas of largely undisturbed habitat, represented primarily by sand and rock, with occasional small areas of permanent vegetation. As a result, the most ecosystem and vegetation degradation is found where water (oases, wetlands, or fringe ecosystems and ecotones) is present and these habitats may be therefore heavily altered by human activities, mainly through vegetation removal.

One of the most widespread and potentially least understood human impacts in dry regions is desert expansion, a phenomenon appropriately called desertification. Additionally, especially in regions where agriculture and/or agro-pastoralism are commonly practiced, the continual loss of fertile land on the outskirts of arid areas is also becoming a troubling development, as is overgrazing, which is the loss of protective cover of plant life, leading to increased wind and water erosion. Another major issue affecting these ecosystems is the continued loss of biodiversity, and especially the declines in the remaining populations of large mammals that are adapted to desert conditions, due to hunting for food or recreation.

A prime example is the addax of the Sahara Desert, which has been hunted to near extinction, as well as many other desert-adapted antelopes that are seriously threatened or endangered due to overhunting. A more recent development, desert groundwater pumping from ancient aquifers with extremely slow recharge rates for irrigated agriculture, is becoming increasingly necessary and desirable in countries such as Tunisia and Algeria. Such developments not only deplete aquifers that formed thousands of years ago and may lead to subsidence, but they also create a suite of problems such as soil salinization and degradation because of deficient drainage. From a conservation perspective, very small areas of the trade wind deserts are under official protection status. This may be partially because of the low population and impracticality of defining borders over this vast area as, for instance, fewer than 2 million inhabitants reside throughout the entire Sahara Desert

Narcisa G. Pricope

Forest Biomes

Boreal Forests and Taiga

The boreal forest is the world's largest terrestrial biome, characterized by coniferous forests. It makes up 29 percent (about 20 million hectares) of the world's forest cover. It occurs in a broad band (50–70 degrees north latitudes) across northern North America, Europe, and Asia in regions with cool temperatures and adequate moisture. It forms a nearly unbroken belt of trees that circles the Earth. Fifty-eight percent of the Earth's boreal forest is in Russia, 24 percent in Canada, and 11 percent in the United States. Boreal forest is also known as taiga or northern coniferous forest. The term *taiga* is of Russian origin and means "marshy pine forest." The term *boreal forest* is sometimes used to refer to the more southerly part of the biome, while the term *taiga* is often used to describe the more barren areas of the northernmost part.

The boreal forest has a subarctic climate characterized by long, cold winters and short, cool to mild summers. The main seasons in the forest are winter and summer. The spring and autumn are so short that we hardly know their existence. Long, cold winters last up to six months, and short summers last one to three months. There is a very large temperature range between seasons. Summer temperatures are mostly cool, averaging from about 19 degrees F to 70 degrees F (minus 7 degrees C to 21 degrees C), and winter temperatures range from about minus 65 degrees F to 30 degrees F (minus 54 degrees C to minus 1 degree C). Mean annual precipitation is relatively low (average 12–33 inches, or 300–850 millimeters), but low evaporation rates make the region a humid climate, which is sufficient to sustain the dense vegetation growth. The precipitation arrives mostly as rain in summer, but also as snow in winter. Annual snowfall varies from about 15 to 40 inches (40 to 100 centimeters). The average length of the growing season in boreal forests is about 130 days, and the length varies in different regions. For example, growing season varies from 80 to 150 days in the boreal plains in Canada, and from 100 to 140 days in the boreal shield.

As a result of its glacial history, most of the boreal forest's soil is geologically young and nutrient poor. Soil development progresses in the forest are very slow because of cold climates. Fallen leaves and moss can remain on the forest floor for a long time in the cool, moist climate, which limits their organic contribution to the soil; acids from evergreen needles further leach the soil. As a result, plants cannot efficiently use much of this mineral content in the soil. There are only lichens and mosses growing on the forest floor because of the acidic soil. In some parts of the forest, permafrost exists, which makes it difficult for plants to root deep into the soil generally.

Overlying formerly glaciated areas and areas of patchy permafrost, the forest is a mosaic of successional and subclimax plant communities that are sensitive to varying environmental conditions. The boreal is typically dominated by a few species of coniferous trees, such as spruce (genus *Picea*), fir (*Abies*), pine (*Pinus*), and larch (*Larix*). There are two main types of boreal forest. One is called "closed forest," which has trees that grow close together and a shaded, often moss-covered forest floor. The other is called "open, lichen woodland," which has trees spaced far apart, with lichen growing in open areas. Decomposition rates are relatively slow due to cool temperatures. Leaf litter in the boreal decomposes 60 times slower than that in the rainforest. Slow decomposition results in an accumulation of peat and humic acids, which render many soil nutrients unavailable for plant growth and limit diversity and productivity of the few tree species. Often the canopy is not dense, and a well-developed understory of acid-tolerant shrubs, mosses, and lichens may be present in the most mesic sites.

"Despite its global significance, the boreal forest is in great danger today."

Compared with the tropical or temperate biomes, there are not a lot of plant species in the boreal because of the harsh conditions. Flora consist mostly of cold-tolerant evergreen conifers with needlelike leaves, such as pine, fir, spruce, and hemlock. In North America, one or two species of fir and one or two species of spruce are dominant. Across Scandinavia and western Russia, the Scots pine is a common component of the boreal, while the boreal of the Russian Far East and Mongolia is dominated by larch, a deciduous conifer species. It also has some broadleaf deciduous trees such as birch, aspen, and poplar, mostly in areas escaping the most extreme winter cold. Underneath the trees, some shrubs (e.g., blueberries, willows, and alders), herbs, mushrooms, lichens, and mosses grow.

The Boreal Habitat

The boreal forest provides habitat for diverse wildlife, including a large range of birds, mammals, and insects. Fauna include woodpeckers, hawks, moose, bear, weasel, lynx, fox, wolf, deer, hares, chipmunks, shrews, and bats. Over 300 bird species live in the boreal in the summertime, while only 30 species stay for the winter. The rest are summer visitors, and they migrate there to nest and feed because the boreal has millions of insects in summer, providing abundant food for birds. The forest also provides birds with a good place to raise their young, with generally more space and less competition with other birds than in the tropic and temperate forests. In fact, about 30 percent of North American migratory birds depend on the boreal for breeding habitat. The boreal is also rich with large herbivorous mammals (e.g., moose and caribou), carnivorous (e.g., lynx, coyote, foxes, and weasels), and omnivores (e.g., bears and raccoons). Although the boreal does not have as many animal species as the tropical or the temperate biomes, it does have millions of insects in the summertime.

For example, Canada's boreal forest includes an estimated 32,000 species of insects. These insects play an important role in the forest—they decompose litter, supply food for birds and small animals, and eliminate diseased trees. Unfortunately, a number of wildlife species are threatened or endangered with extinction in the boreal, including the Amur (Siberian) tiger, the Amur leopard, woodland caribou, American black bear, grizzly bear, and wolverine. For example, the Amur tiger, the largest of all cats, has all but disappeared from China, Korea, and most of Siberia, and a small population (less than 400 lives) in the far east of Russia. Habitat loss, fragmentation, and extensive poaching for illegal trade are the primary cause of decline of these species.

Mainly covered with forest, the boreal zone also contains a large amount of lakes, rivers, wetlands, bogs, and fens that support beaver, fish, waterbirds, and many other wildlife. Especially in Canada, the boreal contains 25 percent of the world's wetlands, extending over more than 459,460 square miles (1.19 million square kilometers). It has been estimated that Canada's boreal region contains over 1.5 million lakes with a minimum surface area of almost 10 acres (40,000 square meters).

The extensive undammed watered landscape of the boreal serves as last refuge for many of the world's sea-run migratory fish, including half of the remaining populations of North American Atlantic salmon. The watered landscape also makes vast and critical contributions to the global environment by stabilizing climate and feeding the productivity of the world's oceans. The wetlands and peatlands store an estimated 147 billion tonnes of carbon, more than 25 years worth of current man-made emissions. The input of freshwater from boreal rivers to the Arctic and other northern seas is critical to forming sea ice, which cools the atmosphere and provides the basis for much of arctic marine biodiversity.

A Threatened Treasure

Around the world, boreal forests are highly valued by humans. They supply many ecosystem goods and services. They act as a reservoir for maintaining biological and genetic diversity. Boreal forests store carbon, purify air and water, and help regulate regional and global climates. They also provide food, renewable raw materials, and outdoor recreational places for human use. In Canada alone, the boreal have long been central to the natural environment, history, culture, and economy of the country. More than 4 million people live in Canada's boreal, and the boreal sustains over 7,000 forestry businesses and 400,000 jobs. For Canada's aboriginal people, the boreal sustains the traditional lifestyle and provides many plant resources that hold special dietary, medical, economic, and spiritual value.

Boreal forest has been influenced and shaped by natural disturbances such as wildfires and insect

A Norwegian boreal forest in early winter at about 1,100 feet above sea level in the hills near Trondheim. The spruce trees are slim and grow in a spike shape, common in a taiga forest. Characterized by coniferous trees, the boreal forest is the world's largest terrestrial biome. It covers a broad band across northern North America, Europe, and Asia in regions with cool temperatures and adequate moisture, forming a nearly unbroken belt of trees that circle the Earth. (Wikimedia)

infestations, which create openings or clearings in the forest. The forest provides a large amount of fuel for fires. Conifer needles and branches are highly flammable, especially during dry weather. The number and distribution of forest fires and area burned varies annually across the boreal. In Canada, there are about 8,000 forest fires recorded every year, which burn, on average, about 2.1 million hectares of the forest annually. That average annual burn area is equivalent to more than three times the current annual industrial timber harvest. Fire is a recurring event in the boreal. The recurrent cycle of large, damaging fire occurs approximately every 70 to 100 years. Fire is a natural and essential part of the boreal forest ecosystem, and many boreal plants and animals flourish in the aftermath of wildfires. For example, the intense heat of the fire releases seeds from hardy cones of black spruce, jack pine, and lodgepole pine. Moose invade the burned area after a fire to find food (various parts of deciduous trees and shrubs). Insects are big consumers of trees. Repeated attacks by insects can kill a tree or weaken it. Acres of trees may be partly defoliated in a single season by insects such as the spruce budwork or the larch sawfly.

Despite its global significance, the boreal forest is in great danger today. Industrial logging, mining, hydroelectric dams, and global warming are among the largest threats to the boreal forest. The boreal forest is being clear-cut to create building materials and consumer products such as toilet paper, office paper, books, and catalogs. Approximately half of the boreal forest is under lease to forestry companies. In Canada, it was estimated that the annual harvest in the boreal was about 700,000 hectares per year, equivalent to about 0.2 percent of the total Canadian boreal forest. In addition to unsustainable logging, exploration and development of oil and natural gas reserves is another large threat to the boreal. From Alaska to Canada to Russia, it is estimated that vast amounts of petroleum products lie under these forests. The high demand for fossil fuels nowadays is pushing exploration and development in the boreal. Global warming is another serious threat. The boreal biome is expected to experience the greatest increase in temperature as a consequence of global warming, especially during the last three decades. Therefore, the boreal forests should suffer most from the climate change. Scientists have already noted that the impacts of climate change on the forest are expected to be largely detrimental. The warmer and drier weather in the summer is also expected to increase tree mortality and reduce tree growth and reproduction. The effects of climate change are expected to create unprecedented levels of insect infestation, drought, and forest fires in the boreal forest.

Jian Zhang

Mediterranean Forests

Mediterranean forests occur in Mediterranean-type climates characterized by mild and wet winters and dry summers. Being part of the Mediterranean-type biome, they are found in Europe, California and northwest Mexico, and a few small areas in Australia, Chile, and South Africa. In western North America, they are called chaparral. In Spain, the most common name is *matorral*. Farther east in the Mediterranean Basin, they are referred to as *garrigue*. Meanwhile, in South Africa they are recognized as *fynbos*, while Australians refer to at least one form of it as *mallee*. Mediterranean forests are classified into three main types: sclerophyllous forests, broadleaved forests, and coniferous forests. These are shelter for plant types from various origins that have attained extraordinary levels of both diversity and endemism, composing 20 percent of the Earth's plant species.

The fynbos alone feature 8,600 different plants, of which nearly 70 percent are endemic. The Mediterranean Basin harbors about 25,000 species of vascular plants, of which 50 percent are endemic to the region. In the arid Australian southwest, around 2,500 vascular plant species exist that are found nowhere else in the world. In California, there are 3,500 plant species, among which more than 2,000 are endemic. The worldwide cover of these forests is nearly 1.5 percent of the total wooded area. Mediterranean forests include a diversity of formations, with different levels of woody vegetation and open areas. They are also highly diverse in terms of the growth form, morphology, physiology, and phenology of the dominant trees in each region. These forests are dominated mainly by drought resistant, hard-leaved scrub of low-growing woody plants. They consist of populations of broadleaved and evergreen sclerophyllous such as the oak and mixed sclerophyll forests distributed in California and the Mediterranean region, eucalyptus growing in southwest Australia, and the nothofagus populations resident in central Chile. The latter enjoy moist conditions during summertime, as they are often residents of the riparian areas.

More than two-thirds of these forest populations are concentrated in the Mediterranean region, and coniferous forests occur especially around the Mediterranean Basin; the rest are scattered over the other continents. Many of the Mediterranean forests' trees have thick, tough bark that is resistant to fire. Trees and shrubs are typically evergreen, rich in essential oils with small and tough leaves, which conserve both water and nutrients. During winter and early spring when rainfall is more abundant, annual plants are common. The highly diverse plants and animals show several adaptations to drought. To avoid drought and fire, most herbaceous plants grow during the cool, moist season and then die back in summer. As Mediterranean ecosystems are subject to periodic fires, many plants produce seeds that will only germinate after fire, while other plants with fire-resistant roots can quickly resprout because of nutrient reserves in their roots.

Throughout history, high population densities coupled with a long history of human occupation have left eternal prints on these forests and they have had substantial influence on the structure of the landscapes in Mediterranean forests. Anthropogenic factors, including forest clearance for timber and fuelwood, as well as agriculture and setting fires to control woody species and encourage grass, harvesting brush for fuel, and grazing and browsing by domestic livestock, induced changes in the vegetation communities. These factors have left imprints on the vegetation cover with maquis, which is a sort of dense shrub formation including wild olive, myrtle, laurel, and juniper; and garrigue, comprised of aromatic low shrub formations such as lavender, myrtle, rosemary, cistus, marjoram, and thyme and a mix of occasional higher shrubs. It is difficult to define whether these formations are original Mediterranean vegetation or just degraded remnants of better forest types.

Initiatives for the protection of Mediterranean forests and their diversity have been launched. Mediterranean forests are either protected under the forestry law or declared to be nature reserves. Protected areas in these regions are often small reserves or recreational parks. Because of the high diversity and endemism of the flora in particular, these may be of great importance for biodiversity conservation. In South Africa, the recently declared 16,000-hectare Cape Peninsula National Park is home to more than 2,000 vascular plant species, of which 90 occur nowhere else and more than 140 are considered threatened with extinction.

Mediterranean Region

The Mediterranean Basin, or the "Old World," lies around the Mediterranean Sea, stretching over an area of around 772,204 square miles (2 million square kilometers). It is embraced by the Lebanese shores in the east, the European folded mountains in the north, and the African Sahara deserts in the south. The largest among the five Mediterranean regions, it is considered as a biodiversity hot spot, harboring a considerable share of the world's terrestrial biodiversity with high total and local species richness and a high spatial heterogeneity. The forests are highly diverse in their architecture, appearance, and woody plant species composition: there are approximately 290 species of trees in the region, of which more than 200 are endemic. This biodiversity shows high resilience to disturbance, which has been related to their evolutionary history.

Mediterranean forests and maquis cover about 7.5 percent of the countries' total land. Around 60 percent of the forest area is concentrated in five northern countries: Greece, Italy, France, Spain, and Portugal. While northern forests cover 77 percent,

The Backbone Trail in the Santa Monica Mountains, California, winds through brushy, chaparral-covered slopes. The region has a Mediterranean-type climate characterized by mild and wet winters and dry summers. Similar biomes are found in Europe, northwestern Mexico, and a few small areas in Australia, Chile, and South Africa. In western North America, they are called "chaparral," whereas in Spain, they are called "matorral." (National Park Service)

southern and eastern forests cover 8 percent and 15 percent, respectively. The prevalence of forests in the north is due largely to natural conditions that allow faster and better growth. In contrast, forests are scarce in southern and eastern countries, where the climate is very hot and dry. Southern forests are scattered in only three countries: Turkey, Morocco, and Algeria. Mediterranean forests are still more abundant in Mediterranean mountains than in the foothills and lowlands. They are shelter to a surprising number of conifers, including pines, junipers, cypress, cedars, firs, and the Barbary thuja.

The old-growth forests at higher elevations in much of the area probably combine conifers and broadleaved species in intricate mixtures with many species. Oak forests dominate in the lowlands of the basin, but as one goes toward the peaks, chestnut, fir, pine, and juniper take over. In this region, pure forest stands of pines or evergreen oaks are not the result of natural occurrence and dynamics but instead reflect human interventions throughout history.

In the southern countries, evergreen oak and conifers such as Aleppo pine, Calabrian pine, thuya, and junipers are resident tree species occupying 9 million hectares. Half of these forests are found in Morocco and most of the rest in Algeria and Tunisia, where wild olive, carob, and arganier are more prevalent compared with other countries. Sclerophyllous oak is the most important Mediterranean forest type from the economic and environmental viewpoints; it is located mainly in the central and western Mediterranean. Trees and bush forms of kermes oak are mostly widespread; the former form replaces progressively holm oak between southern Greece and the eastern countries. Holm oak is replaced by cork oak in the oceanic sites of the eastern Mediterranean, where morphological and climate conditions tend to be mesophilic, whereas it is replaced by pine forests in Italy, Spain, Greece, and Lebanon and in some mountain sites of Morocco, Tunisia, and Algeria. Mediterranean cypress is sometimes found at high elevations.

Hilly oak forests grow in typically submesophilic zones toward the eastern Mediterranean. These become mixed with deciduous oaks, such as zee, afares, Lebanese tauzin, hornbeam, ash, and, occasionally, beech. Beech and deciduous oaks are mostly found in Spain, Italy, and Turkey, whereas in the mountainous sites of Spain, Italy, and France, beech prevails. Mediterranean junipers prevalent in the western part of the region are replaced by thuya and junipers in Morocco, Algeria, Tunisia, and Turkey. Toward the east, juniper can reach the same altitude as the relic cedar and fir forests. The diversity in birds is very high in the Mediterranean region. Sixteen species of birds have evolved in the Mediterranean forest, including laurel pigeon, Corsican nuthatch, somber tit, spotless starling, and Syrian woodpecker.

"These forests are dominated mainly by drought-resistant, hard-leaved scrub of low-growing woody plants."

Mediterranean forests are recognized for their multipurpose aspects. They are acknowledged for their provisioning services, which are not only restricted to wood forest products and nonwood forest products (NWFPs) but also integrate regulating, supporting, and cultural services. NWFPs are not referred to as only for subsistence but they also contribute to national economies. They include cork, charcoal, game, honey, fruits, and mushrooms. Notable regulating services are protection of watersheds and agricultural soils, as well as conservation and purification of water. As in many other parts of the world, the extraordinary rich "megafauna" of the Mediterranean Basin was reduced through the combined effects of a changing climate and of the various types of pressure induced by prehistoric humans. Throughout history, major anthropogenic factors have left prints not only on the forests' composition but also on the landscapes. During the 19th century, the overuse of timber for the shipbuilding industry and war needs strongly affected the surface area of these forests. The invasions and occupation during the Middle Ages led to forest clearing for agriculture, especially on mountainous steep slopes. Grazing especially by goats played an important role in the areas with a dominant pastoral economy in the southern and eastern countries and Spain. Human-made forest fires always invaded the vegetation cover in the past in the entire area. Urban development and economic growth accompanied by the abandonment of farmlands and the rural depopulation in the northern part of the Mediterranean Basin have had a positive impact on the expansion of forest areas as well as on the development and progression of stands of conifers.

However, in the southern Mediterranean, forest area has gone through a noticeable shrinking. Due to the constant changes in environmental conditions, ecosystems' equilibrium has been extremely altered and has threatened the biodiversity of these Mediterranean forests over the past decades. Many ecological factors like climate change, fire, soil disturbance, and the combination of grazing and human land use have contributed to shaping the current Mediterranean landscape and have modified diversity at all levels of organization. Resource depletion caused by overgrazing activities and long-lasting human land use has led to the loss of genetic diversity. Extensive uses of pesticides and fertilizers in the large agricultural areas have polluted the surrounding ecosystems, and the increasing water demands have caused drainage of natural wetlands to provide both water and land resources.

Climate change is predicted to have a pronounced effect in the Mediterranean Basin. Temperatures will continue to increase because of an increase in the concentrations of greenhouse gases, which will affect species distribution and cause, as predicted, species extinction in some regions. The tree taxa in the Mediterranean forests found with high intraspecific biodiversity are threatened, whereby in southern Europe 60–80 percent of tree taxa will go extinct by 2050, compared to 20–40 percent of northern ones.

Projected extinction rate in hot spots might reach 39–42 percent of flora and fauna richness. Moreover, invasion phenomena have taken place in the altered landscape; many disturbed habitats harbor newly introduced alien species, which affect ecosystems' composition by competing with native species, leading to both ecological and economic damages. Therefore, conservation strategies should not only aim to anticipate climatic changes by protecting species located in north latitudes but also protect the most ancient and genetically highly divergent populations of the Mediterranean forests. Since the 1970s and 1980s, most protected areas have been established. At present there are 972 protected areas, which cover approximately 4 percent of the land area.

Matorral

Located in central Chile, the *matorral* stretches over 57,336 square miles (148,500 square kilometers). It is bounded in the north by the Atacama Desert and in the south by the Valdivian forests, while embraced in the east by the Andes and on the west by the Pacific Ocean. It presents a climatic and physiognomic resemblance with Californian ecosystems. These forests are relatively isolated and therefore contain many endemic species such as the Chilean wine palm. It is home to several plant communities. The Chilean *matorral* is a shrubland plant community containing typical plants like cactus. It is an opened-up area not only due to fire frequency but also due to urban sprawl coupled with growth in population. The region is subject to an increased number of fires per year, leading to some changes in use of land. Thus the ecoregion is threatened by conversion for agriculture and pasture. Also, the introductions of exotic species threaten many local and endemic plants, birds, and mammals.

California Chaparral

The California chaparral has been home to various types of plant communities such as evergreen sclerophyll forests and woodlands and evergreen scrubs. Rainfall is only 10–20 inches (25–50 centimeters), and soils are rocky, shallow, and infertile. Low rainfall allows dry litter to accumulate without decomposition, and dead shrubs persist upright as dry sticks. Much of California chaparral is dominated by 3–9 feet- (1–3 meter-) tall short shrubs. At high altitude, manzanita buckthorn and scrub oak occur. In the north, sagebrush and bitterbrush occur; the eastern, drier climates and higher elevations result in pinyon-juniper woodland instead of chaparral shrubland.

The vegetation is a savanna of pinyon pine—small, slow-growing trees with short needles—and juniper trees that may have the stature of larger shrubs. In Arizona and Mexico, oaks may be important. Fires occur in an area every 30 to 40 years, frequently causing great damages to Californians. Winter rains cause flooding, erosion, and mudslides after a fire because no vegetation remains to hold soil in place. Although the shrubs and trees are fire-adapted and resprout quickly, the main new growth is by annual and perennial herbs. These are present before the fire as seeds, bulbs, rhizomes, or other protected structures, and after the burn they grow vigorously, free of shading by charred shrubs; released minerals from the ash also enrich the soil. A few years after fire, larger shrubs dominate again and herbs are suppressed, perhaps by allelopathy or by recovery of the herbivore population.

Some unique plant communities, like southern California's coastal sage scrub, have been nearly eradicated by agriculture and urbanization. The region has been heavily affected by grazing, logging, dams and water diversions, and intensive agriculture and urbanization, as well as competition by numerous introduced or exotic plant and animal species. Anthropogenic activities in this region have altered natural habitats and animals' species richness. This has caused some species to go extinct. Disturbances like vegetation clearing, human-caused fires, refuse dumping, and land conversion have caused habitat fragmentation, which exists, today, only in small, isolated patches. Subsequently, the wildlife species that depend on these areas are becoming rare, threatened, or endangered.

Fynbos

Situated on the southern tip of Africa, the Western Cape is one of the nine provinces of South Africa. It stretches over an area of 49,985 square miles

(129,462 square kilometers) and has a topographical and a wide climatic diversity with many distinct micro- and macroclimates. This has resulted in the complex diversity of habitat types, flora, and fauna of the region. It is home to many indigenous and endangered species. In the Western Cape, expansions of vegetation are limited by the supply of water as well as suitable soils. Agricultural activities account for nearly 55 percent of all water consumption in the Western Cape. Grazing activities occur in the succulent Karoo regions of the western part. These regions are extensively used and are home to many exported species.

Vegetation formations comprise marsh rose, king protea, pincushion protea, belladonna lily, heathers, and many other plants. Among the bird species found are the Cape sugarbird, orange-breasted sunbird, and Cape francolin.

Mallee

Southwest Australia is a biodiversity hot spot in the Western Australia. Limited by the Indian Ocean, the coastal area has a wet winter, dry summer Mediterranean climate, and is one of five such regions in the world. The region covers 137,729 square miles (356,717 square kilometers), originally heavily forested. Today, vegetation in the region is mainly woody, including forests and woodlands, scrubland, and heath land. Vegetation formations comprise tree species such as Mallee, black paperback and wrinkled hakea, and other plant species such as morning flag, desert baeckea, and silvery phebalium. These are home to bird species such as firetail, skylark, little raven, Gilbert's whistler, and the endemic red-lored whistler.

Elsa Sattout

Montane Forests

Mountains rise like islands from the surrounding landscape, serving as landmarks, recreation areas for humans, and refugia for plants and animals that cannot be found in nearby lowlands. Sometimes mountain ranges serve as migration routes, other times they serve as migration obstacles. Often, the resources they hold are biologically and economically important.

Mountains are valuable because of their uniqueness—they are areas of extensive environmental change over relatively small distances. Despite the distinct nature of individual mountains or mountain ranges, some generalizations may be drawn about the characteristics of the mountain environment. For example, temperatures typically decline with elevation, while precipitation usually increases. Even with the greater precipitation, soil development on mountain slopes may be inhibited because steep slopes inhibit the accumulation of organic matter, and soil moisture may be lacking because steep slopes prevent storage of water. High peaks and ridges may be enshrouded in clouds, yet the intensity of solar radiation increases with elevation as there is less atmosphere with which to filter incoming solar radiation.

Other important environmental variables are not dependent on elevation. Geology may influence plant distributions by influencing the physical and chemical characteristics of the parent material from which the soils that plants grow in are derived. Landform characteristics may influence the pattern and frequency of disturbances such as landslides.

Exposure likewise adds another level of complexity to how plants are distributed on mountains. Different exposures—the direction that a slope faces—are subject to different patterns of diurnal (daily) shading, heating, and cooling, which in turn influence moisture supply. Likewise, sites shaded or sheltered by adjacent slopes or adjacent mountains tend to be wetter. More shade leads to cooler temperatures, and adjacent slopes may also dampen wind velocity—both likewise lead to reduced evapotranspiration, a term combining evaporation from soils and plant surfaces with transpiration, or loss of water from stomates (openings) in leaves. Exposed sites feel the full brunt of sunlight and desiccating winds and thus are usually drier.

One by-product of this seemingly infinite number of combinations of site characteristics is an incredible patchwork of biotic diversity in mountainous areas. Nevertheless, by analyzing the montane landscape in terms of environmental gradients, some reasonable generalizations about how montane forest composition and structure change across mountain landscapes can be drawn.

"Mountains are valuable because of their uniqueness— they are areas of extensive environmental change over relatively small distances."

The most useful environmental gradients to evaluate montane forests are elevation, latitude, and moisture gradients. More properly, these gradients are "complex gradients" as they represent gradients of several environmental factors that change—though not necessarily in the same way—in more or less the same direction. A good example of a complex gradient is the elevation gradient in which both temperature and precipitation change with elevation. On a global scale, changes along one gradient may mimic changes along another. In the case of mountain ecosystems, the elevation gradient mimics the latitude gradient in terms of temperature. According to the bioclimatic law, the temperature effect of a change of 400 feet (122 meters) of elevation is equivalent to a change in 1 degree of latitude.

Elevation

One of the most noticeable characteristics of mountains is how vegetation and other phenomena change with elevation. Temperature inversions, where warmer air overlays cooler air, may occur, but, in general, temperatures decrease with increasing elevation. The change in temperature—a measure of the available solar energy that powers biological systems—with elevation is measured by lapse rates. The rate of temperature change ranges between 5.5 degrees F per 1,000 feet (3 degrees C per 300 meters) in dry air to 3 degrees F per 1,000 feet (1.7 degrees C per 300 meters) in air saturated with water vapor. The decreasing temperature affects other important variables. One is vapor pressure, a measure of the amount of water vapor in the atmosphere. Since temperature is the primary variable influencing the amount of water vapor air can hold, vapor pressure decreases with elevation as well. The decrease is logarithmic, with the rate of change faster at lower elevations than at higher elevations. The drier the air, the greater the evapotranspiration demand. Drier air is likewise more effective at transmitting solar and infrared (heat) radiation, which translates into faster heating and cooling.

As temperature decreases, so does the length of the growing season—at least in temperate locations. The growing season may be year-round in tropical locations. Rather than a hot season and a cold season, tropical mountains have hot days and cold nights. Diurnal (daily) temperature ranges decrease with elevation.

Latitude

One of the main latitudinal relationships is the decrease in temperature and annual incoming solar radiation from the equator to the poles. The equatorial zone, with nearly 12 hours of daylight year-round, gets more than enough energy and, as a result, has on average high average temperatures. Where water is plentiful, lush plant growth—especially forest growth—is possible.

In subpolar and polar latitudes where daylight alternates between nearly 24 hours of daylight in the summer and nearly none in the winter, short growing seasons and limited solar energy inputs result in low temperatures and short growing seasons. The combination drastically limits the type of plant growth possible. Because of the low angle of the sun in the sky, slopes cast longer shadows, thus magnifying topographic effects on temperature and moisture conditions in mountainous environments.

Latitude has a major effect on temperature ranges. In tropical regions, diurnal temperature ranges are greater than annual temperature ranges. Outside the tropics, the reverse is true.

Moisture

For decades, researchers studying vegetation patterns in mountainous environments have used slope position and exposure as a proxy for the moisture gradient. In the Great Smoky Mountains of North Carolina and Tennessee, for example, sheltered (shaded) coves would be at the extreme mesic (moist) end of the moisture gradient. Moving from the mesic to the xeric (dry) end of the gradient, ravines and draws would be next, followed by sheltered slopes, open slopes, and exposed ridges and peaks.

In the Northern Hemisphere, southwest-facing slopes, because they are exposed to direct sunlight during the warmest part of the day, tend to be drier because the higher air temperatures, combined with solar heating, increase loss of water through evapotranspiration. On the other hand, northeast-facing slopes, which are exposed to direct sunlight during the coolest part of the day, tend to be wetter because evapotranspiration demand is lower. Between those extremes, from mesic to xeric, are north-, east-, northwest-, west-, southeast-, and south-facing slopes.

Another possible component of the moisture gradient is slope configuration (shape). At one extreme are convex slopes, which tend to disperse moisture and thus are drier. At the other extreme are concave slopes, which tend to collect water and thus are wetter.

Other Factors

Other factors also influence the characteristics of montane forests. For example, where a long mountain range of sufficient height runs perpendicular to prevailing wind directions, the windward slopes of the range force air upward. As the air rises, it cools, and the water vapor it carries condenses and falls as orographic precipitation. Over the ridge, the air follows the contour of the land, sinking, warming, and, in effect, drying. Such a process produces a lush, wet forest on the windward slopes and a scrub forest or woodland with lowland desert on the leeward slope.

This pattern produces the temperate rainforests characteristic of the Pacific Northwest region of the United States. Mountain ranges, such as the Pacific Coast Ranges, Cascades, and Sierra Nevada, intercept moisture-laden air masses driven off the

Cosby Creek flows through the Great Smoky Mountains. In this area, sheltered (shaded) coves are at the extreme mesic (moist) end of the moisture gradient. In the mountains, vegetation and other phenomena change with elevation. Generally, temperatures, vapor pressure, and length of the growing season decrease with increasing elevation, while solar radiation transmission increases. (National Park Service)

Pacific Ocean by the mid-latitude westerlies. As the winds descend back down the slopes, they produce dryland environments in areas such as California's Central Valley and Nevada's Great Basin.

Geology and soils can create unique patterns on smaller scales. Shales, for example, give rise to fine-grained soils from which plants have difficulty extracting water. Serpentine soils are toxic to many plants. The type of vegetation that typically occurs on shale or serpentine barrens tends to have a composition and structure more typical of drier environments.

The Ridge and Valley Province of Virginia features a number of shale barrens. Whereas most slopes are covered with thick stands of tall oak (*Quercus*), hickory (*Carya*), pine (*Pinus*), or hemlock (*Tsuga*), the shale barrens are open, with stunted, scraggly versions of some of the same species. Serpentine barrens in the mid-Atlantic region of the United States often feature post oak (*Q. stellata*) and blackjack oak (*Q. marilandica*), two species more common in the Cross Timbers, a savanna formation on the plains of Oklahoma and Texas.

Generic Patterns

In tropical or warm temperate climates with adequate precipitation, the lower flanks of a mountain (lower montane belt) are typically covered with broadleaved species. In the tropics, the dominant tree species are broadleaved evergreen. In temperate regions they are typically broadleaved deciduous species. There may be a mixture of coniferous and deciduous species in the montane belt above, but conifers typically dominate. Conifers, particularly spruces (*Picea*), firs (*Abies*), and pines (*Pinus*) again dominate the subalpine zone in the Northern Hemisphere. Other species, such as southern beech (*Nothofagus*) and conifers of the genus *Podocarpus* are the main dominants of the Southern Hemisphere subalpine zone.

Above the subalpine zone is the alpine zone, where tundra—a plant community dominated by herbs, grasses, and low shrubs—replaces forest where climatic conditions prohibit the growth of trees. The forest-tundra ecotone (boundary) is the upper timberline (or tree line). In arid environments, the lowlands are dominated by desert or steppe vegetation because the lack of water likewise inhibits tree growth. Woodlands replace desert or steppe vegetation at higher elevations where precipitation is sufficient for tree growth. The desert/woodland or steppe/woodland ecotone is called the lower treeline.

Tree lines and boundaries between vegetation belts may change depending on local or regional conditions. On any given mountain slope, boundaries between vegetation belts tend to be lower on wetter slopes than on drier slopes. Tree lines on isolated mountains tend to be lower. On a much larger scale, tree lines occur lower the closer one is to the poles.

DAVID M. LAWRENCE

Plantation Forests

The world's population just exceeded 7 billion people, and the demand for products derived from wood for this number of people simply cannot be attained by harvesting from natural forests. To augment our needs, we have turned trees into a crop plant by creating plantation forests around the world. Plantation forests, also called "tree farms," are grown on every continent except Antarctica and have been planted for many centuries. Plantations are grown almost exclusively for economic purposes, while benefits to wildlife and the environment are much less emphasized. As natural forests have shrunk in size, plantations forests have increased in area over time.

Plantation forests share many characteristics with more traditional agricultural crops in North America. Plantation forests are generally monocultures, consisting of a single genotype or species of tree. Occasionally, plantation forests have multiple genotypes of a single species, creating closely related polycultures. Tree species used are generally fast-growing, early successional species such as pine (*Pinus* spp.), poplars (*Populus* spp.), or eucalypts (*Eucalyptus* spp.). Regardless of the number of genotypes within a plantation forest, the treatment of these areas is generally the same. Trees are planted at uniform spacing, which is often determined by the size of the equipment and machines used to plant, apply pesticides, and harvest the stand. Trees almost always receive fertilization and receive irrigation in some drought-prone areas. Pest and weed control is often administered through the use of chemicals, but sometimes this can be accomplished through silvicultural and biological control methods.

"Plantation forests that are present in most countries around the world are considered 'wood-baskets.'"

Plantation forests contrast greatly with natural forests, which can have many different tree species. Natural forests often arise through succession, whereas plantation forests are planted, harvested after several years, and replanted, over and over again. Natural forests have many different-aged trees, while trees in plantation forests are all the same age. Natural forests have a diverse forest floor that supports woody vines and shrubs, herbaceous plants, many different arthropods, and even mammals. Plantation forests are generally kept weed- and insect-free, which in turn greatly reduces the attractiveness to most mammals, with the exception of deer, which may feed in plantation forests when the trees are young. To that end, however, most plantation forest managers implement deer-exclusion fences to eliminate deer browse damage. Natural forests do not receive irrigation or fertilization. Rather, they are dependent on nature to provide necessary water and nutrients for tree growth. Plantation forests, however, may be irrigated and fertilized in an effort to maximize tree growth and production.

Purpose and Land Area

Plantation forests are typically grown for industrial (48 percent as of 2000) and nonindustrial purposes (26 percent), with another 26 percent as unspecified uses. In terms of industrial use, plantation forests focus on roundwood production that is classified as either industrial roundwood (e.g., sawlogs, veneer logs, plywood, and pulpwood) or wood fuel (e.g., wood used for cooking, heating, or power) production. Current worldwide harvests for industrial roundwood production are around 1.8 billion cubic meters (more than 63.5 billion cubic feet). Even though plantation forests constitute about 5 percent of total forests in the world, they may contribute up to two-thirds of wood production in some areas. In recent years, there has been a tremendous push to develop technology to make liquid biofuels (ethanol) using wood residues from plantation forests, as well as to use woody biomass along with coal in cofiring power plants. In the next decade, this technology may lead to widespread use of forest biomass for transportation and energy needs. Nonindustrial use of plantations includes providing ecological services such as wind protection, soil and water conservation, carbon sinks, and enhancement of barren and urban landscapes.

The area occupied by forest plantations varies considerably among continents. In North America, forest plantations cover an area over 72,587 square miles (188,000 square kilometers)—that's an area larger than Kentucky and South Carolina combined. Poplars (*Populus* spp. and hybrids) are grown primarily in the Pacific Northwest and the Great Lakes regions of the United States, with smaller areas in the northeast and southeast. Most plantations are grown on 8–12 year rotations, with the majority of wood going for pulp. However, some plantations in the Pacific Northwest are growing trees longer, with a targeted end use as lumber. Trees are generally planted at 2.5-by-3-meters (8.2-by-9.8-feet) spacing and are not thinned prior to harvest. When harvested, the entire site is clear-cut. Trees for the next rotation in the plantation forest are planted between the rows of harvested stumps, so by the time the second generation is ready to harvest, the old stumps have biodegraded. Willows (*Salix* spp.) are grown in the northeast, especially in New York State. These willow plantations are grown for 2–3 years and harvested up to four times before being replanted—a technique called "coppicing." In this case, the biomass produced is used for energy, not pulp or lumber, and may be cut and made into large bales with modified agricultural equipment. In addition to hardwoods, several coniferous species, including pine and spruce (*Picea* spp.) trees are grown in the southeastern and northeastern United States and in parts of Canada. Conifer species are slower growing than poplars and require 1–2 thinning cuts after initial planting. Thinning cuts allow more space to grow for the trees that are left on site. Trees removed in the thinning are generally smaller and used for pulp, while the larger, older trees (up to 30 years old for *Pinus* spp. and older for *Picea* spp.) are used for lumber.

Plantation Forests in Asia

Asia has the greatest area of forest plantations in the world, with over 250,194 square miles (648,000 square kilometers) planted—an area nearly the size of Texas. This region, along with Australia, has the highest rate of forest plantation establishment in the world, with Japan, China, and India having the largest amount of forest plantation land area. Over 45 percent of Japan is classified as forest plantation, with trees such as Japanese cedar (*Cryptomeria japonica*), Japanese cypress (*Chamaecyparis obtusa*), Japanese larch (*Larix leptolepis*), and pine species. Japan uses its forest plantation trees for pulp and lumber products. Several pine and spruce species are grown in China, as well as hardwood species such as eucalyptus, paulownia (*Paulownia* spp.), birch (*Betula* spp.), locust (*Robinia* spp.), and poplars. Trees are harvested in as little as 6 years, and are given intensive nutrient, weed, and pest management. These trees are used for all types of wood products, including pulp for paper and lumber. China alone accounts for nearly three-fourths of the world's poplar production. India grows teak (*Tectona* spp.), eucalyptus, and poplar in its forest plantations. Teak is planted at wide spacing, with one thinning occurring when trees are around 33 feet (10 meters) tall, and another when trees are 99 feet (20 meters) tall. Branches are often removed on the lower portions of the stem, especially when teak trees are young. Teak is generally harvested by 30 years of age, but some plantations up to 60 years old do exist.

A plantation of eastern cottonwood (Populus deltoides) *in the southeastern United States. Eastern cottonwood is a valuable timber species that is highly suitable for plantation management. (David R. Coyle)*

Plantation Forests in Europe

Europe as a whole has over 106,564 square miles (276,000 square kilometers) of forest plantations, which is an area slightly larger than Colorado. Sweden, Finland, and other Nordic and Baltic countries have many acres of willow plantations. In these coppice systems, the trees are cut back after the first year to promote multiple sprouts. Biomass is harvested every 3–5 years thereafter for another 3–4 rotations. Plantations are fertilized, harvested during dormant periods, and treated much like traditional agricultural crops. Sweden is widely regarded as the world authority on willow coppice systems, as it has conducted extensive research and field trials on silviculture and harvesting methods. Pine and eucalypt plantation forests are found in Spain, while poplars are grown in France. Plantation forests in Europe are primarily used for biomass and energy, with a smaller portion going toward pulp and lumber.

Plantation Forests in Africa

With the exception of South Africa, there is little reliable information regarding forest plantations in Africa. Forest plantations cover an area almost as large as Louisiana (over 50,579 square miles, or 131,000 square kilometers), and are primarily planted with teak, pine, eucalypts, acacia (*Acacia* spp.), and types of rubber tree (*Hevea* spp.). As countries in Africa continue to develop, the demand for wood products is growing. A large portion of forest plantations are found in western Africa, where their primary uses are for fuelwood and environmental stability (e.g., preventing sand dunes). The eastern portion of Africa, unfortunately, has been one of the most unstable regions in the world in the last several decades, and likewise forest plantations in this region have been neglected and are impacted by pests and disease in addition to a general lack of silviculture. Central Africa has a low proportion of forest plantations, mostly pines and eucalypts. Here, the wood is used for fuel and lumber. The majority of forest plantations are in southern Africa, specifically South Africa. Eucalypts and pines are most often planted and used as fuelwood, pulp, or lumber for construction. South Africa is widely recognized as a world leader in forest plantation management and is able to supply the majority of the country's wood needs through its plantation forests.

Plantation Forests in South America

Eucalypts and pines are the most commonly grown tree species in South America, which has over 43,629 square miles (113,000 square kilometers)—an area almost the size of Ohio—planted in forest plantations. Brazil, Chile, Argentina, and Uruguay have the most land area, while other countries have much less area in forest plantations. Brazil is one of the most active producers of pulpwood in the world. South American forest plantations are used for not only pulp and lumber but also for exports of wood chips and logs, as well as for biomass and cofiring plants. It is illegal to clear native forests in Chile, making the renewable wood from forest plantations even more valuable. In addition to pines and eucalypts, poplars and willows are grown on a smaller scale, especially in Argentina.

Plantation Forests in Australia

Australia (and Oceania) has over 14,672 square miles (38,000 square kilometers) of plantation forests, or an area the size of Maryland and Delaware combined. In Australia, plantation forests are mostly located in the southeastern area and Tasmania. The predominant tree species planted are pine, teak, mahogany (*Sweitenia* spp.), and eucalypts, with one native conifer, hoop pine (*Araucaria cunninghamii*) also used. These species are used for pulp and paper products, saw timber for building and furniture, and wood products such as particle board and plywood. Plantation forests are an important component of forestry in this region and account for a great deal of wood products. Forest plantations are managed very intensively, with pest and weed control. Australia and New Zealand are regarded as leaders in the development of plantation forest silviculture, and this is reflected in the high economic impact that this sector has on the local economy.

Criticisms of Plantation Forests

The main criticism of plantation forests is that they are monocultures that are even-aged with trees grown primarily for quick wood and fiber production. Hence, plantation forests have a much lower diversity of plant species and other biota that those of nonplantation forests. As monocultures, the trees may be more susceptible to outbreaks of insects and diseases because the trees may represent a single genetic stock and pests can move easily across the homogenous landscape. Plantation forests typically tend to be dominated by nonnative tree species that may have adverse ecological impacts. For example, eucalyptus trees that are widely used in plantation forests have a higher requirement for water than native species and show allelopathy where they produce compounds that inhibit the growth of other plants.

There have been cases where nonnative trees (e.g., pine trees planted in the Southern Hemisphere) have escaped plantations and have invaded natural ecosystems with significant alterations to native forest structure and composition. Nonnative forest pests have frequently been introduced to plantations, from where they can establish and spread into the natural forests. Because fertilizers and insecticides are often used in plantation forests, this may cause chemical pollution in the general area. Forests have been cleared to create plantations in areas where indigenous people used to live, thus causing their displacement. Further, there is a concern that if significant biofuel technology is developed, it could lead to large-scale replacement of natural forests by plantation forests, especially in tropical areas.

Benefits of Plantation Forests

There could be many ecological benefits to plantation forests, especially if plantations are managed with appropriate silvicultural methods and sustainable forest practices. Plantation forests have the maximum benefit when they are planted on degraded lands due to soil erosion after intensive agriculture and mining activities. In some instances, plantation trees have been used to competitively exclude nonnative weeds from landscapes, so that forests could be regenerated to a natural state over years. Planted trees that are hardier than native tree species may prevent further soil loss, rehabilitate the land, and assist with watershed protection. Afforestation of these areas has led to higher sequestration of carbon stocks, which is one of the reasons why many tropical countries are creating plantation forests. It is also argued that the use of existing plantation forests to the maximum capability may reduce the need to cut down natural forests for fiber production in the future. If managed properly, plantations can continue to provide a renewable source of wood for a long time. Since plantations are more productive than natural forests, more wood and fiber can be produced using less area. Sewage and industrial waste has been sprayed in plantations where trees break it down and absorb it, leading to a more environmentally friendly way of disposing waste. Certain types of trees in forest plantations, especially poplars, can also be used for phytoremediation, or removing pollutants from the soil by drawing them up through the roots and accumulating pollutants in tree tissues. Those trees can then be removed and disposed, taking the pollutants with them.

Plantation forests are excellent areas in which to conduct research. Because of the relatively sterile environment compared to natural forests (e.g., no ground vegetation, high survival of trees, low genetic diversity), plantation forests are well suited for experiments on tree physiology, tree growth and nutrient dynamics, and interactions with pests and diseases. By using plantation forests, scientists can isolate specific issues and adequately test them without lots of confounding effects that are often present in natural forests. Thus, plantation forests may not only produce a high amount of wood quickly on a small land area, but they also provide scientists with sites to conduct research to improve forest productivity. Research and production forest plantations contribute to the local economy, especially in the third world countries through providing livelihoods, poverty alleviation, and food security.

The World's "Wood-Baskets"

Overall, plantation forests that are present in most countries around the world are considered "wood-baskets." The area occupied and the importance of plantation forests are increasing over time. Adverse effects of plantation forests include changes in the original forest structure, composition, and processes, with subsequent impacts on native species. The beneficial effects of plantation forests include production of wood and fiber that could be sustainable and renewable, some ecological values especially in degraded and urban areas, and a potential source of income to the local communities. A major challenge for plantation foresters is how to optimize production and efficiency with minimal impacts by insects and disease, while maintaining and enhancing local biodiversity. Management of plantation forests in the future is likely going to be heavily affected by changing technology, policy, and the value any society places on these highly modified habitats.

David R. Coyle
Kamal J. K. Gandhi

Subtropical Forests

Subtropical forests, also known as tropical moist forests or subtropical moist broadleaf forests (SMBF), occur worldwide within the tropical humid forest biome. Variously sized tracts of subtropical forest remain in Mexico, Latin America, eastern Australia, New Zealand, east Asia, and southeast Africa. Subtropical forests flourish in humid regions with annual rainfall varying from about 315 to 512 inches (about 8,000 to 13,000 millimeters), with critical temperatures occurring in a window below 11 degrees F (below 24 degrees C).

In the scientific and popular literature, subtropical forests are frequently unspecified to type or geographic distribution, instead identified variously as "tropical forest," "lowland forest, "tropical evergreen forest, "tropical humid forest, "tropical moist broadleaf forest, or "tropical rainforest." Inconsistent terminology obscures the characteristic and measurable features differentiating among the three ecosystems comprising the tropical humid forest biome (after L. Holdridge): tropical rainforests, transitional wet forests, and subtropical moist broadleaf forests (SMBF) that transition into seasonal tropical forests. Confusing terminology and mapping contribute not only to the relative lack of scientific and media attention to SMBF but also to their critically endangered status compared to tropical rainforest landscapes. This comparison is exemplified by the conservation status of Brazil's Amazon rainforest versus the seriously threatened moist tropical Mata Atlântica (Atlantic Forest) along Brazil's eastern coast.

The two annual rainfall peaks occurring in subtropical forests are more seasonally differentiated (wet season, dry season) than those in wetter, tropical rainforests, occasioned by characteristic and measurable differences in annual precipitation and temperature between the two forest types. SMBF flourish in humid tropical regions with annual rainfall varying from about 315 to 512 inches (about 8,000 to 13,000 millimeters), with critical temperature occurring in a window below 24 degrees C. On the other hand, annual rainfall in tropical rainforests varies from about 512 to 630 inches (about 13,000 to 16,000 millimeters), with critical temperatures occurring in a window above 24 degrees C. This article not only summarizes information about SMBF available in the scientific literature but also highlights research questions and topics required for assembly of quantitative data based upon modeling and experimentation that will permit differentiation of SMBF from other tropical humid forest ecosystems. Tropical humid forests occurring at high altitude (e.g., tropical "cloud" forest) are not covered in this article.

Ecosystem Variability

P. M. Vitousek and R. L. Sanford discussed the within- and between-ecosystem variability of biomass, production, and nutrient cycling in tropical humid forests, concluding that these variations "nonetheless follow coherent, explicable patterns." Tropical humid forest biomes feature relatively higher productivity compared to other terrestrial ecosystems, and biodiversity of tropical rainforests is second only to temperate rainforests. Supporting these evidence-based generalizations about the positive association between rainfall and productivity in tropical forests, C. B. Jones found a wider range of group sizes for mantled howler monkeys (*Alouatta palliata* Gray) in Central American SMBF (Panama) than for groups of the same species inhabiting Mesoamerican seasonal tropical forest (Costa Rica). Jones concluded that higher standard deviations in group size found in Panama were explained by higher carrying capacity in SMBF, the wetter site.

In SMBF, seasonal patterns in plant (primary producers) and animal (primary and secondary consumers) life-history features (survival and fecundity) are less pronounced than those in tropical rainforests and more pronounced than those in seasonal tropical forests. Temporal patterns of biotic life histories (e.g., flowering, fruiting, dispersal) occurring in response to water stress, and water availability, in addition to light reflectance, are the most significant factors determining tropical tree species life-history trajectories and diversity. Analyses of species compositions when tropical rainforests, wet forests, subtropical forests, and seasonal tropical forests are compared regionally demonstrate overlap for some plants and animals whose wide habitat ranges result from one or more adaptable biochemical, developmental, or phenotypic traits. On the other hand, some organisms are specialized for or limited to the restricted abiotic or biotic features of tropical humid forests, in some cases because of endemism (forest inhabitants found in one ecosystem type), in others because of physical factors limited to the particular parameters associated with the subtropics (e.g., restricted ranges of thermal tolerances). Barring cryptic biochemical or phenotypic plasticity, species constrained by adaptations to tropical humid forest conditions are particularly vulnerable to anthropogenic effects (climate change, deforestation, habitat fragmentation) as a result of their relatively limited resilience to environmental perturbations and other stressors (e.g., disease).

"The key problem facing humanity in the coming century (is) how to bring a better quality of life for 8 billion or more people without wrecking the environment entirely in the attempt."

Testing Whittaker's Ideas

Quantitative and experimental tests of Robert Whittaker's ideas have also been conducted. Investigating the spatial dynamics of species assembly rules with theoretical models, S. Nee and R. M. May modeled metacommunity ("secondary") effects for two competing species while decreasing the amount of available habitat ("patch removal"). These researchers found that patch removal may decrease the distribution and abundance of superior competitors while increasing the same parameters for inferior competitors, counterintuitive effects dependent upon rates of dispersal or colonization and rates of patch extinction. For plants and animals, patterns of movement are of fundamental importance as they may counteract effects of genetic drift by maintaining connectivity among subpopulations and populations, decreasing likelihoods of isolation and extinction. Consistent with Whittaker's formulations about how ecosystem networks are assembled and change, Nee and May showed that patch removal has the potential to modify the makeup of communities that received no simulated alterations. Gradients (microscale networks), then, may form, in essence, from perturbed locations to patches that have experienced minimal "intrinsic" perturbations of their own, and the previous authors' generalized mathematical treatments showed that these nonlinear landscape effects are difficult to predict. Again, the forgoing research has not been tested in SMBF, studies that should be prioritized for clear distinctions of the previous biome from tropical rainforest and wet forest ecosystems.

Studying a disturbance gradient (primary to old-growth secondary to plantation) in experimentally fragmented central African SMBF transects, J. H. Lawton and his collaborators documented an edge to interior gradient in invertebrate and vertebrate species assemblies, consistent with propositions advanced by Whittaker, whereby ecological structures and processes result from networks of relatively independent species functions and effects. As expected, species richness declined with degree of disturbance. However, as Whittaker predicted, responses to different perturbation states by one species were poor indicators of responses by others. By manipulating spatial dynamics, Lawton's group showed that species in assemblages were more likely to respond individualistically than holistically and interdependently. These findings addressed fundamental questions concerning micro- and macroscale transitions from one patch, habitat, or forest type to another in SMBF. In addition, Lawton's group studied "boundary dynamics" and "edge effects" across changing plant communities from seasonal tropical, to subtropical, to rainforest ecosystems. These results are not only of fundamental importance to basic and applied ecology but also the research programs discussed herein contribute to conservation biology databases, including information required to protect subtropical and seasonal tropical forests buffering rainforests, reservoirs of Earth's richest biodiversity assemblages.

Studying the Causes of Gradients in Tropical Tree Diversity

T. J. Givnish's classic theoretical study "on the causes of gradients in tropical tree diversity" reviewed hypotheses found in the ecological literature advanced to explain variations in tropical forest microscale variations (e.g., precipitation, forest architecture, soil fertility). Within regions, density-dependent plant mortality in tropical wet forests promoted greater forest diversity, a characteristic with significant implications for plant life-history "trajectories." Two other factors were strongly associated with diversity in tropical forests: (1) a "shifting balance" among tree mortality, competition, and recruitment, and (2) facilitation of diversity by high productivity. The latter relationship is expected if higher tree mortality yields a higher proportion of young, rapidly growing plants, a "fast" life-history strategy, as detailed by S. C. Stearns.

Givnish modeled the aforementioned principal components, accounting for "trends in woody plant diversity along ecological gradients in the tropics." Since (1) and (2) varied by forest type, precipitation was a principal component in combination with dispersal of seeds by vertebrate consumers. Finally, Givnish was unable to exclude the possibility that random drift accounted for certain of his theoretical results concerning the causes of forest gradients (e.g., "the repeated dominance of particular taxa in separate but ecologically similar sites"). These fundamental studies have not been conducted for the discrete traits of SMBF, highlighting, again, the ecological literature's relative lack of focus on this ecosystem as one separate from other humid tropical forest landscapes.

Variations in Nutrient Cycling

Vitousek and Sanford reviewed variations in worldwide SMBF nutrient cycling, emphasizing physiology and community ecology, as well as plant and animal population biology. These authors reported that, within and between SMBF, variations in nutrient cycling were explained by differences in soil characteristics, in addition to duration and intensity of dry seasons as well as altitudinal gradients. These authors highlighted the importance of evaluating species composition as a source of micro- and macroscale variations in SMBF. However, Vitousek and Sanford noted that these variables are difficult to measure without experimental tests. These researchers found that soil fertility and nutrient cycling are closely linked in SMBF. However, biomass varied little across space even though differences in nutrient concentrations of primary producers varied with "soil fertility." This research program is of particular significance because results describe distinguishing features of SMBF that permit comparative analyses.

Variations in Soil Characteristics

A recent review by A. R. Townsend and his colleagues highlighted the importance of variations in soil characteristics as determinants of tropical forest diversity (e.g., soil age, chemical composition). Feedback processes occurred at microscale from soils to plants, on the one hand, and from plants to soils, on the other, operations including effects from plant diversity to variations in "chemical and structural traits" influencing local productivity (e.g., growth rates), decomposition (e.g., leaf litter conditions), and nutrient cycling. At regional levels in tropical ecosystems, biogeochemical variables are strongly influenced by "landscape dynamics" increasing unpredictability in the limiting resources available to plants, organisms upon which animals directly or indirectly depend. This review discussed advances in remote sensing capable of "capturing" features of tropical forest diversity beyond the power of satellite and other "large-scale estimates," methods with the potential to differentiate one category of tropical humid forest from another.

Soils were also the focus of study by J. N. Price and her colleagues. This empirical work focused on the role of small-scale competition among roots and rhizomes below soil surfaces as determinants of above ground, nonrandom assembly patterns. These authors found that root and rhizome assembly in grassland ecosystems was driven by abiotic factors, data having fundamental import for future investigations in humid forest biomes. Investigating the characteristic dynamics and consequences of inorganic components of materials below SMBF soil surfaces will add an important dimension to the existing body of research documenting critical "soil services" (e.g., regulation and buffering of hydrologic and nutrient cycles as well as detritus).

Studies of Predation

New research questions are also advanced in a paper published by O. Schmitz and his colleagues.

Field experiments demonstrated that predation, a stressful and, possibly, lethal environmental event, caused herbivorous, temperate-forest grasshoppers to switch from a nitrogen-rich to a carbohydrate-rich diet with correlated changes in soil carbon cycles. Price and Schmitz published their papers in 2012, suggesting that novel mechanisms associated with consumer-producer interactions remain to be discovered and investigated. Scientists conducting research within and across humid tropical forest landscapes are certain to intensify efforts to increase our understanding of known biogeochemical mechanisms as well as formulate hypotheses concerning dynamic abiotic and biotic processes characterizing SMBF.

Geochemical and Biophysical Dynamics and the Biodiversity Crisis

A case study based on research conducted in one Belizean SMBF will serve as an approximate model of geochemical and biophysical dynamics in this ecosystem, highlighting their current biodiversity crisis. According to the Forest Resources Assessment Programme, Belize is a biologically rich Mesoamerican nation with up to 60 percent of original tropical humid forest cover still remaining. Belize's population density is among the lowest in the world, with approximately 250,000 inhabitants per square kilometer. Black howlers (*Alouatta pigra*) and Central American spider monkeys (*Ateles geoffroyi yucatanensis*), both members of the herbivorous primate family Atelidae, are the only nonhuman primates inhabiting Belizean forests, although populations may be fragmented, locally endangered, or locally extinct. In Belize, large tracts of forest and their attendant biogeochemical components are under private oversight (e.g., nongovernmental organizations, farmers). Following R. H. Horwich and J. Lyon, clay and fertile limestone soils are most characteristic of Belizean SMBF, supporting ecosystem gradients from more productive to less productive habitats, respectively. Most SMBF occur as a mosaic of variably sized tracts, from relatively intact stretches of landscape to fragments and patches of habitats formed by anthropogenic effects, particularly selective cutting, deforestation, rural poverty, external markets, unsustainable patterns of resource exploitation, poor management, and "milpa" ("slash-and-burn") agricultural practices.

Figure 1 exhibits approximate species distributions of Belizean black howler and spider monkeys. One suggestive pattern of results is that, according to geographic surveys, black howlers, folivore-frugivores, entered Belize from the north, moving southward across the Maya Mountains and Cockscomb Range in southern Belize. On the other hand, the mountains appear to have been a geographic barrier for the primarily frugivorous spider monkeys spreading from south to north. The ability to process leaves (black howlers) as a nutritional source is thought to provide a "fallback" dietary source permitting flexible switching of food sources, enhancing capacities for dispersal and colonization.

As M. L. Cody pointed out, frugivores (spider monkeys), on the other hand, are more sensitive and vulnerable to environmental stress, including variations in the predictability of fruit in time and space, conditions that may limit or retard the geographical spread of species. Colonizing abilities and frugivory are generally unrelated, unless fruit is evenly rather than patchily dispersed. A satellite image of forest cover created an overlay for the species distribution map (Figure 1), showing that, consistent with expectation, monkeys were least likely to occur along Belize's coastline where deforestation is most severe. *A. pigra* persist in more deforested areas, reflecting robustness due to their tolerance of leaves. *A. geoffroyi yucatanensis* was more likely to occur in larger tracts of SMBF, similar to distribution patterns observed by C. A. Jost for howler and spider monkeys in Costa Rica.

Terminology and Definitions

Conventional terminology and definitions of tropical humid ecosystems have not been standardized in the scientific literature. However, in 2002 D. M. Olson and E. Dinerstein (World Wildlife Fund) published a global classification of 238 "ecoregions," defined as "regional-scale (continental-scale) units of biodiversity." These authors determined an ecoregion's vulnerability to extirpation based on qualitative measures of species richness and endemism, uniqueness of abiotic and biotic components, degree of fragmentation, as well as other factors. Of 238 ecoregions, 50 occurred in tropical or SMBF. Most of the latter ecoregions were found in the Paleotropics (Old World: 38/50 ecoregions), and, SMBF were most commonly represented overall (27/50). Among terrestrial ecoregions worldwide, species diversity of the SMBF, Atlantic Forest, was second only to Amazon rainforests.

Despite the conceptual utility of Olson and Dinerstein's ecoregion analysis, their model has not become the scientific standard, possibly because it is based on qualitative factors rather than quantitative parameters. Nonetheless, this paper provides important details about what we do not know about humid tropical forests that would permit conservation biologists to develop quantitative, predictive models to facilitate the preservation of biodiversity in SMBF. Increased knowledge of SMBF will advance what E. O. Wilson has called "the key problem facing humanity in the coming century, how to bring a better quality of life for 8 billion or more people without wrecking the environment entirely in the attempt."

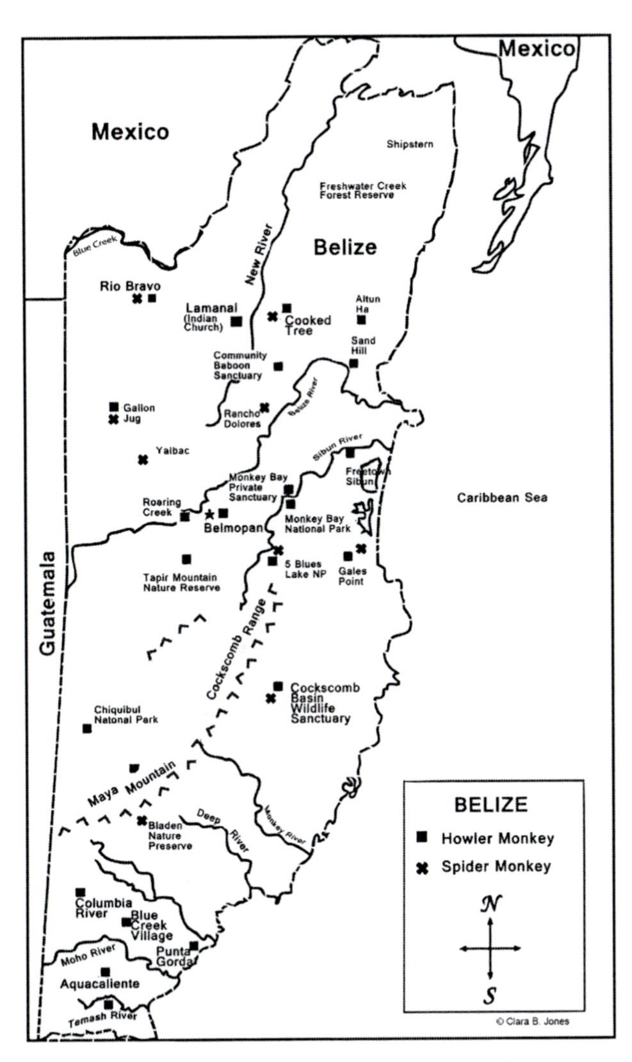

Figure 1. Approximate species distribution and range limits of the two primate species in Belize (black howler monkeys and Central American spider monkeys) based upon surveys conducted in the 1980s and 1990s by Community Conservation, Inc. in SMBF. Some data points represent reports by local residents. (© Clara B. Jones)

Early Subtropical Forest Research

Biome and ecosystem concepts were developed initially by several investigators central to the establishment of ecology as a scientific discipline distinguishable from natural history. Since the 19th century, biomes and ecosystems have been understood as assemblages of organisms inhabiting characteristic environments. However, early researchers held different perspectives about what abiotic (geochemical) and biotic (organismal) regulatory mechanisms accounted for ecosystem functions, complexity, interspecific associations, and stability. As synthesized by S. A. Levin, broadly speaking, one perspective emphasized a reductionistic approach whereby biomes and ecosystems might be assembled and reassembled from basic abiotic and biotic components. Other ecologists thought of biomes and ecosystems holistically, whereby components interacted synergistically, a viewpoint emphasizing component interdependency more than the independent function and, possibly, changeability of component parts.

Though these formulations are not mutually exclusive in any strict sense, they characterize different schools of thought in contemporary ecology. The contributions of F. E. Clements, H. A. Gleason, A. G. Tansley, Charles Elton, Albert Lotka, R. Lindeman, and Eugene P. Odum represent classic treatments of ecosystem and biome concepts. Later, Whittaker and Holdridge independently introduced the first comprehensive maps and classification systems of terrestrial habitats, ecosystems, and biomes throughout the world.

A gradient of temperature and precipitation occurs across tropical humid forests. R. H. Whittaker introduced the gradient analysis concept to explain changes in the distribution and abundance of individual plant species along correlation curves in an attempt to document gradual and individualistic (i.e., species-typical) patterns of change along topological gradients. Whittaker's schema was supported in Argentina by G. E. Zunino's research team studying six ecological factors along biogeochemical gradients in five different ecosystems, including SMBF ("paranaense"). In 2004, J. B. Bastow and his colleagues reevaluated Whittaker's database, finding his methods and results unconvincing, primarily because of inappropriate sampling of plants. Despite their criticisms of Whittaker's methods and results, Bastow et al., consistent with other reports, empirically confirmed that Whittaker's gradient analysis paradigm is, effectively, correct.

Analyses of gradients are essential to landscape and metapopulation ecology, in particular, the causes and consequences of species' spatial dispersion. The aforementioned research questions emphasize patch and subpopulation dynamics, especially the ways in which these mechanisms, functions, and effects influence dispersal, colonization, and other plant and animal population and subpopulation phenomena. The "patch" network view of physical and biotic environments has received increased attention in recent years due to researchers' attempts to document the effects of habitat fragmentation and other landscape perturbations on populations. T. A. Guisan and his colleagues, among other theoretical ecologists, have developed conceptual and simulation models for the analysis of gradients and landscapes, methods in need of application to the distinctive quantitative traits differentiating SMBF from other tropical humid forests.

Questions related to spatial dynamics in tropical and temperate forests have been addressed by numerous ecologists, notably R. H. MacArthur, R. C. Lewontin, J. H. Connell, S. Harrison, S. Nee, R. M. May, J. H. Brown, J. Clobert, A. A. Dhondt, E. Danchin, S. A. Levin, I. A. Hanski, M. E. Gilpin, and other scientific associates. Though research on spatial dynamics in tropical forests is less common in SMBF compared to tropical rainforests and seasonal tropical forests, descriptive research has been conducted in SMBF by L. Poorter (trees, Bolivia), G. H. Adler (rodents, Panama), J. H. Lawton (beetles, Central Amazon), and J. E. Fa (mammals, western and central Africa), among others. Despite theoretical and empirical support for the robustness of gradient analysis at the level of microclimates, Vitousek and Sanford cautioned that attempts to apply the paradigm as a continuum across macroclimates (across forest types) are suspect because of significant variability and discontinuity of tropical forest characteristics, termed "state factors" by H. Jenny (biotic and substrate components, climate, topography, and time).

South America's Atlantic Forest

One of the world's most ecologically important, critically threatened SMBF is South America's Atlantic Forest (Mata Atlântica), approximately 330 million acres stretching from northeastern Brazil to northeast Argentina and eastern Paraguay. According to separate reports from the Nature Conservancy and World Wildlife Fund, more than 85 percent of Brazil's original Atlantic Forest landscape has been deforested or fragmented. The evolution of biotic diversity (e.g., species richness), community assembly patterns, as well as abiotic, biogeochemical gradients (e.g., variations in below- and aboveground soil compositions) have been facilitated by Pleistocene forest shrinkage patterns in addition to factors discussed elsewhere in this article. This United Nations Educational, Scientific and Cultural Organization World Heritage Site exhibits biodiversity rivaling that of the Brazilian Amazon, including many threatened or endangered endemic plant and animal species, as well as two indigenous human tribal groups, the Tupi and the Guarani. One source quoted the Brazilian conservation biologist, A. R. Mendes Pontes: "Species in the Mata Atlântica are the living dead."

Clara B. Jones

Temperate Coniferous Forests

Trees with needlelike leaves, known as "conifers," are an ancient plant group, with fossils that go back 300 million years to the late Carboniferous Period when much of Earth's coal was formed. Coniferous trees are generally more tolerant of drought and fire than are deciduous trees, although some species do inhabit moist forest ecosystems. Coniferous forests of the temperate zone mainly grow in North America, where they occur in the Pacific Northwest, the Rocky Mountains, as stands of eastern hemlock in the northeast (currently disappearing due to an invasive insect pest), and in the Coastal Plain of the southeast. Very small coniferous temperate forests also grow in Japan, Argentina, and Chile. At the highest elevations in the southern Appalachian Mountains in North Carolina coniferous spruce-fir forests occur, but these are ecologically more similar to northern boreal forests than to temperate coniferous forests. With the exception of cypress, which grows primarily in southeastern swamps, the tree species of coniferous forests are almost always evergreen as opposed to deciduous trees, which lose their leaves every autumn. Fire is an important feature of coniferous forest ecosystems, and without fire most of these ecosystems would eventually develop, or "succeed," into different forest types. Other important environmental determinants of these forests are the amount of soil moisture and, in the Rocky Mountains, elevation. As winters become warmer and growing seasons are lengthened in duration, insect pests and disease pathogens are increasing the mortality of many temperate conifer tree species.

Pacific Northwest Coniferous Forests

The beautiful temperate evergreen forests of the Pacific Northwest hug the coast of the Pacific Ocean, where climates are mild and very moist. These forests grow within 37 to 74 miles (60 to 120 kilometers) of the coast and receive enough rainfall to be called "temperate rainforests." Massive evergreen conifers are important for ecosystem functioning, wildlife, recreation, and local and regional economies. Giant trees include Douglas fir, western hemlock, western red cedar, Sitka spruce, and coast redwood. These trees occur in diverse tree communities with multiple dominant species and are generally very large and very long-lived. Some of these species, notably the famous redwood (*Sequoia sempervirens*), include the oldest trees in the world. These moist, mild forests with very long intervals between natural fire events (several hundred years) are unusual on the planet, as most such ecosystems are dominated by deciduous tree species, not conifers. The humid, moist conditions result in exuberant growth of ferns as well as mosses and lichen on tree trunks and branches, leading to names like "Hall of Mosses" and "Elvin Forests" for these temperate rainforest ecosystems.

Major environmental changes over the landscape ("gradients") that explain differences between forest types in the Pacific Northwest track changes in latitude, distance from the coast, and elevation. The two most important environmental factors explaining differences in forest types are soil moisture and temperature. In coastal regions, temperatures are mild and there are many cloudy days. Differences between day and night temperatures, and differences in temperatures from one day to the next, are mild, making these climates very moist and stable. Most of the precipitation occurs from October through March. As the landscape position moves from south to north, increasing in latitude, the temperature generally decreases and the precipitation increases.

The main forest type in the Pacific Northwest is dominated by Douglas fir and western hemlock and often includes western red cedar and other species of primarily coniferous trees. Forests of these massive trees range from sea level up to 2,296–3,280 feet (700–1,000 meters), with Douglas fir trees that may be 1,000 years old.

Rocky Mountain Coniferous Forests

The eastern slope of the Rocky Mountains rises dramatically from the relatively flat expanse of the Great Plains, reaching peaks of 14,436 feet (4,400 meters) in elevation. Not surprisingly, the elevation where an ecosystem occurs most strongly determines what type of ecosystem it will be. As elevation increases, temperatures decline and precipitation generally increases, as do solar radiation, wind, and snow depth. Although there are important differences in temperature and moisture between, for example, north- and south-facing slopes at a given elevation, lower-elevation climates are generally warmer and drier, whereas higher-elevations sites are colder and moister. The lowest elevations of the Rockies are dominated by grasslands and mixed oak deciduous forests, whereas the highest elevations are evergreen boreal forests grading into alpine meadows. Temperate coniferous forests dominate large expanses of the midslopes of these mountains. Woodlands comprised of juniper and pinyon pine tree species occur upslope from the oak forests, and heavy grazing by livestock has increased the abundance of juniper trees over this landscape by eating grass and reducing the grass-competitive shading of juniper seedlings. Juniper-pinyon grade upslope to Ponderosa pine woodlands and forests and then further upslope to forests dominated by Douglas fir, an important tree for use as a Christmas tree in the United States, that grow at about 5,410 to 8,900 feet (1,650 to 2,700 meters) in elevation. The highest elevation temperate deciduous forest is dominated by lodgepole pine, with grades into spruce-fir boreal forests at higher elevations.

Eastern Hemlock Forests of Northeastern North America

Eastern hemlocks are "climax" species in forests, meaning that they do not require a disturbance event such as fire or the death of canopy trees creating a light gap in order to successfully reproduce and replace themselves in the forest canopy. In fact, hemlocks produce such dense shade that few if any other tree species are able to successfully regenerate under hemlock stands; instead, seedlings of other tree species generally die within a few months or years of germinating. Hence, despite having a dense canopy of evergreen needles, there is little plant ground cover or undergrowth in dark hemlock stands.

Hemlock stands do, however, have some unique hydrological features that enable them to form temporary freshwater "vernal pools" in the spring that are essential for populations of amphibians including frogs, toads, and salamanders. Because these pools are temporary, they do not support populations of fish, which could otherwise eat tadpoles and other offspring that inhabit the pool. And dense, evergreen hemlock forests provide essential wildlife habitat in the winter months, helping to protect white-tailed deer, wild turkey, and other animals from cold winter winds and snow. Hemlock trees are very slow growing and extremely long-lived, but unfortunately they are currently being decimated by an introduced species called the hemlock woolly adelgid.

Southeastern Coastal Plain Coniferous Woodlands and Forests

The Coastal Plain is a flat, sandy region of the southeastern United States with a large amount of biodiversity and also one of the fastest-growing human populations in the nation. Three main types of temperate coniferous forests occur in the southeastern Coastal Plain, and differences among them depend largely on soil moisture and associated fire frequency. The first is the longleaf pine forest, which once accounted for 36 million hectares (90 million acres) of the landscape from North Carolina to Florida to eastern Texas, and occurs in various vegetative structures from quite wet to very dry soil moisture conditions. The vast majority of the former longleaf pine system has been converted to agriculture and forest plantations, making natural longleaf pine forests one of the most endangered ecosystems in the United States, with only 2 percent still remaining.

Wet longleaf pine woodlands contain some of the highest numbers of species ever recorded at small spatial scales, including wild orchids and carnivorous plants like sundews, pitcher plants, and Venus flytraps. Like other coniferous forest systems, longleaf pine forests and savannas depend on regular fire events to remain healthy. The second type of temperate coniferous forest in the southeast is very wet bald cypress and pond cypress swamp forests, or "bayous," which can have standing water several meters deep. These wet forests are less prone to fire, but during especially dry periods or when humans alter groundwater aquifers, these forests, too, can burn. Unlike most other conifer trees, cypress trees lose their needles in the autumn, and these swamp forests are important for freshwater fisheries, including crayfish and shrimp. The third type of temperate coniferous forest in the southeast is the sand pine woodland, or "scrub," that occurs on very deep, very infertile, and extremely dry white sands in central Florida. Many of these woodlands have been converted to citrus plantations. Sand pine scrub woodlands are the habitat of rare animal species, such as the gopher tortoise and scrub jay, as well as 12 types of endangered plant species, including wildflowers.

Fire in Coniferous Forest Ecosystems

Fire, set either by natural lightening strikes or by human land managers, is an important feature of almost all coniferous forests, with the exception of those dominated by eastern hemlock. Evidence of past fires is found in fire scars on tree trunks and in charcoal layers in soils. There is a major difference between ground- or surface-level fires, which burn less intensely and generally do not kill large trees, and crown or canopy fires, which do. When fires are suppressed by people, these coniferous forests can develop into deciduous hardwood-dominated stands or change into overly crowded tree conditions that burn much more intensely with more extreme damage than natural fires when a burn finally does occur. The latter happened in the Yellowstone fire event of 1988, which formed the largest recorded fire in the history of Yellowstone National Park.

Fires had been suppressed for decades, and normally widely spaced trees became crowded by smaller, younger trees that would have otherwise been burned as seedlings when ground-level fires occurred naturally. Instead, due to fire elimination, these younger trees were allowed to grow taller and when fires did ignite, they formed a "ladder effect" and the flames burned up into the

The Hall of Mosses in Olympic National Park's Hoh Rainforest in Washington. The humid, moist conditions encourage the exuberant growth of ferns, mosses, and lichen on tree trunks and branches. The climate in the temperate evergreen forests of the Pacific Northwest is mild and very moist; they even receive enough rainfall to be called temperate rainforests. There are long intervals between fire events, making these forests very unique. (Wikimedia/Konrad Roeder)

large trees, resulting in a canopy fire. Thus land managers now set controlled, or "prescribed," fires to mimic natural fire frequencies and promote the health of these ecosystems.

In the wet and foggy Pacific Northwest, forests have fire intervals of 500–800 years as evidenced by fire scars and soil charcoal. When a fire does occur, it is a catastrophic fire burning up the giant canopy trees as fuel. In contrast, other temperate coniferous forests typically have more frequent and less severe fire events. Douglas fir forests of the Rocky Mountains have typical fire frequencies of 50–100 years, while the more arid Ponderosa pine forests burn more frequently every 40 years or so. Natural fire frequencies can be as often as every few years in moist longleaf pine forests and woodlands of the southeastern Coastal Plain. Florida sand pine scrub systems are so dry that plant cover is extremely limited and fuel that forms from dead plants and leaves takes a long time to accumulate. Hence, fire intervals in these systems may be 20–50 years, whereas nearby moist longleaf pine woodlands could burn every few years. The cones containing the seeds of sand pine are serotinous, meaning they require the high heat of being burned in a fire before they can open and release the seeds to produce the next generation of trees.

"Coniferous trees are generally more tolerant of drought and fire than are deciduous trees, although some species do inhabit moist forest ecosystems."

Frequencies and sizes of natural and accidental arson-caused fires have been going up in the United States over the last several decades, largely as a result of earlier spring snowmelt and higher spring and summer temperatures. For example, the National Interagency Fire Center finds that the size of the area burned by an average fire in the United States was about 47 acres between 1987 and 2000, and increased to 100 acres during 2001–07.

Insect Pests and Disease Pathogens

Like fire, climate is another important controller of the insect pests and pathogens that shape the health of all ecosystems, including temperate coniferous forests. With warming winter temperatures, disease pathogens and insects like pine bark beetles as well as human pets' outdoor flea populations are better able to survive and even increase their distribution over regions like mountaintops that were formerly too cold. This is occurring now as the mountain pine bark beetle that is causing over 90 percent tree mortality rates in western lodgepole pine stands has been able to cross over the Rocky Mountain Continental Divide and now is killing jack pine trees in eastern North America. The near synchronous death of these standing trees forms fuel for extreme canopy-level fires, further hastening change in these forest ecosystems and having major ecological and economic consequences. Longer growing seasons also enable insects and pathogens to produce more generations per year, which exacerbates negative effects on tree health. For example, the hemlock woolly adelgid beetle, which invaded the east coast of North America from Asia, is wiping out trees and stands of eastern hemlock. In New England, this insect generally produces two generations per year and takes 15 years or so to kill the trees. In the warmer climates of the southern Appalachian Mountains, this insect can produce three generations per year and typically kills mature trees in just four years.

JACQUELINE E. MOHAN

Temperate Deciduous Forests

Temperate deciduous forests exist in the midlatitude regions around the globe, including North America, Europe, Asia, and southern South America. Four seasons characterized by cold, relatively dry winters and warm, wetter summers are found in these forests. The cycle of seasons influences the shedding of leaves by deciduous tree species. These forests provide important resources to civilizations around the world, but many modern activities and urbanization threaten to decrease the extent of these forests. Conservation efforts focusing on increasing management and awareness of temperate deciduous forests are working around the world to preserve these ecosystems in the face of global changes in climate and land use.

Temperate deciduous forest regions are characterized by relatively warm summers and cold winters with growing seasons ranging from 140 to 200 days with four to six frost-free months. Annual temperatures in temperate deciduous forests range from minus 22 degrees F to 86 degrees F (minus 30 degrees C to 30 degrees C) with an average of 50 degrees F (10 degrees C), and precipitation ranges from 29 to 59 inches (750 to 1,500 millimeters) per year and mostly occurs during warm months. Temperate forest species typically thrive on neutral to slightly basic soils, and soils are typically very fertile due to the continuous decomposition of newly deposited litter such as falling leaves and woody debris. The four distinct seasons that occur in all temperate deciduous forests trigger the characteristic shedding of leaves by deciduous, broadleaved species and subsequent regrowth of leaves during spring. The cyclical adaptation of shedding leaves during the cold, drier periods allows plants to survive harsh winters in a rather dormant stage and reanimate during more favorable spring conditions.

Distinguishing Characteristics

The four distinct seasons and the changes that broadleaved trees experience during the seasonal cycle are the most defining characteristics for the temperate deciduous forests. The changes experienced physiologically by trees are unique across the seasons and represent a significant adaptation to the harsher conditions experienced by trees. For comparison, the trees in warm, wet climates such as the tropics never experience a dormant period in which leaves are shed. The shedding of leaves by trees in the autumn decreases water loss due to transpiration, a side effect of photosynthesis. When leaves are shed, water loss is decreased and photosynthesis is in turn decreased. Since new carbohydrates are no longer produced by photosynthesis, trees enter a stage of torpor, or dormancy, in which most physiological functioning ceases. Growth of woody material stops as well.

Leaves experience a change of color in autumn triggered by a change in the allocation of sugars and chemical components and a withdrawal of chlorophyll from the leaves. Trees are able to draw nutrients from the leaves back into woody tissues for storage over winter. For example, nitrogen, an often limiting nutrient in many forests, can be translocated, or transferred, from the leaf back into the tree prior to leaf shedding. Once leaves are shed, trees enter dormancy for the winter, and transpiration and photosynthesis cease for the most part. Thick exterior bark allows trees to withstand cold winter temperatures. Leaves and flowers reemerge in the spring when growing conditions become more favorable for growth, and photosynthesis resumes. In the spring, large conducting elements comprising the tree's vascular system (i.e., phloem and xylem) reestablish as well, allowing for transport of water from the roots to the treetops and transfer of sugars and carbohydrates generated by photosynthesis from the leaves to the rest of the tree for use in growth and storage. These conducting elements comprise the annual growth rings of trees, and thus a new, annual ring is added each year to deciduous species as trees exit dormancy and resume growth and photosynthesis each spring.

The most common deciduous tree species that exist in temperate forests belong to the oak (*Quercus*), ash (*Fraxinus*), maple (*Acer*), beech (*Fagus*), hickory (*Carya*), and chestnut (*Castanea*) families. Biodiversity is high; along with canopy-dominant tree species, understory shrubs, herbs, and flowering plants are also important for these ecosystems. Closed-canopy forests typically contain shade-tolerant tree species reestablishing in the forest understory. When forest stands reach maturity, canopies are referred to as "closed," signifying a relative closure in leaf space and a decrease, but not total elimination, of the sunlight penetrating the canopy. There are also several different kinds of plants that exist in the shady conditions of the forest floor where only small amounts of sunlight penetrate, such as shrubs, flowers, and mosses.

Global Distribution and Structure

In terms of global distribution, temperate deciduous forests are found in the eastern half of North America in latitudes between 35 and 48 degrees north and in central Europe in latitudes between 45 and 60 degrees north. Deciduous forests reach farther north in Europe than in other regions of the Northern Hemisphere as a result of the Atlantic conveyor belt carrying warmer waters from the mid-Atlantic and Caribbean up around Great Britain, resulting in warmer, temperate climates in higher, northern latitudes. The deciduous forests of Asia are located in southwest Russia, Japan, Korea, and eastern China. In South America, southern Chile and the mid-coastal regions of Paraguay contain deciduous forests. Also in the Southern Hemisphere, New Zealand and southeastern Australia have extensive deciduous forests. While all these regions are characterized by four distinct seasons over the year, there are unique characteristics of each region attributed to changes in soil texture, nutrients, topography, and common species of flora and fauna. For example, the deciduous forests of the eastern United States contain interspersed coniferous trees, especially in northern regions and areas of high elevation such as the Appalachian Mountains, with species indigenous only to North America.

"The four distinct seasons and the changes that broad-leaved trees experience during the seasonal cycle are the most defining characteristics for the temperate deciduous forests."

Temperate deciduous forests cover large areas in the Northern and, to a lesser degree, Southern Hemispheres. Further categorization of these forest communities can be done by assessing the repeating units of dominant tree species in a stand. For example, some of the most common forest communities in the United States can be described by eight categories: mixed mesophytic, Appalachian oak, hemlock-white pine-northern hardwoods, oak-hickory, maple-basswood, beech-maple, oak-pine, and southern pine. Mesophytic signifies an environment receiving a moderate amount of moisture.

The forest structure in temperate deciduous regions is separated into five zones based on vegetation height. The tree stratum zone comprises the first zone and ranges in height between 60 and 100 feet (10 to 30 meters). The tree stratum zone contains trees such as oak, beech, maple, chestnut hickory, elm, basswood, linden, walnut, and sweet gum trees. Smaller tree and juvenile saplings are in the second zone. The shrub zone, or third zone, consists of short shrubs and flowering plants such as rhododendrons, primrose, bluebells, painted trillium, azaleas, mountain laurel, and huckleberries. The herb zone is the fourth zone, and it contains plants close to the ground such as herbal plants like ferns and perennial forbs that blossom in spring. The ground zone contains low-to-the-ground growth forms such as lichens, true mosses, and club mosses. The canopy in most deciduous temperate forests contains openings and gaps created by fallen trees and low tree densities that allow some light to penetrate to the forest floor, thus allowing for a great diversity in flora, which in turn supports a great diversity in fauna.

Animals found in temperate deciduous forests must adapt to the seasonal cycles, so many hibernate or migrate when freezing temperatures arrive. Some animals are adapted to the harsher winter months and can withstand the high risk of predation and cold temperatures during the winter. For example, black bears are adapted to temperate conditions. They have sharp claws for climbing trees and, as omnivores, they eat other vegetation and prey on other forest creatures. While their thick fur protects them from the cold temperatures of winter, they do hibernate during the time of hardest freeze to avoid food shortage. Other common forest animals are deer, gray squirrels, opossums, mice, and raccoons. Amphibians such as salamanders and frogs, reptiles such as snakes and turtles, and a multitude of insect species including beetles and mosquitoes also contribute to the diversity of fauna found in deciduous temperate forests. A huge variety of bird species including robins, eagles, cardinals, chickadees, broad-winged hawks, various owls and woodpeckers, hummingbirds, and wild turkeys also are found in these forests.

Timber and Resource Management

Temperate deciduous forests have acted as important sources of timber historically and continue to provide important resources presently; thus, these areas have often been settled and used for resources. Less than one-quarter of old-growth deciduous forests still exist globally due not only to timber harvesting but also due to land clearing for agriculture. For example, many European colonists in the northeastern United States cleared the deciduous forests for agriculture. Upon finding more arable, or easier to plow, soils toward the western United States, colonists allowed many former agricultural fields in the northeast to return to forested states. Prior to European arrival, forests in the northeastern United States contained higher species diversity and greater natural variation in forest stands, which were influenced by natural and human disturbances such as hurricanes and clearing of small areas for native villages. This led to trees of varying age, density, size, and species across a wide range of forested sites.

With the arrival of Europeans, forests diversity decreased and tree ages and densities became more homogenous, or had decreases in variety, due in part to the influences of larger-scale hunting, trapping, and clear-cutting by Europeans beginning around 1820. Through the passage of about 30 years, European agricultural efforts in New England notably declined and many of the fields returned to forested states. Early forest regrowth was primarily represented by white pine, but conifers have since been replaced by the diversity of broadleaf species that currently comprise eastern forests. Forest regrowth in the northeast United States has continued for the most part, and the deciduous forests of New England represent secondary succession forest growth, or rather the forest has regrown after the original forests were cleared for timber or agriculture. In another example, forests in China have been harvested for over 4,000 years, thus most of the forests existing in the country are human-made.

Secondary succession and current forest management of temperate deciduous forests create some ecosystem problems; because most trees are of the same age, or even-aged, stands are denser than historic stands, and species are often selected for their commercial value instead of to maintain historic biodiversity with a wide variety of species. To contrast current temperate deciduous forests, it is important to recognize some of the valuable characteristics of old-growth deciduous temperate forests. Old-growth forests are more complex forest structures characterized by trees representing a variety of ages, large fallen trees and woody debris on the forest floor, and multiple layers in the canopy representing differences in tree size. Forest regrowth typically decreases the complexity of forest communities by decreasing the variety of age, layers, species, and habitats available.

The ecosystem services provided by temperate deciduous forests, or the commercial or aesthetic value that is placed on nature, are not limited to timber products such as wood or paper. Forests provide oxygen as a by-product of photosynthesis, decrease atmospheric carbon dioxide through sequestration, provide food including mushrooms and game, create recreational spaces and natural beauty, and purify the water that many communities around the globe rely upon. The change in the autumn colors of foliage in deciduous trees leads to great beauty. This can be recognized among the wide array of tours and recreational opportunities in the fall in New England. Tourists flock to these regions to view the autumn colors among the maples, and Arcadian National Park in Maine is flooded with guests enjoying the reds, yellows, and oranges that flame through the leaves during the fall. Not only does this represent a great aesthetic value, but also an important economic aspect for the region.

Forests are often managed to maintain high densities of valuable tree species such as red oak, and stands are often even-aged because most trees within a stand would have been established at the same time following forest clear-cutting. High densities of trees make stands more vulnerable to the negative impacts of drought as competition between neighboring trees creates more demand for limited water resources. This makes trees more vulnerable to drought-related mortality or secondary stresses such as insect or disease pathogen attacks. An example of this can be observed in the oak-hickory stands of northwest Arkansas. An oak decline event in the early 20th century that resulted in high mortality of red oak species was linked to increased vulnerability of trees to insect infestation following a cycle of droughts. The mortality event was also related to the unnaturally high densities of red oak species that existed in the area. As a commercially valuable species, the even-aged stands were managed for high timber production, and these high densities increased moisture stress during drought. The increased stress may have made trees more vulnerable to insect attacks as more resources had to be allocated to accessing water (i.e., roots) than to defensive chemicals. Decreased diversity of species also poses a threat to forest ecosystems.

For example, eastern U.S. forests were dominated by chestnut trees in the early 1900s until the chestnut blight, an introduced fungal pathogen, mortally impacted this key species and left the canopy in many forest stands open to other species such as maples and sycamores. A pathogen decimating an entire forest species is rare, but as introduced pathogens become more common, it is vital for continued ecosystem function to maintain a wide variety of codominant tree species within forest canopies. The management practices that create even-aged, high-density forest stands with relatively low diversity found in many temperate deciduous forests make these regions more vulnerable to the impacts of disturbances such as hurricanes, windfall events, and disease pathogens.

Black bears are adapted to temperate conditions with sharp claws for climbing trees. As omnivores, they eat vegetation and other forest creatures. Thick fur protects them against cold temperatures. (U.S. Forest Service)

Environmental Challenges and Forest Conservation

Temperate deciduous forest habitats are also threatened by timber harvesting, acid rain, and invasive plants and animals worldwide. Nonnative, invasive species may compete for food and habitats, potentially threatening the native plant and animal species. Because temperate deciduous forests provide rich areas of flora and fauna diversity along with globally important ecosystem services, conservation of these regions is important. Recycling programs and awareness programs sponsored by organizations such as the U.S. Environmental Protection Agency (EPA) and the European Union (EU) European Commission have led to increased recycling rates over the past several decades. Acid rain impacts leaf photosynthesis and soil nutrient balances and has been recognized as a serious problem in many areas such as central Europe. Evidence of leaf bleaching and slowed growth resulting from acid rain deposition in forest stands has been linked to vehicle fossil-fuel emissions through decades of research. Programs in the early 1990s through the EPA and the EU to decrease fossil-fuel emissions of sulfur by-products, which cause sulfuric acid deposition in rain, have helped decrease the impacts of acid rain on forest communities up to 40 percent since the 1970s. Another threat to temperate deciduous forests is land use change. Continued clearing for agriculture and commercial mining threatens forests globally. For example, in Germany, much of the historic deciduous forest land has been strip-mined for metal and minerals, decreasing the range of forests in the region.

The large amount of standing biomass existing in temperate deciduous forests represents a significant contribution to the carbon balance of the Earth. The carbon stored in the woody biomass, roots, and soil of these forests is a large component of the global carbon balance, which is a biogeochemical cycle of exchanges among the biological components, soil, rocks, atmosphere, and water cycles of the Earth. The carbon cycle is vital for recycling and reuse throughout the biosphere and all organisms globally.

Forests in general store up to 86 percent of global carbon stocks. Disturbance of soils along with burning or decay of woody biomass releases this stored carbon back to the atmosphere. Because trees convert carbon dioxide into carbohydrates during photosynthesis, releasing oxygen, they are an important component of global atmospheric carbon dioxide levels. The drawdown of global atmospheric carbon dioxide levels can be observed in deciduous forests during leafing-out in the spring. Given that deciduous forests are most extensive in the Northern Hemisphere due to larger land masses, the Northern Hemisphere spring dominates the drawdown trend, most famously observed by the Keeling curve, which measures global carbon dioxide levels.

Related to changes in global carbon dioxide levels, the recognized increases in global temperatures also threaten forest biomes through a decrease in moisture availability and increases in secondary stress factors such as invasive insects and pathogens. Climate change impacts may also alter the species habitats, composition, and adaptations of forest stands, leading to changes in forest structure or function. For example, the impacts of global climate change have already been noted in hardwood forests in Ohio, where decreases in annual snowfall, important for soil water recharge, and increased temperatures in winters have placed carbohydrate and water constraints on some forest tree species. With global increases in disturbances and climate impacts, conservation efforts in forests are an ever-increasing concern for many U.S. national organizations such as the Sierra Club and Audubon Society and for international organizations such as the Commonwealth Forestry Association.

Alexis S. Reed

Tropical Rainforests

Tropical rainforests are forest biomes with an average rainfall between 67 inches (170 centimeters) and 400 inches (1,016 centimeters). The World Wildlife Fund (WWF) defines them as a type of tropical wet forest, tropical moist broadleaf forest, or lowland equatorial evergreen rainforest. Tropical rainforests are found in the equatorial zone between the Tropics of Cancer and Capricorn, roughly between latitudes 28 degrees north or south of the equator, where the direct angle of the sun rays causes high rates of evaporation. As water vapor evaporates from the forests, it rises, cools, and falls about an eighth-inch per day, giving these forests their name. Average humidity is usually around 77–88 percent. Mean monthly temperatures exceed 64 degrees F (18 degrees C) throughout the entire year. Tropical rainforests are found in five major geographical areas: Central and South America in the Amazon River Basin; Africa in the Zaire Basin; a small amount in West Africa and East Madagascar; in Indo-Malaysia on the west coast of India, Assam, southeast Asia; as well as New Guinea and Queensland, Australia.

There are four main types of rainforest; lowland equatorial evergreen rainforest, moist deciduous and semi-evergreen seasonal forests, montane rainforests, and flooded forests. Lowland equatorial evergreen rainforests have high rainfall throughout the year. They form a belt around the equator of the globe, covering the Amazon Basin of South America, the Congo Basin of central Africa, Indonesia, and New Guinea. Moist deciduous and semi-evergreen seasonal forests have high rainfall overall but have a distinct warm summer wet season and a cooler winter dry season. Some species of trees in this type of rainforest even drop their leaves during the winter dry season to conserve water. These forests make up parts of South America, Central America and the Caribbean, coastal west Africa, India, and Indochina. Montane rainforests are also sometimes called cloud forests because of their almost constant shroud of fog. These forests are found in cooler-climate mountain areas, depending on the latitude, between elevations of about 4,920 and 18,827 feet (1,500 and 3,300 meters). Flooded forests are caused by seasonal flooding of tropical rivers. There are seven types of flooded forests recognized in the Tambopata Reserve in Amazonian Peru alone.

It is estimated that more than half of all plant and animal species are found in this biome and there are likely millions of species yet to be discovered. Many of these species are, or could be, important sources of food and medicine. Tropical rainforests produce about 40 percent of the Earth's oxygen and sequester carbon from the atmosphere, which helps to mediate climate change. Due to heavy logging and clearing for agriculture throughout the 20th century, tropical rainforests now cover less than 6 percent of the Earth's land surface. They are threatened globally by deforestation and fragmentation for agriculture and urbanization, mining, drilling, and climate change.

Layers of the Tropical Rainforest

Tropical rainforests are typically made up of four layers or zones typified by the structure of the vegetation. These different zones form habitats for a variety of animal species. Though this stratification is not always clear-cut, these four layers are generally found at different heights in the forest: starting at the forest floor, the lower canopy, the upper canopy, and ending at the emergent layer.

The forest floor of the tropical rainforest is completely shaded from the dense forest canopy except for clearings, formed by fallen trees, where the sun can reach through the opening. Only about 1 percent of the sunlight that strikes the upper canopy reaches the forest floor. Except for dense growth near riverbanks, swamps, and clearings, most areas are so shaded that very little can grow, leaving a relatively open forest floor that allows for easy movement of large animals. Litter falls to the ground from the trees above and then is broken down by decomposers like termites, earthworms, and fungi. The constant heat and humidity accelerates the decomposition process. This leads to a very short nutrient cycle wherein organic matter is consumed almost as quickly as it is deposited, leaving only a thin layer of poor quality soil.

Also called the lower canopy, the understory is composed of 60-foot trees, the trunks of upper canopy trees, shrubs, and plants. Epiphytes, mosses, and lichens grow on the trunks and branches of trees. This area is dense and humid, with very little air movement. This area receives slightly more light than the forest floor, 5 percent of the light that strikes the upper canopy, but is mostly still in constant shade.

The canopy, found above the understory, is also called the upper canopy and it is made up of 60- to 130-foot trees. The trees in this layer compete and grow taller to reach the sunlight and block light access to the layers below. They form the roof of the rainforest. Leaves on these tallest trees often have "drip spouts" that allow rainwater to run off them to keep them dry and light and to prevent mold growth. Most of the animals in the rainforest live in this layer, where there is lots of food available from fruiting trees. Many animals that live in the upper canopy never have to go to the ground, getting most of their water from their food or rain pooled in the trees.

The emergent layer is the only one truly unique to the tropical rainforest, as the other layers are found in temperate rainforests as well. This layer is made of widely spaced trees that grow above the upper canopy. These trees, called emergents, are 100 to 240 feet (30 to 73 meters) tall with umbrella-shaped crowns. Emergent trees are exposed to extreme drying winds and heat. Many have adapted small pointed leaves that they sometimes drop during the dry season in monsoon areas to conserve water. Due to the shallow soil, these giant trees have shallow roots and must grow large buttress structures that can grow up to 35 feet wide for support. The emergent layer is home to animals like the crowned eagle (*Stephanoaetus coronatus*), king colobus (*Colobus polykomos*), and large flying fox (*Pteropus vampyrus*).

"It is estimated that more than half of all plant and animal species are found in this biome and there are likely millions of species yet to be discovered."

Disturbance and Succession

Succession is the ecological process that changes biotic community structure over time. This generally trends toward a more stable, diverse community structure after the initial disturbance. Disturbances can be a natural phenomenon or human-caused. Natural phenomena include hurricanes, volcanoes, river movements, or fallen trees. Such natural disturbances are well documented in the fossil record to encourage speciation and endemism. Fallen trees, often caused by windstorms, are especially important disturbances in tropical rainforests. They open clearings in the dense canopy, allowing light to reach the forest floor where seedlings are waiting to spring into growth to take advantage of the gaps. These saplings might otherwise never be able to establish themselves and mature. About 70 percent of seedlings depend on such gaps for germination or growth past seedling size.

Soil Ecology

As previously mentioned, soil quality in tropical rainforests is typically poor due to the short nutrient cycle of the ecosystem. High rates of precipitation also contribute to the leaching or loss of nutrients from the system. Soil types throughout rainforests are, however, variable and some areas are very fertile. Each area's soil type is influenced by that region's climate, vegetation, topographic position, parent material, and age. There are two main classes of soil: ultisols and oxisols. Ultisols are defined as well-weathered, acidic red clay soils. They have low levels of calcium and potassium. Their high clay content helps them to retain water. Oxisols are also acidic, reddish soils but are old and highly weathered and leached because of good drainage, leaving them nutrient-poor. The red color in both of these soil types originates from the high heat and moisture environment, which forms oxides of iron and aluminum. These oxides are then insoluble in water and not readily taken up by rainforest plants.

The soils of the eastern and central Amazon rainforests and those of southeast Asia are generally old and mineral-poor. The nutrient flood plains of the western Amazon rainforests in Ecuador and Peru and volcanic areas of Costa Rica, southeast Asia, Africa, and Central America have young, mineral-rich soil. These soil characteristics influence the ecological processes of the regions in question. For example, primary productivity, or wood production, is higher in the western Amazon region than in the eastern Amazon oxisols. Aboveground productivity is closely linked to soil types and mineral contents. While the physical properties of a forest, like disturbance regimes, control tree turnover rates, the chemical properties such as nitrogen and phosphorous levels control the growth rates of trees. Phosphorous, potassium, calcium, and magnesium are all needed by growing trees and generally come from the weathering of rocks. Increased rates of decomposition result from increased phosphorous content in the soils. Soils that are limited in phosphorous content limit the growth of trees and subsequently the ability of a forest to uptake and sequester carbon.

The Amazon River Basin contains more types of plant and animal life than any other biome. The second-most biodiverse rainforest is found in southeast Asia. African tropical rainforests have the lowest variety of all the rainforests. In areas of high diversity, tropical rainforests can harbor between 40 and 100 species in just two and a half acres.

Plant Biodiversity

Seventy percent of all plants in tropical rainforests are trees. There are more kinds of trees in this biome than in any other area. For example, one area in South America was found to have between 100 and 300 tree species in just two and a half acres (one hectare) of forest. Trees in the rainforest are often straight and unbranched due to the need to grow fast and tall to compete for light and space. They tend to have smooth, thin bark because they do not need protection from cold temperatures or to retain water. This smooth bark also deters parasitic epiphytic plants. Due to this similarity in adaptations of rainforest trees, it is often difficult to identify species by their bark and leaves and they can only be differentiated by their flowers, which are often highly specialized to different pollinators.

Many plants are adapted to help them deal with the intense water inputs to the system. They have developed structures like drip tips and grooved leaves with oily coatings that shed water so that it does not weigh them down, which might cause them to break. The trees give off water through their pores, called stomata, which further increases the humidity of these forests. This transpiration can account for as much as 50 percent of the precipitation in a rainforest. Some upper canopy trees have small, dark green, leathery leaves to reduce water loss in the direct sun but the same trees may produce large leaves lower in the canopy to funnel off water and increase their light-catching surface.

Competition for sunlight and space has also led to many plant adaptations. Plants and trees in the lower canopy have large leaves to absorb as much sunlight as possible. Some leaves even turn to track the sun as its angle changes throughout the day. There are plants that specialize in growing on tall trees in order to gain access to the light. These plants are called epiphytes and derive their water and nutrients from the air and rain as well as from debris that accumulates around them. Epiphytes include orchids and bromeliads.

There are over 2,500 species of vine that grow in the tropical rainforests. Lianas are long-stemmed, woody vines that start out as small shrubs and then send out tendrils to grab onto sapling trees. As the tree grows upward, the liana vine grows with it to reach the sunlight of the canopy. Lianas will even grow from one tree to another to reach the best source of light and can make up about 40 percent of the plant mass in the rainforest canopy. The rattan vine similarly has spikes on its underside that point backward to grab onto saplings. Strangler vines use trees as support and as they grow thicker they can actually strangle their host tree, which will die and rot away, leaving the "hollow" tree form of the vine.

Tall rainforest trees develop wide buttress roots or a system of stilt roots to support them in shallow soil, but these structures have other benefits for the trees as well. Buttress roots aid in the competition for and efficient uptake of nutrients and water. The roots help to collect leaf litter and funnel nutrient-rich rain water as it drains down the trunk and holds it in pools that can be absorbed as needed. The large root structures increase the surface area of the tree, which increases gas exchange. They also help to reduce soil erosion that would wash away needed nutrients.

Throughout tropical rainforests, tree species are not often found in monoculture, or large areas of only one species. This spacing is due to high seed dispersal from fruit-eating animals as well as the highly competitive nature of tree establishment in the forest. This well-spaced biodiversity ultimately helps to prevent the spread of infections among the trees and decreases mass die-offs. Many tree species also have staggered blooming and fruiting periods, which helps to provide a year-round source of food to rainforest animals.

Animal Biodiversity

Tropical rainforests are home to a huge diversity of animal life. Some forests support ungulates like okapi (*Okapia johnstoni*), tapir (*Tapirus* spp.), or Sumatran rhinoceros (*Dicerorhinus sumatrensis*). Others have apes like the western lowland gorilla (*Gorilla gorilla*). There are many predators in rainforests, such as leopards (*Panthera pardus*), pira-

The forest anole Anolis nitens lives in leaf litter under the canopy of the Amazon rainforest. Tropical rainforests host an enormous diversity of animal life, many of which also live in the canopy. (National Science Foundation)

nha, poison dart frogs, ring-tailed coati (*Nasua nasua*), boa constrictor (*Boa constrictor*), and many insects, which make up the largest group of animal species in tropical rainforests. Many species are adapted for a specialist life in the treetops and have bright colors, sharp patterns, loud vocalizations, fruit-heavy diets, and structures to aid in climbing. For example, New World monkeys such as the spider monkey (*Ateles* spp.) all have prehensile tails that help them to climb in their canopy home. Other treetop specialists include the three-toed sloth (*Bradypus* spp.) which may only climb to the forest floor once a month to defecate, the kinkajou (*Potos flavus*), harpy eagle (*Harpia harpyja*), macaws, and other parrots.

Tropical rainforests have existed for hundreds of millions of years. Most that remain today are fragments of forest that covered the Mesozoic era Gondwana supercontinent. When this land mass separated, it resulted in a great loss of amphibian diversity but the consequent drier climate led to the diversification of reptiles. Many amphibians (frogs, toads, salamanders, newts, and wormlike caecilians) and reptiles (snakes, lizards, turtles, tortoises, and crocodiles) are found in current-day tropical rainforests. They form an important part of the rainforest ecosystem, helping to cycle nutrients through the food web. The most common types of amphibian in the rainforest are frogs. More than 1,000 species of frog are found in the Amazon Basin alone. Many rainforest frogs are quite different from temperate frogs. The high humidity of the rainforest allows them to stay moist outside bodies of water and many are adapted to live high in the trees. These species also often lay their eggs in moist leaf litter or puddles of water formed in bromeliads high in the canopy.

Theories for the Origin of Biodiversity

There are two main theories for the origin of the great biodiversity of the tropical rainforests: the interspecific competition hypothesis and the Pleistocene refugia hypothesis. The interspecific competition hypothesis states that in tropical rainforests there is a high density of species with similar niches, or the way that they respond to distributions of resources and competitors within an ecosystem. Since there are limited resources available, if two species share a niche and are therefore directly competing for resources, then one of those species must eventually lose the competition and find a new niche or go extinct. This pressure to find a new niche is called niche partitioning. Niche partitioning can involve utilizing a different habitat, food source, or behavior, such as eating the same food but at a different time of day.

The Pleistocene refugia hypothesis was formulated by Jurgen Haffer in 1969. He theorized that the present diversity was a product of rainforest patches being separated by stretches on nonrainforest vegetation during the last glacial period. These remaining rainforest patches served as refuges for the species of the rainforest and over time the species in each of these patches changed, causing speciation, or the creation of new species. At the end of the glacial period when humidity increased and the rainforest patches reconnected, the new rainforest stretch was populated by a larger diversity of species. This theory has been met with much debate and skepticism based on evidence that places rainforest speciation before the Pleistocene glacial period.

Importance to Humans: Carbon Sequestration, Medicine, and Food

Tropical rainforests provide many important services to humans, including oxygen production, carbon sequestration, and sources of food and medicine. It is estimated that more than half of all plant and animal species are found in tropical rainforests and there are likely millions of species yet to be discovered. Many of these species are, or could be, important sources of food and medicine. Tropical rainforests produce about 40 percent of the Earth's oxygen and sequester carbon from the atmosphere, which helps to mediate climate change.

Carbon flux is the exchange of carbon dioxide between the atmosphere and a reservoir, also known as a sink. The net primary productivity is the amount of carbon that is retained in plant biomass over time. It is also the gross primary productivity minus carbon that is released through autotrophic respiration. This metric is used to measure the amount of carbon in a tropical rainforest. Tropical rainforests are important as a carbon sink because of their above- and belowground biomass, the amount of carbon they store, and the rate of fixation by photosynthesis. Tropical rainforests absorb about 4.8 billion tons of carbon dioxide per year.

One-quarter of all medicines originate from tropical rainforest plants. For example, curare is made from a tropical vine (*Chondrodendron tomentosum*) that is used as an anesthetic and muscle relaxer, and quinine comes from the chinchona tree (*Cinchona* spp.) and is used to treat malaria. More than 1,400 plants are thought to contain properties that could potentially cure cancer. Rainforest frogs have been especially important for pharmaceutical discoveries because of the unique chemicals found in their skin. These chemicals, which the frogs use for defense or healing, have been found to act as powerful painkillers, muscle relaxants, and offer possible cures for cancer.

Many cultivated foods and spices originate from the tropical rainforest biome. Over 250 types of fruit have been harvested from the rainforests, including banana, mango, and papaya as well as yam, coffee, chocolate, macadamia, and sugarcane. In New Guinea alone, 251 edible fruit trees have been found and most of these in the last few decades. These crops are now mostly grown on plantations but much of their genetic diversity is maintained through pollination from wild stock. This introduction of new genes helps the crops to avoid disease and pest damage.

Various Threats to Tropical Rainforests

Most of the original extant of tropical rainforests has been lost due to clearing of forests for logging, mining, drilling, and agriculture. The ecosystem of the forest biome has also been affected by poaching by humans, the introduction of invasive species and diseases, and climate change.

Deposits of precious metals such as gold and diamonds and fossil fuels such as oil and natural gas have instigated forest clearing to allow access to these resources. Such mining and drilling is often given priority over environmental concerns to ensure economic growth in developing countries. For example, after decades of mining, Ghana has only 12 percent of its original tropical rainforest left intact. Mining and drilling activities are also often associated with air and water pollution that contaminate rainforest ecosystems.

Much of the tropical rainforests worldwide have been converted to open farm land. Then thin, infertile soil is quickly leached of all minerals from farming and is quickly eroded by heavy rains. This poor productivity has led to a technique of slash-and-burn agriculture. Once one patch of land has been used up, farmers push farther in to the forest, which is cleared and then burned to add the organic matter to the soil.

The abandoned patch of forest is often completely eroded of soil, leaving only hard red clay behind, making it difficult for new trees to grow. However, this fallow period after agriculture can allow secondary forest to regrow and slowly replenish the soil layer. Fertilizers and pesticides used by farmers to boost the nutrient content of the rainforest soil often run off and contaminate soil and rivers.

The delicate ecosystem of the tropical rainforest biome can be affected by many anthropogenic activities other than forest clearing. Poaching can remove species that are important to the system's functioning. Staggering amounts of wildlife are taken from rainforests every year for food and medicinal uses. In the Brazilian Amazon alone, 9.6 to 23.5 million mammals, birds, and reptiles are harvested annually. It is estimated that this number might be six times higher in Africa due to the demand for meat in poor and rural areas.

Invasive species are plants and animals that, when introduced to a new region, adversely affect the ecosystems therein. Invasive species compete with native species and can bring with them new parasites and diseases. For example, the brown tree snake (*Boiga irregularis*) was accidentally brought to Guam in a ship's cargo and has since extirpated, or caused the local extinction of, most of the native forest vertebrate species. Invasive diseases can be detrimental to tropical rainforest ecosystems as well. Such a disease is caused by the chytrid fungus (*Batrachochytrium dendrobatidis*), which is thought to have caused the extinction of over 100 species of frogs in the past few decades. Frog species in the tropical rainforests are especially hard hit because of the constant moisture and temperatures.

Tropical rainforest ecology is very dependent on climate, with many of its processes driven by the characteristic constant high heat, humidity, and rainfall. Therefore, this biome is very sensitive to changes in precipitation and temperature regimes such as those that might accompany global climate change. The climate in these equatorial forests is controlled by bands of clouds called the intertropical convergence zone (ITCZ) that forms near the equator. This zone is created by the convergence of trade winds from the Northern and Southern Hemispheres. Changes in the position of this convergence zone throughout the year cause slight seasonality in tropical rainforests, leading to wet and dry seasons in some forests.

Some regions have already experienced strong warming trends with an increase in average temperature of almost half a degree Fahrenheit (0.26 degree C). These regions have also experienced a decrease in rates of precipitation. For example, tropical rainforests in Asia have suffered from increased dry season intensity, with longer periods of drought conditions. However, there has thus far been little change in the tropical rainforests of Amazonia. El Niño Southern Oscillation events also drive rainforest climate variability between years. These events are likely to increase in duration and intensity with anthropogenic warming. These altered climate regimes may cause stress and increased mortality of trees within the tropical rainforest biome.

Hannah Bement

Tropical Seasonal Forests

When most people think of a tropical forest they envision a jungle that is hot, humid, dark, noisy, and filled with strange, possibly dangerous animals and plants. In fact, many different types of forests can be found in zones surrounding the equator, and these forests vary in their physical characteristics (average rainfall and soil types) and biological characteristics (plant and animal species). Although all forest types exhibit some degree of temporal variation, tropical rainforests are characterized by less severe cyclic variations compared to seasonal forests (Table 1). In tropical seasonal forests (Figure 1), temporal patterns of temperature and rainfall, as well as plant and animal activities, are correlated within and between years, requiring organisms to shift their responses to changing conditions, to escape harsh conditions, or to minimize potential threats to survival and reproduction with intermediate traits.

For example, the root systems of plants inhabiting seasonal environments must have the capacity to regulate water and nutrient absorption in both wet and dry conditions. Animals also must accommodate to seasonal changes since resources required for survival and reproduction, particularly food and mating sites, are not continuously available. If a population of a given species is sufficiently diverse, evolutionary processes are expected to favor traits minimizing negative consequences to life history (patterns of mortality and reproduction) occasioned by variability in resource dispersion and quality over time and space.

Scientists utilize a variety of research methods to evaluate the nature of variability in tropical seasonal forests and the effects of these variations on plant and animal populations. One of these techniques utilizes a multiyear series of rainfall data yielding an unbroken sequence of points, each representing total monthly rainfall. A quantitative test, time-series analysis, is used to produce a model summarizing past, current, and estimated future rainfall events. This statistical approach describes features of the rainfall series that may be similar within and between years and that may serve as predictable environmental stimuli eliciting compensatory physiological and behavioral responses in organisms.

Thus, individuals in populations located in tropical seasonal regimes survive or die depending upon the success or failure of their phenotypic traits. In the real world, however, reoccurring cycles of rainfall or other environmental stimuli will not be perfectly correlated over time within and between cycles, exposing organisms to some degree of unpredictability and risk. Since the response of plants and animals inhabiting tropical seasonal forests will not reliably fit many challenges presented by changing conditions, the survival and reproductive success of individuals in populations will depend upon phenotypic flexibility. In tropical seasonal forests, then, developmental, physiological, and behavioral features are expected to represent "best-of-a-bad-job," rather than optimal, responses.

Forests, their canopies, and other characteristics are fundamental to the processes, functions, stability, maintenance, and persistence of ecosystems. Locally and globally, the physical and biological elements of forests filter and cool the atmosphere, absorb and deflect sound and light, provide refuge for animals and plants, and buffer the potentially damaging effects of temperature (desiccation), rainfall (erosion), and wind (tornadoes may destroy forests and extirpate species). Detritus (falling leaves) and decomposing trees produce humus, the primary building block of soil. Trees also produce large quantities of oxygen and absorb carbon dioxide, processes critical for regulation of the negative consequences of global climate and other anthropogenic effects.

The varieties of tree species grouped into different tropical seasonal forest types (habitats) are determined by the range of local (food) and global (climate) conditions in given environments that change across time and space. Local (niche) and global factors responsible for habitat characteristics may be classified as physical (abiotic) or biological (biotic). Abiotic factors include all inorganic properties of forest landscapes such as altitude, geological and geographic formations, and waterways. Biotic factors include all organic features of forest landscapes, for example, distribution, abundance, type, and composition of plant and animal species.

Central American Tropical Dry Forests

The tropical dry forests of Central America can be used as prototypes for tropical seasonal forests worldwide. These landscapes are characterized by clearly defined wet and dry seasons of about six months duration each and by two easily distinguishable habitat types, deciduous and riparian (riverine). Gordon Frankie reported that Costa Rican riparian and deciduous forests exhibit flower and fruit activity primarily during the dry season, November through April. In the deciduous forest, leaf fall is synchronized for most trees during the early to mid-dry season. In both habitats, soil, altitudinal gradients, and local disturbances (e.g., tree-fall gaps) are important determinants of plant growth, distribution, and abundance patterns. Climate, particularly rainfall, is the most significant predictor of seasonality.

Most trees in the riparian forest retain their leaves throughout the year, displaying a phenological (tree activity) pattern similar to wet forest sites in Costa Rica. Riparian habitat, with higher humidity and greater proportion of evergreen vegetation, is most likely characterized by a higher level of primary productivity compared to deciduous habitat, though few empirical studies exist to test this assumption. Riparian habitat is also likely to exhibit greater complexity and stability resulting in greater resilience when confronted with environmental perturbations (for example, drought and hurricanes).

Fruiting Patterns

For more than 30 years, Theodore Fleming, a biologist at the University of Miami, has conducted research on plant-animal interactions in many Latin American tropical seasonal forests, including those in Costa Rica. During that time, he has served as a faculty member and administrator with the Organization for Tropical Studies (OTS), an internationally recognized consortium of universities in the United States and Costa Rica whose primary goals are to facilitate scientific initiatives in tropical ecosystems and to train future tropical ecologists. Comparing fruiting patterns over time and space in New World (Central and South America) and Old World (Asia and Africa) tropics, Fleming and his colleagues demonstrated that seasonal unpredictability of fruiting patterns is greater in Old World forests. This finding has important implications for differential physical (soil composition and temperature) and biological (species composition, abundance, survival, and reproductive strategies) features. The differential environmental patterns documented by Fleming's group are expected to have significant consequences for patterns of mortality and survivorship of individuals, populations, and species inhabiting tropical forests west and east of the Atlantic Ocean.

The presence of temporal variation in tropical seasonal forests is generally problematic from the organisms' point of view, since their biological processes, from molecular and cellular to phenotypic levels, must be sufficiently flexible to endure both wet and dry conditions. Although plants and animals experience disadvantages associated with inhabiting tropical seasonal forests, a major advantage is that time-varying patterns associated with seasonality yield environmental predictability, providing clear cues and signals about changing conditions (for example, changes in sunlight, humidity, or rainfall patterns). The discrete stimuli signaling wet and dry seasons in time-varying environments modify physiological and behavioral processes of organisms, allowing adjustments to the

Table 1. What is a tropical seasonal forest?

Component	Definition
Tree	A long-lived woody plant occurring in a variety of types with a minimum height of 5 meters and distinctive characteristics (such as leaves and average circumference and height)
Forest	Land with tree-crown (canopy) layer greater than 10 percent in an area greater than 0.5 hectare
Natural forest	Forest composed of indigenous (native) trees that have not been deliberately planted or replanted
Seasonal forest	Forest in changing environments (for example, summer, fall, winter, spring, or wet season, dry season)
Tropics	The tropics include regions between 20 degrees north and 20 degrees south of the equator with average temperatures varying more over the short-term than average monthly temperatures vary over the long-term. Sea-level temperatures in the tropical zone are usually greater than 20 degrees C (60 degrees F), and forests in the tropical zone are characterized by variations in rainfall.
Tropical seasonal forest	Tropical seasonal forest environments are distributed worldwide in North America, Central and South America (Latin America), Asia, Africa, and Australia. They are generally found between 10 degrees north and 10 degrees south of the equator and differ from other tropical forests by rainfall patterns.

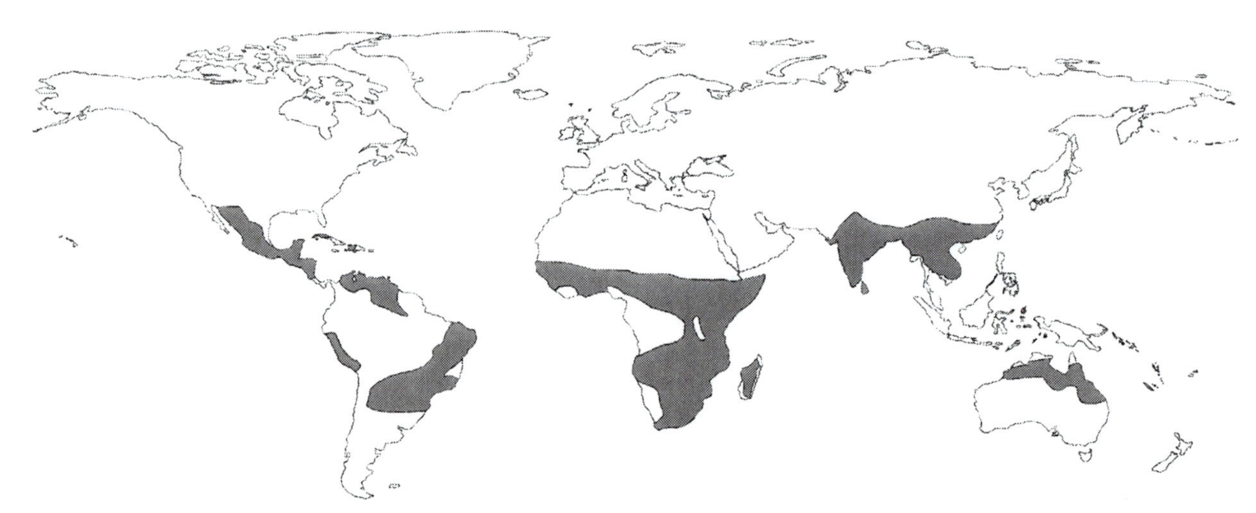

Figure 1. Worldwide distribution of the seasonal tropical forest biome (shaded areas). The geographic ranges of these forests have decreased significantly over the past 100 years, due in particular to anthropogenic effects such as habitat destruction and fragmentation, slash-and-burn or "milpa" agricultural practices, and climate change. (© Clara B. Jones)

stress of different temporal features (for example, abundance and distribution of nutrient and water resources).

Ecologists studying plants and animals inhabiting tropical seasonal forests measure patterns of mortality to evaluate the success of organisms' responses to temporally varying regimes. On the whole, these studies show that many organisms display a trade-off between adult mortality and mortality of immature age-sex classes, and adult survivorship is commonly favored in seasonal forests if survival and reproduction exhibit certain features (extended life span). Ecological studies have shown that this life-history trade-off results from uncertainty and risk associated with reproducing in seasonally changing habitats in which, compared to wet tropical forests, cues and signals predicting the initiation and termination of seasons are highly variable. These challenges may stress developmental, physiological, and behavioral processes influencing lifetime reproductive success of individuals in a population. Statistical analyses of time-varying features are predictable on average, but the onset, duration, and termination of temporal patterns within and between years are usually difficult for the perceptual mechanisms of organisms to assess. Ambiguity and error resulting from between-cycle variability usually leads to significant mortality of one or more age-sex classes in a population.

Modes of Reproduction

All other things being equal, plant and animal populations inhabiting seasonal environments may respond to cycle variability with one of six modes of reproduction, and it is important to remember that each life-history strategy entails both costs and benefits for the organisms displaying it. One mode is characterized by brood parasitism whereby some bird species in tropical seasonal forests are parasitized by other species of birds that lay clutches in nests of the parasitized (host) species. The parasitic species destroys some proportion of the host's clutch, resulting in partial or complete loss of the host's seasonal reproductive investment. Parasite-host associations are highly specialized with respect to the species involved, and parasites gain an advantage over their hosts by laying their eggs prior to hatching of host eggs. Though this life-history pattern appears to be a reproductive dead-end for hosts, the lifetime reproductive benefits to hosts must have been higher than alternative patterns available over evolutionary time. In New World (neotropical) tropical seasonal forests, ground cuckoos (*Tapera naevia*, *Dromococcyx pavoninus*, *D. phasianellus*) are common brood parasites.

A second reproductive pattern is displayed when plants and animals specialize to the most common type or subtype of tropical seasonal forests, such as riparian or deciduous habitat. Specialists are usually characterized by a diet limited in food types and a tolerance for a narrow range of environmental conditions. Of approximately 230 primate species, my research suggests that 18 species are extremely specialized. Of these 18, however, most are exclusive to rainforests, two prefer wet forests (*Hylobates hoolock*, *H. syndactylus*), while the geographic distributions of five species are limited to or include large areas of seasonal forests (*Propithecus diadema*, *Nasalis larvatus*, *Presbytis pileata*, *P. potenziani*, *Gorilla gorilla*). It is interesting to note that the diets of the previous seven species include significant proportions of leaves, an observation worthy of further study since a folivorous diet is often thought to buffer primates from environmental stressors because, all other things being equal, leaves are continuously available. The neotropical plant *Tabebuia chrysantha* (araguaney) displays a life-history mode specialized to optimize lifetime reproductive success in deciduous habitat, and, like other specialist plants, produces flowers and fruits in dry season, the period most beneficial to the survival of new growth. A specialist strategy may be beneficial when conditions favor the specialized traits; however, specialists may be highly vulnerable if environmental stimuli change significantly (changes in rainfall patterns incompatible with a root system's particular functions).

"On one point there is no argument: tropical forests are, indeed, in trouble."

In a third case, some species exhibit a phenotypic compromise or mix of traits advantageous to an intermediate fit across changing conditions (across both deciduous and riparian habitats in Central American environments). These generalist phenotypes are the most common ones displayed by plants and animals in tropical seasonal environments. All other things being equal, generalists thrive in a wide range of environmental conditions and utilize a variety of resources (for example, food items, resting sites). Compared to other life-history modes, generalist strategies may seem intuitively to be least costly to lifetime reproductive success; however, generalists may be spread too thin in time and space to respond successfully to many stressful environmental events, such as climate change or habitat destruction. Recent scientific evidence supports this proposition. The mantled howler monkey (*Alouatta palliata*) inhabits a broad range of forest types, including tropical seasonal forests, throughout the forests of Middle America and the Pacific coast of northern South America. Mantled howlers prefer a diet of flowers, fruit, and new leaves. Kathryn Stoner and other investigators have shown that mature leaves serve not only as a fallback food when preferred items are unavailable to howler monkeys but also as a buffer against environmental stress if distribution, abundance, or quality of preferred plant dispersion changes in time or space. The *Alouatta* genus is extremely successful biogeographically as the most widely distributed neotropical primate genus. However, the most well-known and widely ranging generalist species among mammals is the omnivorous *Homo sapiens*.

A fourth reproductive pattern displayed by some tropical seasonal forest species entails specialization on one of the environment's morphs (for example, dry or wet conditions). A number of these taxa are migrating species such as cockatoos (*Calyptorhynchus* spp.) that move to and from rainforests and tropical seasonal forests to maintain a relatively unchanging exposure to a tolerable range of humidity. This strategy permits the species to escape the negative effects of the least favorable suite of conditions to the more favorable suite of conditions (favorable for breeding or preferred food items).

A fifth life-history strategy observed in tropical seasonal forests is mast fruiting, whereby individuals of a given plant species coordinate flower and fruit production, usually in accordance with the highest availability of pollinators. In the tropical seasonal forests of Central America, *Manilkara* spp. display mast fruiting. A sixth life-history pattern observed by a relatively few tropical seasonal forest plant species such as *Pithecellobium saman* entails the production of flowers and fruit biannually. This mode of reproductive allocation may represent a lifetime energy-conservation strategy for some large trees. For each of the six life-history modes, the evolutionary and ecological strategies characteristic of specific populations depend upon a suite of factors. Among the most important factors are ancestral and derived traits of a species, as well as differential life-history metrics describing populations, such as generation time (T), rate of intrinsic increase (r), and net reproductive rate (R_0).

Threats to Tropical Seasonal Forests

The tropical seasonal forest biome buffers the tropical rainforest biome and the tropical savanna biome. Thus, tropical seasonal forests are of critical importance for the preservation of biodiversity in tropical zones worldwide. Habitat destruction and other anthropogenic effects threaten evolved (for example, cooperative) and by-product (for example, competitive) associations within and among species. Local (patch or niche) and global (climate) effects interact, potentially inducing ecosystem perturbation and instability, in some cases leading to a cascade of network changes, including loss of species and ecosystem collapse.

Climate change and other factors varying in frequency, rate, duration, intensity, and quality contaminate, modify, and extirpate biogeochemical environmental components and attendant processes in tropical seasonal forests. The severity of anthropogenic outcomes continues to increase, with the potential to drive inherently resilient ecosystems beyond their capacities to accommodate and adapt to changing regimes. Jean Carlos Santos and his colleagues studied the research and conservation status of the Brazilian Caatinga, a tropical seasonal forest. These investigators showed that tropical seasonal forests are typically resilient and, thus, of particular scientific interest for a general understanding of factors associated with developmental, physiological, and phenotypic as well as biogeochemical buffering in complex systems. Santos's group highlighted the ecological, social, and economic importance of tropical seasonal forests. However, using tropical dry forests as an assay, these authors found that only 1 percent of dry forests in Central and South America, and only 5 percent of these forests worldwide, are protected.

These Brazilian scientists also found that tropical seasonal forests have been relatively ignored in the scientific literature, with tropical humid forests receiving the most research and conservation focus. Tropical seasonal forests are frequently located in areas having few academic resources, are poorly represented in the scientific literature, and are rarely highlighted as hot spots. Tropical seasonal forests require attention from professional, nonprofessional, and student conservationists, who can apply their knowledge, skills, and labor in ongoing initiatives from community to international levels in order to understand, address, and manage crises attendant to changes in tropical ecosystems. As two high-profile ecologists stated in the early 1990s: "On one point there is no argument: tropical forests are, indeed, in trouble." The worldwide status of tropical seasonal forests deserves treatment equal to that received by humid tropical ecosystems.

Clara B. Jones

Grassland, Tundra, and Human Biomes

Agroecosystems

The term *ecosystem* was first coined by a British ecologist, Sir Arthur George Tansley, to denote complex systems comprising both abiotic and biotic components. Agroecosystems, or agricultural systems, are specialized, human-managed ecosystems in which communities of plants and animals interact with their physical and chemical environments modified for the production of food, fuel, and fiber for human consumption and processing. Agroecosystems are more open than natural systems (forests).

They rely on the external inputs of energy and nutrients, and most of the production is removed out of the system. In general, agroecosystems are not self-sustaining and are driven by population, market, and policy needs and regulated by environmental feedback mechanisms. Agroecosystems are the basic unit of agroecology studies. Specifically, agroecology refers to the science of understanding the ecology of agroecosystems. However, in recent times agroecology has denoted an integrated discipline that includes elements of agronomy, ecology, sociology, and economics.

Agroecosystem dynamics can be studied using key properties such as productivity, sustainability, stability, equitability, and autonomy. These are also called *system properties* or *emergent properties.* The term *productivity* relates to the quantity of food, fuel, or fiber that an agroecosystem can produce for human use. The term *stability* refers to consistency of production, while *sustainability* refers to the maintenance of a specified level of production over the long term. *Equitability* refers to sharing agricultural production fairly, whereas *autonomy* refers to agroecosystem self-sufficiency. These properties can be used to compare the performance of agroecosystems in diverse landscapes and at varied spatial scales.

However, assessing agroecosystem properties is more complicated as the properties themselves have multiple dimensions and meanings. The definitions also vary based on the scale at which the properties are analyzed. For example, the productivity of a shifting cultivation field may be high per unit area of land on the cultivated field itself, but the productivity may be low in terms of the total land area occupied by the entire shifting cultivation cycle including forest fallow.

Structure and Function

Significant structural and functional differences in agroecosystems versus natural ecosystems have been noted by several researchers. The structural composition of an agroecosystem varies based on the location and scale at which it is studied. Agroecosystem structure can also change over a period of time due to management practices. The structure of agroecosystems can be perceived in terms of hierarchy theory.

Agroecosystems may be thought of as occurring in nested hierarchies, that is, agricultural farms nest in communities, which are nested in watersheds, which are further nested in larger regions, and so on, up to the globe. Each level in the nested hierarchy has its own unique properties (ecological, cultural, social, or economic features) and contributes to the nature of, and is affected by, levels above and below it. Across each level, agroecosystems may exhibit different interactions. For example, at a farm level, the input elements of agroecosystems include land, labor, and capital, which are interlinked together and interact with external attributes such as markets, policies, etc. To understand the complex interactions among different levels of agroecosytsems, researchers have identified three different dimensions, that is, environmental (for example, soil and other biophysical attributes), economic (for example, market) and human dimensions (for example, rural community). The division of agroecosystems into individual dimensions such as those described above facilitates understanding in a more systematic manner.

In comparison to natural ecosystems such as forests, conventional agricultural ecosystems exhibit a simplified, trophic structure due to few selected crops or animal types. Agroecosystems consist of three intermingled and strongly interacting subsystems: the managed fields, referred to as the productive subsystem; the semi-natural or natural habitats surrounding them; and the human subsystem composed of settlements and infrastructures. Agroecosystem function is a consequence of agroecosystem structure, which can vary widely based on location and management practices. Agroecosystem functions involve movement of material, energy, and information from one agroecosystem to another or movement in and out of the agroecosystem. Compared to natural ecosystems, agroecosystems are characterized by relatively high net productivity, simple trophic chains, low genetic and species diversity, low habitat heterogeneity, etc. Further, the multiple facets of agroecosystems and the services they offer have been well recognized. Agroecosystems offer a variety of functions. The functions refer to how agroecosystems operate in generating different services, which include production services, ecosystem regulation, and cultural services. Furthermore, these ecosystem services are strongly dependent upon the management practices.

The production services of agroecosystems include food, fuel, and fiber. Agroecosystems also serve as repositories of agrobiodiversity and genetic resources. The ecosystem services include providing habitat for biodiversity, primary production, biogeochemical cycling, soil formation and retention, carbon sequestration, water cycling, energy flow, etc. Regulation services include soil erosion control, climate mitigation, water purification, pollinating mechanisms, etc. The cultural services include agroecosystems serving as knowledge systems for education and inspiration, spiritual and religious values, recreation/aesthetic importance, etc.

In contrast to the above-mentioned positive ecosystem services, agroecosystems can also impact the environment in a negative way depending on management practices such as nutrient runoff due to excess nutrient application, greenhouse gas emissions, sedimentation, pesticide poisoning of humans, etc.

Livestock

Livestock are an important component of traditional agroecosystems and are considered a secondary food production system. Livestock provide draught power for farm operations including plough operations. Livestock are also the source of organic manure for crops and the main source of protein for human beings through the supply of meat and milk. In addition, livestock provide a source of income through the sale of animals. Compared to the specialized agroecosystems, such as single-cropping production systems or livestock-alone systems, integrated crop-livestock agroecosystems are considered highly beneficial in terms of agricultural productivity, environmental quality, operational efficiency, and economic performance. The synergies brought about by integrating crop and livestock systems can result in a positive feedback. For example, the crop residues that are left after harvesting can serve as a feed, while the livestock excreta can serve as manure to soils.

Integrated crop-livestock agroecosystems have also been associated with positive environmental effects. Increased organic carbon and nitrogen in soils of crop-livestock systems due to livestock manure have been associated with greater aggregate stability of soils, thereby conferring resistance to soil erosion. Another improvement obtained by diversifying solely from cropping systems or livestock systems and choosing integrated cropping-livestock systems is mitigation of losses from disturbances in either of the individual systems. In essence, crop-livestock agroecosystems can achieve both profitability and environmental benefits.

An entomologist and farm manager work together to eradicate pests on this agroecosystem, an apple orchard in Washington State. (USDA Agricultural Research Service)

Agroecosystem Health

The concept of agroecosystem health has been thoroughly debated in literature. Some researchers focus on symptoms of ecosystem stress, whereas others focus on ecosystem organization and change in relation to stress. In recent times, the health concept is generally judged to be well suited to describe the state of agroecosystems given the many ecosystem services that agroecosystems provide and their importance to human livelihood. In general, agroecosystem health has been closely linked to agroecosystem sustainability. Researchers preferring the agroecosystem health concept argue that sustainability is a broader concept while agroecosystem health is a more focused term. To describe agroecosystem health, different indicators have been proposed by earlier researchers, which vary from the farm to the landscape level. For example, agroecosystem health might be evaluated on the basis of structural criteria (resource availability, accessibility, diversity, equitability, and equity), functional criteria (productivity, efficiency, effectiveness), organizational criteria (integrity/coherence, self-organization, autonomy, self-dependence/self-reliance), or dynamic criteria (stability, resilience, capacity, and time to respond).

Alternatively, agroecosystem health evaluation could integrate different ecological processes such as cybernetics, water cycle, mineral cycle, and community dynamics in addition to human and biophysical phenomena and processes at different spatial scales. Several authors tried to quantify agroecosystem health, some using a conceptual framework, others using driving forces, and others by integrating biophysical and socioeconomic datasets using spatial tools and a multicriteria decision-making framework.

Scale

There is a significant need to understand scaling aspects in agroecosystems. Agroecosystems exist and function at plot level (less than a hectare) to landscape scale (many square kilometers). The information that is generated at the plot or field level cannot be generalized to the regional, national, or global level. While most of the decisions made by policy makers are at a regional or national scale, those decisions may not be congenial to implement at the plot or farm level. Thus, interdisciplinary approaches are needed to link plot level data and extrapolate it to landscape scale. To implement any such approaches, the data requirements for analyzing agroecosystem properties and functions at multiple scales can vary. Extensive georeferenced data might be needed to quantify some of the agroecosystem properties, and such data are rapidly becoming available. With advances in remote sensing technology and availability of improved satellite data, crop-type mapping, including field size, is plausible. Further, geospatial information technologies facilitate storage analysis, retrieval, and display of spatial and nonspatial information useful for agroecosystem research. Both local and federal government agencies are rapidly developing geospatial technologies to establish databases and make them available to the scientific community. Therefore, characterizing agroecosystem properties and functions at different scales will become more feasible over time.

Sustainable Agroecosystems

There is a growing recognition in the scientific community of the importance of adopting more sustainable and integrated practices for agricultural production in diverse landscapes that depend less on chemical and energy-based inputs. Relating to the same, the term *sustainable agriculture* is most commonly used to synthesize a variety of concepts and perspectives associated with agricultural practices that differ from conventional production. Conventional agricultural production relies heavily on external inputs. For several years the conventional practices emphasized short-term economic and production gains at the expense of long-term economic, environmental, and community interests. For improved agricultural management, the adoption of more holistic systems-based approaches have been proposed as an alternate path. The goal is to develop sustainable agroecosystems that have high production efficiency, economic viability, environmental compatibility, and social acceptability with less impact to the environment. Although designing such agroecosystems is highly challenging, efforts are underway through farming systems research.

Recent studies suggest organic farming as highly sustainable in the long term. Organic farming relies on management practices that enhance organic matter in the soil and limits the use of synthetic fertilizers, pesticides, plant growth regulators, livestock antibiotics, or genetically modified organisms. It also relies on improved crop rotations such as involving legumes, green manure, compost, and biological pest control. As defined by N. H. Lampkin in 1994, the aim of organic farming is "to create integrated, humane, environmentally and economically sustainable production systems, which maximize reliance on farm-derived renewable resources and the management of ecological and biological processes and interactions, so as to provide acceptable levels of crop, livestock and human nutrition, protection from pests and disease, and an appropriate return to the human and other resources."

Though organic agriculture is closely linked to sustainability, it can also have negative impacts to the environment such as the leaching of nitrates through legumes, or volatilization of ammonia from livestock waste. However, several researchers agree that the negative impacts caused by organic farming are low compared to conventional systems.

Four different attributes are most commonly used to assess agroecosystem sustainability in farming systems. They include diversity, cycling, stability, and capacity. Maintaining diversity in agroecosystems is important for risk minimization, genetic conservation, efficient resource use, and biological pest control. *Cycling* refers to effective flows of nutrients. Open nutrient cycles lead to losses, whereas closed cycles ensure that nutrients remain in the system. Thus, agroecosystems having closed or tighter nutrient cycles are better performing systems and are much more sustainable. *Stability* is also called *homeostasis* or *resilience* by different researchers. These terms refer to low variability or resistance to change and the ability of a system to maintain its productivity when subject to disturbances. *Capacity* refers to quality of the soil and water resource base and their ability to produce and sustain biomass. These attributes together with surrogate variables can be used to quantify the sustainability of farming systems.

Economic development and sustainability of agroecosystems will depend largely on our ability to manage these agroecosystems in a systematic manner. Maintaining crop diversity in agroecosystems is considered one of the important factors contributing to ecological sustainability. However, crop diversity can contribute to ecological sustainability only if the different crops fill various functions (for example, nitrogen fixation, production of organic matter to maintain soil quality, and provision of ground cover to prevent erosion) necessary for maintaining a productive agroecosystem. In addition, a diverse cropping system can reinforce monetary productivity and stability by allowing farmers to adjust the areas they plant to different crops in response to market opportunities.

"Conventional agricultural ecosystems exhibit a simplified, trophic structure due to few selected crops or animal types."

Agroecosystems and Role in Climate Change

Climate change is mostly attributed to the increase in greenhouse gases in the atmosphere. The changes in greenhouse gas concentrations are projected to lead to regional and global changes in climate and related parameters such as temperature, precipitation, or soil moisture. These changes can also affect agroecosystems and their productivity. Further, agroecosystems act as both sources as well as sinks for greenhouse gases.

Agriculture itself accounts for approximately one-fifth of the annual greenhouse gas emissions through the release of carbon dioxide (CO_2), methane (CH_4), nitrous oxide (N_2O), etc. Thus, agroecosystems have a major role to play in the mitigation of these gases. CO_2 is mostly released through deforestation for agriculture expansion. CO_2 is also released during the burning of agricultural crop waste, for example, rice/wheat residues. In Asia, agricultural residue burning is common as it is an easy way to dispose of residues. Fossil fuel combustion during agricultural management practices also releases CO_2. The largest source of CH_4 from agriculture is through paddy fields. Because the paddy fields during most of their cropping period are submerged in water, CH_4 is released due to anaerobic decomposition. Other sources of CH_4 include animal husbandry and agricultural residue burning. Most of the N_2O emissions from agricultural systems are from nitrogen fertilizers, leguminous crops, animal waste, and agricultural residue burning. Most of the above-mentioned greenhouse gas emissions can be mitigated through management practices both in the crop and livestock systems.

Agricultural soils can also act as a CO_2 sink mechanism. "Carbon sequestration" refers to removal of CO_2 from the atmosphere into a long-lived stable form, which can be beneath the soils. Soil carbon sequestration is considered an effective tool for mitigating CO_2. The most common management practices that are followed in the cropping systems for soil carbon sequestration include improved tillage practices, reduced cropping intensity, and incorporating organic inputs through fertilization. Tillage and soil carbon are negatively related as soil tillage accelerates organic carbon oxidation releasing high amounts of CO_2 to the atmosphere. Thus, management practices that reduce tillage result in increased soil carbon. Cropping intensity may positively enhance soil carbon as more biomass is incorporated into the soil through crop residues. Carbon sequestration can also be maximized through fertilization options such as organic compost, livestock manure, etc., and other organic amendments.

One of the most important challenges for agroecosystems will be adapting to future climate change and unfavorable weather conditions. Developing cultivars that require either longer or shorter growing seasons based on location, that can tolerate drought conditions, and heat resistance, and, most important, that can produce higher yields, will be the key adaptation strategy.

Greater emphasis may also be placed on the moisture-conserving managing practices such as minimum tillage, conservation tillage, or no tillage as climate change predictions in different regions suggest moisture shortage. Practices such as intercropping, multicropping, and relay cropping may be beneficial in terms of achieving overall production per unit area and soil moisture conservation. The importance of no-till agriculture is gaining significance. It is a way of growing crops without disturbing the soil through a tillage mechanism. This practice has been shown to increase the amount of water and organic matter in the soil, increase nutrient retention, and reduce soil erosion. Cover-cropping is another option that has been shown to enhance soil fertility and soil quality, suppress weeds, and help control pests and diseases. In essence, the key for the future will be to design and manage agroecosystems that have high production efficiency as well as being economically and environmentally efficient over long-term periods.

Krishna Prasad Vadrevu

Montane Meadows

Montane meadows are geographically limited to the high elevation mountain ranges. For example, in the Colorado Plateau the meadows typically start between 7,500 and 8,500 feet (2,286 and 2,591 meters) in elevation, with some variation depending upon location. These meadows are nestled within mountains around the globe, including the Alps and Pyrenees in Europe, the Himalayas in Asia, the Andes in South America, and the western mountain ranges in the United States. Despite the disparate geographic locations, they do have some climatological similarities.

Temperatures in montane meadows are generally cool, and precipitation is generally high. In Arizona montane meadows, the temperatures can range from minus 46 degrees F (8 degrees C) for a low in January to 80 degrees F (27 degrees C) for a high in July. Montane meadows, on average, are wetter than many other grassland biomes. These meadows receive multiple sources of water, including the melt of the winter snow from the surrounding mountains as well as rain from the annual summer monsoons. The monsoons start in July and last through October, helping to maintain a higher than average annual precipitation. This precipitation can range anywhere from 20 to 35 inches (508 to 889 millimeters), which allows for a rich floral community with substantial ground cover in the wetter areas.

The floral and faunal composition of montane meadows is unique and these meadows are often surrounded by encroaching pine forests. Vegetation in montane meadows will vary depending on geographic location. However, the patterns are similar and consist of grasses, forbs, and shrubs. Grasslands can include brome grasses and fescues. Forbs can include fleabane, scarlet gilia, and yarrow. The shrubs include rabbitbrush and sagebrush. Conifer forests that surround these meadows vary in species composition based on geographic location. Forests in Arizona include ponderosa pines, aspen, and mixed conifer forests.

The conifer forests that surround the meadows are constantly moving in. The main force that limits their takeover is the presence of an active fire regime. Fire drives the nontolerant tree seedlings and forest herbs back to the established tree line. Research in 2009 by the Joint Fire Science Program suggests that the meadows are shrinking in size due to conifer encroachment. In fact, meadow species tend to get replaced by forest herbs within just a few decades of tree invasion. Fire is very important to meadow maintenance, and research suggests that it takes a careful mixture of prescribed burns and tree removal to help maintain meadows. Interestingly, burning itself does not seem to impact meadow species, but burning does drive back forest herbs. Alternatively, it is the tree removal that seems to directly benefit and maintain the meadow species.

The Ak-kem River, in the Altai Mountains in southern Siberia, runs through a montane meadow. With some variation depending upon location, these meadows are limited to high-elevation mountain ranges. Generally, temperatures are cool and precipitation is high. Pine forests often encroach these meadows of grasses, forbs, and shrubs. (ThinkStock)

Montane meadows are fragile from a conservation perspective, which comes in part from an inadequate seed bank. Normally, a stable seed bank will lead to species maintenance, or in the case of disturbance, recovery. Almost 75 percent of the species that define montane meadows are absent from the soil seed bank. This means that once an area has been disturbed, the damage is permanent unless an active recovery effort takes place. Typical natural disturbances include water erosion, drought, and ungulate grazing.

Anthropogenic disturbances include climate change that will invariably impact moisture levels. Despite these higher than average precipitation levels, montane meadows can vary in moisture levels between wetter and drier conditions. Because of this variation, these meadows will be impacted differentially by climate change. Research by Diane Debinski of Iowa State University, using data collected from 1997 to 2007, suggests that long-term drought will negatively impact montane meadows by decreasing the number of flowering plants and their associated pollinators. When examining the reaction of wet to dry meadows to extensive drought, researchers have discovered that meadows with middle-of-the road moisture levels were the most susceptible to drought. Debinski concluded that dry meadows tended to just get drier and wet meadows tended to still have moisture that kept some consistency of the vegetation. However, meadows that depended upon some moisture (medium amounts) had a vegetation shift to more drought-tolerant species or a disappearance of cover overall, which will eventually impact pollinators of the region.

Vegetation in the montane meadows ultimately drives the faunal composition. Grasslands in the United States attract grazers such as mule deer and elk. Floral plants and their fruit attract pollinators such as white-lined sphinx hawkmoths (*Hyles lineata*) and Rufous hummingbirds (*Selasphorus rufus*) and frugivores such as *Heliothis* spp. caterpillars. Grazing by ungulates is a natural process in the montane meadows and in fact has an important consequence in shaping plant life histories.

Scarlet gilia (*Ipomopsis* spp.) has a unique life-history strategy whereby plants in certain populations actually benefit from being browsed. Scarlet gilia initially germinates to form a rosette. The plants can stay in this form anywhere from two to 10 years. If the plants are not browsed, they will send up a single aboveground shoot that will flower and set seed. However, if plants are browsed at this stage, some populations will send up multiple stalks that produce on average twice as many flowers, set seed, and reproduce. This will increase the overall seed production and plant fitness and is known as overcompensation or tolerance.

Not all plants are able to overcompensate, and browsing and grazing can be detrimental to this ecosystem if left unchecked. Constant grazing by cattle can change the vegetation composition, cause water tables to decline, and increase tree density. Grazing is among one of many disturbances currently impacting the range and quality of montane meadows. Combined with tree encroachment, changing fire regimes, and climate change, the future of montane meadows remains questionable.

LYNN CARPENTER

Pasture Lands

Pasture land is an anthropogenic biome represented by lands where predominant plants are grasses, grasslike plants, forbs, or shrubs. These lands are managed as grazing systems designed for the production of domestic animals for consumption, including production for meat, milk, and other major animal products. Since historical times these sites have been predominantly located in natural grasslands, savannas, steppes, wetlands, open and dense shrublands, tropical evergreen forests/woodlands, boreal evergreen forests/woodlands, boreal deciduous forests/woodlands, temperate needle-leaf evergreen forests/woodlands, some deserts, and tundra. For millennia, the widespread presence of human populations and their activities altered these ecosystems around the world, both intentionally and unintentionally. These effects started with low-impact practices like hunting and gathering, but through the centuries humans developed and controlled many tools that allowed them the permanent use of lands for agriculture, forestry, livestock, and settlements. These activities changed many natural processes and with them the composition and structure of flora and fauna, as well as the hydrologic and biochemical cycles at local, regional, and global scales.

Development of Pasture Lands

Historically, people managed livestock to produce food, directly as meat and dairy products and indirectly as animal power that increased cropland production. In the 18th century livestock were managed at local scales because technology and transport facilities were poor, and because livestock depended on the availability of food resources in areas where disease constraints allowed these activities. For this reason, pasture lands were kept geographically close to human settlements and barely distributed around the world. However, during the following centuries there was a considerable increase in human population around the world, as well as a growth in technology development. Due to these factors, the demand for livestock products expanded pasture distribution in almost every corner of the planet in order to produce enough commodities to sustain the growing population. As a result, the location and extension of pasture lands were undergoing important shifts. The increase of this biome exerted an enormous pressure in all wild lands, principally in the drier biomes of the Americas, Australia, central Asia, and southern Africa, but also in the moister wooded biomes of sub-Saharan Africa, Central America, and Eurasia. As a result of the increase of human use of natural resources and appropriation of the net primary production, the global extension of pasture lands increased considerably since the Industrial Revolution, growing from 3 percent of the ice-free land cover before 1700 to 26 percent in 2000.

Because of these developments, pasture lands have now become one of the most extensive anthropogenic biomes, occupying more than 26 percent of the ice-free land in the world in an area around 13,513,576 square miles (35 million square kilometers), including large areas where there previously was little or no livestock grazing in North

America, South America, and Australia. In many areas, pasture lands have expanded to occupy virtually all the land that can be grazed and for which there is no other activity. Except in bare areas located in dry or cold deserts, or in dense forests, pasture lands are present to some extent in all regions around the world. In North America a large extension of the continent has been considered as a pasture, except in areas such as the northern extreme in Canada and Greenland, where the extreme temperatures impedes this land use. Also the rough mountain systems of the Rocky Mountains in western North America, both the Sierra Madre Oriental and Occidental in Mexico, the Andes Range and Patagonia along the southwestern coast of South America, and the Brazilian highlands have been difficult places to sustain livestock production. The Amazon rainforest is also considered one of the few places in the world that has not been transformed to a pasture land. In Africa, the scenario is similar; most of the continent has been grazed except the Sahara Desert in the north and the tropical forest in central Africa. In Eurasia, the extreme weather conditions and complexity of the mountain system such as the Himalaya Range, the Gobi Desert, the Scandinavian and Siberian regions, as well as the tropical rainforests at the southeastern extreme of the Asiatic continent, have disabled these areas to use the land as pasture, while the rest of the world has virtually been transformed to pasture lands.

Pasture land productivity differs in different sites on the planet and is mainly defined by climatic and soil conditions. Managed grazing is preferentially employed in areas that are much drier and stabler than the biome mean, while regions with large interannual precipitation and temperature variation strongly affect vegetal production, grazing capacity, and human living conditions. Savannas, grasslands, shrublands, and deserts support the largest extent of managed pastoral systems; these dry-land biomes cover about 51 percent of the total land area of the Earth (around 25,888,845 square miles or 67 million square kilometers), and they support 78 percent of the global pasture lands of the Earth. Other biomes also support managed grazing systems but in less proportion. For example, temperate deciduous and temperate evergreen broadleaf forests and woodlands support grazing systems, and about 656,374 square miles (1.7 million square kilometers) of tropical evergreen broadleaf forests (corresponding to 10 percent of all tropical forests on the planet) have been cleared for managed grazing. Nowadays pasture land area is expanding annually in wild regions such as the Amazon Basin, Congo, and southeast Asia rainforest.

"Today, we live at a time in which it is necessary to reconcile conservation strategies with livestock management strategies."

Besides savanna, shrubland and desert are the most-used biomes for grazing systems, although they are limited in their soil type and the fertility associated with it. For instance, many types of grassland are supported in soils composed of aridisols, entisols, and alfisols that are not fertile enough to sustain a grazing system for a long period of time. In the colder boreal biomes, grazing takes place preferentially on alfisols and spodosols but not on frozen gelisols, which are the most common soil order found in these regions, the reason why these lands are still under low grazing pressure. In humid tropical regions ultisols represent the best place to maintain a grazing system, while oxisols that are also well represented in the Amazon Basin, Congo, and southeastern Asia are nutrient-poor soils for managed grazing systems.

The five countries with the most pasture lands are Australia (1, 698,850 square miles or 4.4 million square kilometers), China (1,544,409 square miles or 4 million square kilometers), United States (926,645 square miles or 2.4 million square kilometers), Brazil (656,374 square miles or 1.7 million square kilometers), and Argentina (540,543 square miles or 1.4 million square kilometers). However, based on the proportion of total land area that a country uses for grazing, Mongolia, Botswana, and Uruguay are the most-affected countries, with 80 percent, 76 percent, and 76 percent of pasture land coverage, respectively. Countries with the highest stocking rates are Malaysia with 320 animal units per square kilometer, India with 272 animal units per square kilometer, North Korea with 213 per square kilometer, and Vietnam with 184 per square kilometer.

Consequences of Pasture Land Expansion and Management

Although managed grazing systems have been developed throughout the world for thousands of years, the spatial extension and intensity of grazing systems have considerably increased in the last three centuries. Pasture land development is influenced not only by a complex range of environmental conditions but also by many social and economic factors that have a strong influence on extension and pressure. Throughout human history, pastures have been managed under different conditions leaving the land in a broad number of states. Pasture lands have been managed in different ways according to property and access rights. Historically, three main types of land tenure have been recognized: private (owned by an individual or a company), communal, and public property. The interaction of this management generally is noxious for the ecosystem as a result of a complex and inadequate set of rules imposed by state and local officials who pretend to control the use of natural resources. This complexity generally leads to conflicts among stakeholders who claim access to pasture lands.

However, not all grazing systems or practices lead to these negative impacts. Domestic cattle (cows, horses, and goats) may have positive, neutral, or negative impacts on vegetation structure and composition, depending on the grazing intensity and the site. In areas where grazing intensity is light to moderate, the habitat of wild ungulates, the effect can be positive in stimulating the growth of some species and increasing diversity in the landscape. For the most part, however, where livestock is raised, pasture land is aimed toward the goal of producing the greatest possible number of animals. At these sites, generally bounded by fencing, cattle cause plants to recover. In these cases the rate of extinction of species tolerant of grazing increases dramatically, but there are also increases in the rate of colonization of grazing-tolerant species or those that are not consumed by livestock. Gradually the original vegetation (before introducing livestock) is replaced, and there comes a point where even when grazing decreases, the vegetation generally does not return quickly to its original state and can take many decades to return, or even not return to its original state. The impacts on soil structure are also notable. Much of the nutrients are either recycled naturally in the ecosystem, or removed and not redeposited in the place of origin, and the extensive areas without plant cover created by overgrazing promote soil erosion. After many years of inadequate, intense, chronic, and uncontrolled management, many pasture lands are now exposed to three main processes that have negative effects on ecosystems: desertification, woody encroachment, and deforestation.

The bioclimatic and edaphic conditions under which managed grazing occur have, to some extent, contributed to the development of these three processes. Desertification has occurred in arid regions of the southwestern United States, Australia, South Africa, and Argentina as a result of chronic grazing and pronounced climatic variability with special influence of drought periods. Woody encroachment has occurred in semi-arid to mesic environments as a result of large-scale grazing, fire suppression, and climatic variability. Deforestation continues to expand principally in the humid tropics in South America, Africa, and Asia in part because of grazing development on infertile soils that often cannot maintain large-scale livestock production.

As the human population grows and land scarcity increases, intensification and improvements in traditional animal production should be implemented to satisfy human demand for livestock products. Unfortunately, most pasture productivity is declining in many arid regions, and many grazing systems are disappearing in tropical forests. These are being replaced by more concentrated grazing systems that can lead to greater degradation of pasture land. In the northeastern United States, northwestern Europe, and densely populated areas of Asia, animal production has become mechanized and dependent on external fertilizer and feed inputs. Industrial meat production is growing rapidly, and in general in these systems animal numbers exceed the carrying capacity of the land, and waste is saturating the surrounding environment causing habitat fragmentation and eutrophication of freshwater and marine ecosystems.

Intensification of animal production and grazing systems is likely to continue, and it will require the expansion of pasture lands and improve livestock production through a supplemental feeding. In both cases, however, expansive management will cause greater degradation processes and will intensify environmental impacts over the land surface. Even more, the emergence of large economies such as China, India, Brazil, Russia, Mexico, South Korea, Indonesia, and Turkey as new centers of demand and production will potentially promote the expansion of pasture lands globally. Humid tropical ecosystems represent the only viable way to expand global grazing systems beyond their current geographic extent, but this expansion into marginal areas has already more or less reached the limits imposed by climate and soil factors, not to mention the degradation of some of the most biodiversity-rich areas in the world.

The success of pasture expansion into forest will depend mainly on local and regional policies in concerned areas. But since the prospect of expansion on pasture lands is limited, the intensification of pasture production on the most suitable land, and the loss of marginal pastures, is likely to continue. Climate change is also likely to alter pasture land systems. The impact on pastures will be greater than in other anthropogenic biomes. While the cropland biome can be more easily manipulated by irrigation or use of fertilizers, pasture lands depend largely on weather conditions. On dry pasture lands the impact is projected to be dramatic, reducing forage yields by as much as 25 percent by 2030, but on the other hand, pastures located in cold areas are expected to benefit from rising temperatures.

The grazing activity of this herd of roughly 3,000 sheep, pasturing in the shadow of the Centennial Mountains in Dubois, Idaho, helps create a mosaic of different densities and ages of sagebrush crown. This in turn helps wildlife, such as the sage grouse, thrive. A wide variety of pasture lands are managed as grazing systems for livestock—natural grasslands, savannas, steppes, shrublands,and even some deserts and tundras, to name a few. (U.S. Agricultural Research Service)

The management and long-term conservation of pasture lands requires a broad understanding of the patterns and processes that have determined their structure and composition, as well as the impact that humans have had on these lands for centuries. Today, we live at a time in which it is necessary to reconcile conservation strategies with livestock management strategies, and this will only be possible by integrating the knowledge of many disciplines. For this integration to occur it is essential to promote social and governmental participation in the design, dissemination, and implementation of best-management practices involving the conservation of ecological processes in the ecosystems in order to maintain their goods and services. It is necessary that management policies be based on sound scientific data to ensure better management of ecosystems, and that these policies integrate environmental and social needs focused on restoring and preserving the health of ecosystems in the long term.

José F. González–Maya
Eduardo Ponce

Temperate Grasslands

Several characteristics of the temperate grassland biome have contributed to its historic and current importance in shaping people's livelihoods and customs, as well as local, regional, and now global economies. In particular, the location of temperate grasslands at mid-latitude regions ranging from 30 degrees to 60 degrees north and south of the equator is conducive to human settlement. Large swaths of these grasslands cover parts of North America, South America, Eurasia, and southern Africa, while less extensive grasslands occur in Australia and New Zealand. An agreeable climate, highly fertile soils, and the availability of forage for livestock provided incentive for a diversity of people from these regions to settle in the temperate grassland zone. The many names used to describe temperate grasslands, including *prairie* in North America, *steppe* in Eurasia, *pampas* in South America, and *veld* in South Africa, reflect how different cultures identify portions of this expansive biome. Given the history of human settlement in temperate grasslands, one of the main threats to their integrity is land-use change. Prudent stewardship is required so that the healthy functioning of this biome, which is highly productive and of great economic importance, can be maintained over the long term.

Temperate grasslands are largely defined by their lack of woody vegetation. They are typically dominated by perennial grasses and sedges (graminoids) that are well adapted to semi-arid to subhumid climates with marked extremes in temperature (both among seasons and over the course of a day) and seasonal precipitation. These grasslands also contain a variety of nongraiminoid herbaceous plants called forbs, which, although less common than the dominant graminoids, are the main contributors to plant diversity in these systems. Forbs often flower in synchrony following monsoonal moisture events, giving rise to a colorful palette of blooms that cover vast expanses of land. In some cases small shrubs are present, for example, where the local topography and soil texture favor them. More generally, plant community composition varies in relationship to many factors other than climate including edaphic (soil) conditions, fire regimes, and grazing by wild and domesticated ungulates (hoofed mammals).

Temperate grasslands are broadly considered to be water-limited in the sense that woody vegetation is typically not supported. However, there is substantial variation in the amount and timing of precipitation among regions, which gives rise to distinct differences in grass canopy height and species composition. For example, in North America, as one moves from the subhumid central lowlands in the east to the semi-arid western extent of the Great Plains, precipitation becomes more limiting. This gradient in precipitation is mirrored by a transition from tall-grass prairie in the east, to short-grass prairie in the west, with midgrass prairie occupying the area in between. In contrast, the gradient of precipitation availability in the Eurasian steppe runs from north to south, with grasses and forbs becoming shorter, less abundant, and less diverse as one moves south and precipitation becomes more limiting.

Photosynthetic Grass Pathways

An important distinction among grasses is whether they have a C_3 (when carbon dioxide breaks up into a three-carbon compound) or C_4 (when carbon dioxide breaks up into a four-carbon compound) photosynthetic pathway. The designation as a C_3 or C_4 grass indicates whether the first product of carbon fixation during photosynthesis is a three-carbon or four-carbon molecule. Several differences in the C_3 and C_4 photosynthetic pathways result in plants of each type achieving optimal photosynthetic rates under different climatic conditions. As a result, the distribution of C_3 and C_4 grasses varies strongly in relationship to water availability and temperature among regions. The C_3 grasses, often referred to as "cool season" grasses, dominate in areas with relatively cool temperatures and where precipitation is high during the winter months. These grasses germinate in early spring and achieve maximum growth when temperatures are still relatively low (approximately 20 degrees C or 68 degrees F).

In contrast, the C_4, or "warm season" grasses, dominate in areas where both temperature and average annual rainfall are higher than where C_3 grasses dominate, and where the majority of precipitation occurs during the summer. Germination of C_4 grasses is delayed until the summer months, and maximum growth occurs at quite high temperatures (approximately 35 degrees C or 95 degrees F). The distribution of C_3 and C_4 plants in relation to their climatic requirements results in gradients in their relative abundance at large spatial scales. For example, temperate grasslands in the southeastern United States are dominated by C_4 grasses, while C_3 grasses become increasingly dominant to the north and west. Differences in the chemical composition of C_3 and C_4 grasses (for example, the ratio of carbon to nitrogen in their tissues) affect ecosystem-level properties such as rates of litter decomposition and nutrient cycling.

Grasses also exhibit a number of morphological (structural) characteristics that adapt them to the precipitation and temperature extremes to which they are exposed, as well as to fire and grazing by herbivores. In particular, grasses have several characteristics that help them cope with stressful conditions, including vegetative (asexual) reproduction; buds (new growth) that occur at or below the soil surface; cell division (new growth) that occurs at the base of leaf blades rather than at the tip; and fibrous root systems that are concentrated in the shallow layers of the soil profile. Vegetative reproduction occurs when individual plants send out specialized structures such as stolons or rhizomes in order to establish a new individual. Eventually the new individual breaks off from the mother plant, forming an independent, genetically identical individual (or clone). This type of reproduction is beneficial to plants growing in harsh environments because it requires less time and energy than sexual reproduction, which depends on pollination and the formation of seeds. Once grasses have established, new buds are often produced at or below the soil surface, which protects this young and delicate tissue from fire and grazing herbivores. Additionally, cell division typically occurs at the base of leaf blades, rather than at the tip, as is common in other types of plants. This allows for quick tissue regeneration, even if the majority of a leaf is lost to herbivory. Grasses also have fibrous root systems that extend horizontally from the plant through the shallow layers of the soil profile. The fibrous roots, which are finely branched and have a high surface area, allow grasses to take maximum advantage of periodic monsoonal rains that are short in duration and infiltrate only the shallow soil layers.

Soil Diversity in Temperate Grasslands

There is a great diversity of soils that underlie temperate grasslands. This diversity reflects variation in the many factors that shape soil properties, including climate, parent material (for example, bedrock), topography, and the plant and microbial (that is, bacterial and fungal) communities present in a given area. Temperate grassland soils do share some basic characteristics, however. Importantly, they tend to be extremely fertile, and because of this they have largely shaped patterns of human settlement following the discovery of agriculture. They are typically high in dissolved nutrients (for example, nitrogen, phosphorous, and potassium) and organic matter, making them amenable to crop production. The top soil layers are often dark and rich in color due to a well-developed layer of humus (organic matter). High evaporation rates and capillary action prevent the leaching of nutrients to deep soil layers, where they would be inaccessible by plant roots. Additionally, the dense fibrous root systems of native grasses effectively bind nutrients, which keep them from leaving the system, and they prevent soil erosion by anchoring the soil and trapping moisture. The texture of grassland soils, which is defined by the particle-size distribution, varies from clays to loams to sands (with clays being composed of the smallest particles and sands of the largest particles). Soil texture is important because it influences water infiltration and percolation rates, as well as water-storage capacity, all of which directly affect plant distributions.

All of the attributes discussed so far, including abiotic factors such as temperature, precipitation, and soil characteristics, and biotic factors such as plant community composition and plant canopy height, affect nutrient cycling dynamics in grasslands. In addition, microbes, which are the main decomposers of organic matter, are important in shaping nutrient cycles. Nutrients such as carbon and nitrogen move through different "compartments" of the ecosystem, including the atmosphere; living organisms such as plants, animals, and microbes; and the soil. They exist in different forms depending on which compartment they are in. Often they enter the system as inorganic molecules (for example, gaseous carbon dioxide [CO_2] or nitrogen gas [N_2] from the atmosphere) and are transformed into usable (organic) forms that can be assimilated by plants and animals. For example, during photosynthesis, plants provide the crucial service of "fixing" CO_2 into organic forms that can be used by herbivores. In turn, most plants rely on microbes that transform inorganic nitrogen into organic forms that can be used to make the proteins that catalyze photosynthesis. Given that a diversity of species shape nutrient cycling dynamics in natural grasslands, modifying native plant and soil communities (for example, by converting native grasslands to single-species crops) can have marked impacts on nutrient cycling regimes.

Fauna of Temperate Grasslands

The vegetation of temperate grasslands in large part determines the abundance and composition of the associated fauna. Grasses support several characteristic mammal assemblages, most notably herds of grazing ungulates as well as burrowing rodents and lagomorphs (for example, rabbits and hares). Bison (*Bison bison americana*) and pronghorns (*Antilocapra americana*) represent the dominant large ungulates of the North American Great Plains, while wild horses (*Equus przewalskii* and *E. gmelini*), asses (*E. hemionus* and *E. kiang*), and saiga antelope (*Saiga tatarica*) were prevalent on the Eurasian steppe. Other examples of grassland ungulates include the pampas deer (*Ozotoceros bezoarticus celer*) of South America and the sable antelope (*Hippotragus niger*) of the South African veld. Vast herds of these animals historically populated grassland ecosystems worldwide, acting as keystone species that shaped the abundance, distribution, and adaptive characteristics of the plants on which they grazed. While some of these species, such as the pronghorn, are still widespread, most of them have been hunted to near extinction or widely displaced due to habitat loss. For example, in a few short decades, European settlers of the American west hunted an estimated 70 million bison to near extinction.

Many reasons underlie this tragic slaughter including the attractiveness of the skins for commercial sale, the perception that bison competed with cattle for forage, and the desire of the U.S. government to pressure Native American tribes who relied on bison for food and clothing. In many cases, individuals of these once prolific keystone ungulates exist only in nature reserves or as managed semi-natural populations (as with the American bison and the pampas deer). Instead, domesticated ungulates such as cattle or sheep number in the millions in many temperate grasslands and have in some places virtually replaced

The Badlands National Park in South Dakota is home to bison, one of the quintessential symbols of the North American plains. Temperate grasslands like this attract human settlements for abundant hunting or livestock grazing. (National Park Service)

the native herbivore assemblages. Unfortunately, this shift also has negative implications for many of the large carnivores that historically fed on native ungulates.

Fossorial rodents and lagomorphs have also been important in shaping temperate grasslands, due in large part to the effect that their burrowing has on soil characteristics and plant distributions. The natural disturbances created by these ground-dwelling mammals, including the black-tailed prairie dog (*Cynomys ludovicianus*) of North America and marmots (*Marmota sibirica*) and social voles (*Microtus socialis*) of the Eurasian steppe, help to maintain plant and animal biodiversity in grassland systems. In lieu of other protective structures such as trees or caves, burrows provide shelter to a large variety of these relatively small, herbivorous animals. Burrow excavation and maintenance act to mix the soil layers, often times redistributing deeper soils to the surface and reducing soil compaction. Burrows also serve as habitat "microsites" that are different from the surrounding intact grassland. For example, some forb species recruit almost exclusively on disturbed soil created by burrowing, while insects and spiders often prefer the cooler, more humid habitat provided by underground burrows. In many cases, burrowing rodents live socially, in groups of related individuals that share tasks such as raising offspring and keeping watch for predators. Social living concentrates the number of individuals occupying a given area, leading to pronounced effects on habitat structure and function. For example, the preference of black-tailed prairie dogs for native grasses gives co-occurring forbs a competitive advantage, eventually leading to pronounced differences between plant community composition on and off prairie dog colonies.

Impact of Human Activities

The history and future of temperate grasslands cannot be fully considered without understanding how human activities have altered this biome over time. Because of intensive land-use change, temperate grasslands today have been mostly altered from their pristine state. One of the earliest impacts humans had on grassland ecosystems occurred in the veld of southern Africa, where people used fire as a tool in hunting game. Fires were set to corner large herbivores, thereby facilitating their capture, or to create flushes of nutritious new plant growth that would attract herds of grazers. However, because periodic fires are a natural occurrence in most temperate grasslands, the use of fire by early peoples may not have had a large negative impact (although human-caused fires likely differed in frequency and intensity from natural fires). Indeed, in wetter regions, periodic fire, along with sustained grazing, may actually be required to prevent the encroachment of woody vegetation.

A more challenging issue in terms of land-use change, at least in regard to the maintenance of native biodiversity and ecosystem health, is the conversion of millions of acres of native grassland to cropping systems or for use as intensive livestock operations. The spread and intensification of agriculture, in large part due to its mechanization over the last 200 years, is associated with marked increases in human populations as well as a more sedentary lifestyle. Because agriculture-based societies have access to a stable food supply, they have largely circumvented the need to migrate in search of the plants and animals they historically used for food or clothing. Unfortunately, the modern agricultural practices that became widespread in the early 20th century tend to be largely unsustainable, having led to the degradation of grassland habitats worldwide. For example, soil erosion due to poor farming practices and overgrazing of livestock has threatened the integrity of temperate grasslands in North America, Eurasia, and South America.

A classic example of the hardship associated with unsustainable farming practices is provided by the 1930s Dust Bowl era of the American west. After the Civil War, the government provided incentive for large numbers of settlers from Europe and the eastern United States to homestead the Great Plains. During this time, when precipitation levels happened to be much higher than usual, agriculture became widespread and intense, eventually replacing millions of acres of native grasses with cultivated crops. Many circumstances, including overplowing and severe drought, eventually led to extreme soil erosion across the American and Canadian Great Plains. The unprecedented loss of topsoil rendered the land useless for farming, forcing countless farmers to abandon their homesteads for lack of food and livelihood. The devastating ecological and social impacts that resulted are captured by pictures of denuded landscapes and massive dust clouds rolling across the plains and engulfing abandoned homesteads. Some of the dust clouds traveled far enough to darken the skies of far-away cities like Chicago and Boston.

"Vast herds of these animals historically populated grassland ecosystems worldwide, acting as keystone species that shaped the abundance, distribution, and adaptive characteristics of the plants on which they grazed."

The issue of soil erosion is not unique to the North American prairie. For example, as sedentary agriculture replaced nomadic grazing in northern China, overgrazing led to erosion and salinization (an increase in salt content) of the soil. The same issue arose in the Argentine pampas following the introduction of horses, cattle, and sheep and the advent of sedentary agriculture in areas that experienced periodic drought. Fortunately, an increased awareness of the biological properties of these systems has led to the implementation of sound policy, which has to some extent reversed the degradation associated with early intensive agriculture.

Indeed, the concept of sustainable agriculture is gaining traction today as different stakeholders acknowledge that maximizing agricultural production at the expense of ecosystem functioning and socioeconomic justice is shortsighted. However, the movement toward more sustainable practices must be balanced with the obligation to feed and cloth an ever-growing human population that has come to rely in large part on industrial-scale agriculture. Several organizations have begun to advocate a responsible transition from industrial agriculture back to smaller family-owned farms. They argue several benefits for such a transition including a reinvigoration of local farming economies; an incentive to maintain the natural resources (for example, water and healthy soil) that maintain profitable farms over the long term; a reduction in harmful pollutants due to large-scale application of chemical pesticides and fertilizers; and healthier foods derived from organic farming practices. Such organizations include the National Sustainable Agriculture Coalition in the United States and the European Initiative for Sustainable Development in Agriculture. There are also agencies that specifically advocate sustainable ranching practices, for example, by adjusting stocking rates (that is, the number of animals per unit area) to levels that are sustainable given the various ecological characteristics of the land. In fact, agricultural land is coming to be viewed as an important resource in the face of increasing urbanization worldwide. Given that much of the world's agricultural land is concentrated in the temperate grassland biome, long-term solutions for its sustainable use are needed.

CHRISTINA ALBA

Tropical and Subtropical Grasslands

Grasslands can be defined as ecosystems dominated by grasses (Poaceae family) or graminoids (grasslike plants, usually monocots, for example, from the Sedge family), with few or no trees. There are two principal ways that can be used to classify the vegetation of a defined area: in terms of floristic composition or in terms of vegetation physiognomy. The floristic classification is based on the presence of a set of species (that is, composition), which will be used to define this area as a floristic unit. The physiognomic classification, in contrast, takes into account parameters of the entire species set at any given point in time, such as height, estimated cover, and importance of different growth forms (that is, vegetation structure). The definition of grasslands presented above contains both floristic (large predominance of one botanical family, Poaceae) and physiognomic (comparatively low vegetation height due to dominance of herbaceous species, high grass/tree ratio, and growth form, considering the grassy habit as predominant) elements.

However, it is widely accepted that the worldwide distribution of ecosystems, including grasslands, is mainly related to climate conditions, and this can be used for a classification of grass-dominated ecosystems. David J. Gibson, for example, uses the Köppen climate classification system to define different grassland ecosystems around the globe.

Grasslands occur in tropical wet, dry, and desert climates, as well as in subtropical, temperate, and alpine climates. These different climates encompass a large variation in factors such as soil type, mean temperature, rainfall, and moisture. Considering that different species have different adaptations and strategies to endure different environmental conditions, this climatic variation is reflected in differences of composition and structure among different grassland ecosystems.

The frequency and abundance of tree species is an important characteristic for grassland classification—especially in tropical and subtropical grasslands. This is chiefly due to the conceptual overlapping between two ecosystem types in which grasses dominate: grasslands and savannas. Following the classical definition by Heinrich Walter, savanna ecosystems are tropical formations with the presence of grasses and woody plants, but do not include open grasslands without a woody component. Pragmatically, savannas can be considered as (tropical) grassland ecosystems with more than 10 percent tree cover, usually with tussock grasses and shrubs shaping the lower layer of the vegetation. This tree cover may be relatively uniform and dense (as in some parts of the Brazilian Cerrado) or clumped, shaping a parklike landscape (as in some parts of Africa). Grasslands, on the other hand, present no or less than 10 percent of tree or shrub cover. In many regions around the world, savannas and grasslands form mosaiclike patterns in the landscape, and the limits between both formations are neither clear nor stable. In this article we will focus on grassland ecosystems. However, the grassland component of savanna ecosystems will be briefly considered.

Natural Grasslands Worldwide: A Brief Evolutionary History

Grasslands cover 15.84 million to 21.62 million square miles (41 to 56 million square kilometers), which represents 31 to 43 percent of the world's surface (excluding permafrost areas). During past geological eras, under different climates and land mass distribution, grassland-covered areas where even larger. Although the first grasses grew under forest cover or near forest borders, current evidence indicates that expansion of grasslands and diversification of grasses took place under conditions of increased aridity. The emergence of considerable shrub or tree layers in grasslands (as we see in tropical savannas today) occurred much later, when shifts in temperature and moisture allowed the establishment of such growth forms.

Fossil records provide evidence that large grazing animals and fire were important factors in the evolutionary history of grasslands. Altogether, climatic and, in some cases, edaphic conditions, grazing, and fire are constraining forces that have shaped grassland ecosystems as we see them today. In consequence, many species in the extant grassland flora show adaptations to avoid or resist drought and disturbances such as grazing and fire.

"Today, large areas of natural grasslands in the tropics and subtropics still support herds of cattle, sheep, and horses."

Extant grasslands may be natural or secondary, that is, the consequence of habitat conversion by human populations. Most grassland ecosystems present in highland montane climates are natural and largely determined by climate. Temperate grasslands in central and western Europe are considered secondary, being mostly linked to historical forest cleaning and subsequent mowing, burning, and grazing. Eurasian steppes and South American and African grasslands are considered natural (or climatogenic). In places where present climatic conditions are suitable for forest establishment, open grasslands gradually gave place to savannas (for example, Africa and central Brazil), shrublands or forests (southern Brazil and Africa). Some grasslands in tropical regions are clearly secondary, for example, cultivated pastures used for cattle grazing in tropical rainforest regions, originating from destruction of natural forests. This type of grassland will not be discussed in this article.

Grasses: The Determinants of Grassland's Physiognomy

The grass family does not only give the name to the vegetation type but also dominates most grassland systems. Current estimations of species richness in Poaceae range from 7,500 to 11,000 species, distributed in 600 to 700 genera in all continents and environments. The importance of this family to humankind is unquestionable, since all important cereal crops and sources of forage are grasses. Moreover, almost half of the Earth's surface is covered by landscapes (natural or human-driven) chiefly characterized by grasses.

Grass species are known to thrive under a wide amplitude of ecological conditions; grasses can be found from wet tropical climates to deserts. Many grasses can resist disturbance events such as complete burning or removal by grazing, or manage to resprout immediately. These abilities are consequences of evolutionary adaptations within the family, such as reduction in many vegetative and reproductive structures, growth forms adapted to disturbance (buds protected by leaf sheaths), or high importance of vegetative propagation in many grass species. Further, different photosynthetic pathways allow for good performance under different climatic conditions.

Reduction and simplification in aboveground structures directly influence effectiveness of biomass allocation: Grasses spend less energy and grow faster than trees, for example. The trade-off for this reduction seems to be clear: Grasses are apparently more fragile than trees, because they lack features such as secondary thickening in cell walls. However, grasses compensate for this with astonishing resistance to drought, resprouting ability after disturbances or unfavorable seasons, and ability to reach faraway sites by wind dispersal.

A considerable share of the resistance and resilience of most grasses can be attributed to their growth form. Meristematic tissues of grass shoots ("tillers") in most grasses are located near, at, or below the surface of the soil, providing protection to disturbances. Each tiller comprises photosynthetic leaves and (potentially) reproductive structures of the species. Most grass species can produce many tillers in a single growing season. This feature leads to the formation of dense tufts and multiple flowering culms in some species. Growing tillers may shape various patterns depending on the species, but there are two generally recognized growing forms: caespitose growth (erect tussocks) and rhizomatous growth (prostrate). The production of new tillers, which may be nearly unlimited in perennial species, allows the species to expand its cover in the landscape, "moving" to new sites with potentially better resources without the need of a full seed-dependent cycle of colonization and establishment.

Grasses can be divided according to their photosynthetic pathways: C_3 (when carbon dioxide breaks up into a three-carbon compound) species are cool season grasses, usually with short life cycles, and C_4 (when carbon dioxide breaks up into a four-carbon compound) species are warm season grasses, usually with long life cycles. The two pathways are named after the first stable product of photosynthesis, consisting of three carbons in C_3 grasses and four in C_4 grasses. The C_4 pathway is optimized considering carbon assimilation and enhances the plant's water use efficiency, which is advantageous in hot and dry environmental conditions. As a consequence, there are ecological differences between C_3 and C_4 grasses: The first have more efficient physiological processes under temperate conditions, whereas the latter are more efficient under tropical conditions, in which temperatures are higher throughout the year.

The Role of Disturbance in Grassland Ecosystems

Disturbance can be defined as the partial or total removal of biomass from an individual plant. On the other hand, on the level of the plant community, grazing or fire can be considered normal or, in case of climate favorable to forest development, even necessary factors for the existence of this vegetation type. The plant species present in a community subject to recurrent disturbances show a number of adaptations to these processes, such as high regeneration ability. Fire may cause local removal of individuals and even species, but prevents shrub encroachment and keeps at bay the process of forest expansion over grasslands that takes place in ecosystems under moister conditions. Grazing may negatively affect grassland diversity when management is not adequate, and overgrazing may lead to decreased soil cover, erosion, and species removal. However, well-managed cattle-raising may be one of the most ecologically sustainable economic activities in grassland ecosystems, since native grass and legume species may be maintained and used as the primary source of forage.

Differences Between Tropical and Temperate Grasslands

In comparison with their tropical and subtropical counterparts, temperate grasslands suffer colder winters and milder summers, larger amplitudes in temperatures around the year, a shorter period of vegetation growth, and much more frequent frost events. Annual rainfall is usually lower in temperate grasslands than in tropical and subtropical grasslands. However, many (but not all) tropical and subtropical grasslands show great seasonality in rainfall: a pronounced dry season, when mean rainfall may be less than 2.36 inches (60 millimeters), followed by a similarly pronounced wet season. Soils in temperate grasslands are usually deep and fertile, but plant growth is nutrient-limited because much of the soil nitrogen is inaccessible to the roots.

In addition, tropical and subtropical grasslands receive much more solar radiation throughout the year, which implies consequences concerning species composition. In tropical grasslands, C_4 grasses are dominant, and the contribution of C_3 grasses is usually reduced. However, the further south or north of the equator (that is, passing first to subtropical and then to temperate grasslands), the contribution of C_3 species enhances, considering both composition (more C_3 species) and structure (higher abundances of C_3 species).

With few exceptions (for example, the Indian subcontinent), natural grassland vegetation can be found—even though not always as a dominant vegetation type—in tropical and subtropical regions around the world. In the following we will briefly present the different vegetation types, beginning with those under a tropical climate.

South America

The South American llanos (Spanish for "plains") follow the valleys of the Orinoco River in Venezuela and Colombia. The region is dominated by savannas, but patches of typical grasslands, with almost no trees, also occur. They can be divided into different local vegetation physiognomies, classified mostly according to soil water availability. The lower layer of the vegetation is dominated by arrow-grasses (*Trachypogon* spp.) in some places, and three-awn grasses are also common (*Aristida* spp.). Legume species are abundant, enhancing forage value. Grasslands and savannas are grazed by cattle and the native *Hydrochoerus hydrochaeris* (capybara or chigüiro), the largest living rodent in the world. The llanos are threatened by land conversion to crops and monospecific plantations of exotic trees.

The Brazilian Cerrado is a biome that covers about 772,204 square miles (2 million square kilometers) in South America. Although the biome as a whole can be considered a tropical savanna, its landscape actually consists of mosaics of tree-free grasslands, of different savanna physiognomies (from scattered to dense tree cover), and of gallery forests. The disturbance regime (fire and grazing) and, to some extent, soil properties shape the distribution between savanna and grasslands. Typical grasslands are locally classified according to the presence of woody elements: *campos sujos* contain some scattered shrubs and trees, whereas *campos limpos* are dominated by grasses (common genera are *Echinolaena, Elyonurus, Paspalum, Trachypogon*, and *Tristachya*), without woody species. The Cerrado is considered to be one of the biodiversity hot spots of the world; the biome as a whole and the grassland areas are threatened by land conversion to establish pastures with nonnative species, uncontrollable fires in tree-dense savannas, and invasion of exotic species (*Brachiaria* spp.)

Further to the south, the flooding grasslands of the Pantanal cover 54,054 square miles (140,000 square kilometers) of Brazilian, Bolivian, and Paraguayan territory. Large areas of grasslands and forests that grow along river courses are yearly flooded. Natural grasslands are mostly used for cattle-raising, sometimes with native grasses as the main source of forage. Among these grasses, especially *Paspalum almum* and *P. plicatulum* present high forage value. The fauna of the Panta-

Impalas graze on the new grass sprouting under fire-blackened woody vegetation in Kruger National Park, South Africa. In August 2011, one year after a prescribed burn to kill off woody shrubs and trees that had been encroaching on the area where wildlife graze, the tropical grassland area had completely regenerated. Fire disturbance is sometimes necessary for the existence of grassland species, which show a number of adaptations to this process. (NASA)

nal is outstandingly rich (for example, 650 birds, 1,100 butterflies, 80 mammals, and more than 250 fishes). Hunting is a major threat for this ecosystem since it has historically depleted populations of large predators such as the jaguar (*Panthera onca*). The invasion of the exotic tree *Vochysia divergens* (cambará) threatens the grasslands, and is kept at bay with fire (which is also considered a threat, mostly to animals) and selective logging.

The Campos of southern Brazil are under subtropical climates, and harbor approximately 2,500 nonwoody plant species, many of which are endemic. At higher altitudes (greater than 700 miles or 1,127 kilometers), these grasslands occur in mosaics with Atlantic and Araucaria forests, and *Andropogon lateralis* (caninha grass) is the dominant grass over large areas. At lower altitudes, grassland vegetation is dominant, and forests are mostly restricted to water courses. In these lower altitudes many grasses with high forage values are found, such as *Paspalum notatum, P. nicorae, P. pumilum,* and *Axonopus* spp. Different classifications for local grassland physiognomies exist, mostly related to soil and climatic factors. The southernmost areas of the Campos stretch south to temperate areas and the pampa biome (or Rio de la Plata Grasslands) of southern Brazil, Uruguay, and Argentina. Extensive cattle-raising is historically the most important economic activity in the Campos, and land conversion to crops and exotic tree plantations are the most pressing threats to this ecosystem.

Africa

Grasslands and savannas are the dominant landscapes in large areas of Africa. Atlantic forests mostly follow the equator line and the Zaire River, in central Africa, and are surrounded by woody savannas. Further north toward the Sahel and then the Sahara Desert, and east toward the Horn of Africa, trees become rarer and more scattered in the landscape, and the woody savanna gives way to typical grassland. Transitional ecosystems such as arid grasslands, shrublands, and open-canopy woodlands also occur, and vegetation physiognomy is largely determined by water availability. African grasslands and savannas are known for their large migratory herds of wildebeest (*Connochaetes taurinus*), zebra (*Equus burchelli*), and eland (*Taurotragus oryx*); by the Thompson's gazelle (*Gazella thomsonii*), buffalo (*Syncerus caffer*), and topi (*Damaliscus korrigum*); and by their famous predators, the lion (*Panthera leo*) and the cheetah (*Acinonyx jubatus*). There are also giraffes (*Giraffa camelopardalis*) and African elephants (*Loxodonta africana*), mostly in *Acacia* savannas and in the Serengeti volcanic grasslands. Threats to African grasslands include overgrazing, conversion to croplands, and uncontrollable wildfires. Trophy-hunting is also a problem, and has greatly depleted populations of mammals such as the white and black rhinos (*Ceratotherium simum* and *Diceros bicornis*).

The velds ("fields" in Dutch) are grasslands and savannas from the plateaus of South Africa and Zimbabwe. Locally, these ecosystems are divided into high velds (above 1,400 miles, or 2,253 kilometers in altitude, cooler, and with high rainfall) and low velds (below 700 miles or 1,127 kilometers in altitude, hot and dry). The dominant grasses are *Cymbopogon plurinodis, Diheteropogon filifolius, Heteropogon contortus, Themeda triandra, Brachiaria serrata, Digitaria eriantha,* and *Setaria flabellata.* In heavily grazed areas, this composition changes with more forbs, replacement of some dominant grasses, and tree encroaching.

Asia, Eurasia, and Australia

Most tropical and subtropical grasslands in southern and southeastern Asia are secondary, being the result of abandoned crop areas that were originally rainforests. Large areas of midaltitudes (300 to 700 miles, or 483 to 1,127 kilometers) are dominated by *Imperata cylindrica*, whereas *Arundo madagascariensis* is the dominant grass in higher areas (greater than 900 miles or 1,148 kilometers). The larger continuous grass-dominated ecosystem in the region is the *Sehima nervosum–Dichanthium annulatum* grasslands in India. In some tropical areas trees (for example, *Pinus merkusii, Eucalyptus alba,* and *Casuarina junghuhniana*) form a rather continuous layer, characterizing savanna-like landscapes.

In Australia, climate conditions and soil fertility define different grassland ecosystems. A mosaic of savannas and grasslands are distributed in a belt around the arid center of the continent. Typical savannas, with a continuous tree layer, are usually dominated by eucalypts (*Eucalyptus* spp.) or wattles (*Acacia* spp.). In low-fertility soils, there are tall-grass grasslands dominated by *Themeda triandra*, *Schizachyrium fragile*, and *Sorghum* spp. Additionally, tropical and subtropical tall-grass formations with *Heteropogon contortus* and *H. triticeus*, as well as midgrass grasslands with *Aristida* spp., *Dichanthium sericeum*, *Bothriochloa decipiens*, *B. bladhii*, and *Chloris* spp., can be found. Fire is a natural disturbance and is also used as a management practice to prevent forest expansion. In some areas, absence of fire for long periods leads to closed eucalyptus forests.

Natural grasslands can also be found in the tropics and subtropics as "islands" of often relict vegetation scattered in forest ecosystems. These grasslands usually occur at higher altitudes, where temperatures are lower throughout the year, and are mostly maintained in the present climate by fire and grazing. Such formations are common in the Brazilian Atlantic Forest, in hilltops of mountain ranges above 800 miles altitude (1,287 kilometers), and are locally known as *campos rupestres*. Similar altitude grasslands occur in mountain ranges of Africa and Asia, for example, on so-called inselbergs.

Land Use and Impacts on Conservation

Most grasslands originally occupied landscapes with a rather gentle topography (except in altitude grasslands), and in many cases were associated with deep and fertile soils (although grasslands may also occur on shallow, litholic soils). Also, it is easier to remove the original vegetation and plow the land in grasslands than in forests. The combination of these factors led to the conversion of many grassland ecosystems to crop production throughout the tropics and subtropics, which has led to species and habitat loss. In the past decades, natural grasslands in the tropics and subtropics have been increasingly converted to planted forests, often monocultures of exotic species, mainly to produce paper, resins, and firewood, again leading to pronounced biodiversity losses and landscape changes.

Grasslands are historically related to pastoral activities, dating back to the Neolithic. Today, large areas of natural grasslands in the tropics and subtropics still support herds of cattle, sheep, and horses. In many cases, the raising of grazing animals relies solely (or mainly) on native forage, which usually maintains the local biodiversity (both vegetal and animal) in good conservation status.

Abandonment, or nonmanagement, may have negative effects on natural grasslands. The absence of fire and grazing, the main nonclimatic factors that have influenced the origin, evolution, and maintenance of grassland ecosystems worldwide, may promote the dominance of tall, coarse, caespitose grasses, causing diversity losses in the plant community. This process may be followed by shrub encroachment, which leads to out shading of herbaceous plants. Under tropical and subtropical climates, where moisture conditions are suitable to forest establishment, this shrub encroachment may be followed by forest expansion and complete loss of the original natural grassland ecosystem.

Future of Tropical and Subtropical Grasslands

As human population grows, so does the demand for food. In many regions of the world, grasslands are the primary targets for conversion into croplands. The conversion rates exceed legal protection in grasslands more than in any other inland ecosystem: approximately 1 hectare of natural grassland is protected for every 10 hectares lost.

A less obvious but nonetheless serious threat to grassland biodiversity is the suppression of management and disturbance. In many tropical and subtropical grasslands, current climate conditions are suitable for development or expansion of shrubland or forest vegetation, and grazing and fire impede these successional processes. If they are suppressed, the grassland community will, in the long run, be converted into a community dominated by woody species, losing its original biodiversity and ecological properties. Conservation efforts that focus on natural grasslands therefore must take management into account.

Because of human-induced climatic change, both the global temperature and atmospheric carbon dioxide concentration are expected to rise at unprecedented rates. These changes most likely will directly affect tropical and subtropical grasslands, where C_3 (temperate grasses and most forbs, shrubs, and trees) and C_4 species (tropical grasses) coexist. The effects of climate change on grassland distribution and on the distribution and abundance of C_3 and C_4 grasses has become an important research topic.

As in many other ecosystems, invasion by alien species also threatens grasslands. In African and Australian grasslands and savannas, invasive non-native species increase the amount of flammable biomass in natural grasslands, enhancing the intensity of fire events and thus having an impact in the community at the local scale (for example, local extinctions).

Also, many grass species that are first introduced outside their native range as forage may become dominant and may lead to losses in biodiversity and productivity. In southern Brazil, the African grasses *Eragrostis plana* and *Brachiaria* spp. that present no invasive behavior in their natural environment become immensely aggressive, outcompeting and excluding virtually any native grassland species.

No vegetation type is stable, and changes are not necessarily negative, both from conservationist and agricultural perspectives. However, as natural grasslands constitute the basis for livestock production and harbor important—and often neglected—parts of the world's biodiversity, however, management decisions should consider possible changes and aim at the stabilization of the important ecosystem functions and services provided by grassland vegetation.

Pedro Maria de Abreu Ferreira
Gerhard Ernst Overbeck

Tropical Savannas

Tropical regions contain higher species diversity than any other biomes of the world. This is attributed to their extensive area that creates many habitat types, their comparatively stable and warm climate with higher primary productivity, their long growing season, and the long time since they underwent major perturbations. Tropical savannas denote communities with an expanse of grass layer and scattered woody vegetation (shrubs and trees). They are found in tropical and subtropical regions, shaped by a succession of dry and wet seasons that differentiate them from other biomes. Tropical savannas cover about 10,656,420 square miles (27.6 million square kilometers), approximately 12.5 percent of the global land mass. They comprise roughly 40 percent of Africa, and they are also represented in parts of Australia, South America, and Asia.

Thus, savanna ecology is influenced by both grasses and woody plants, generally warmer temperatures throughout the year, and two distinct dry seasons. Summers are wet, hot, and humid with most rainfall during this period 15–25 inches (381–638 millimeters), while the dry season lasts for a longer period of the year (up to eight months) with warmer temperatures. However, the amount of rainfall and humidity in the savannas are not sufficient to support forest growth. Furthermore, high evaporative demands as a result of intense illumination cause savannas to experience a net water deficit for most of the year. Tropical savannas are maintained by complex and dynamic interactions among climatic factors, topography, edaphic factors, geomorphology, herbivory (grazing and browsing), fire, and human activities.

Savannas are generally categorized based on their canopy cover, spatial layout of woody plants, and stature. One of the resultant categories is *savanna grassland* consisting of sparsely scattered trees or shrubs. On the other hand, *savanna parkland* comprises discrete patches of woody vegetation interspersed over a continuous herbaceous plant layer. Due to various abiotic influences such as soil, altitude, and topography, savannas are interrupted by bands or areas of woodlands, forests (riparian, groundwater, or montane forests) or wetlands forming what are popularly called *savanna landscapes*.

Tropical savannas are known to be subclimax communities that are maintained by the soil characteristics (edaphic subclimax savannas), grazing (biotic subclimax savannas), and fire (pyric subclimax savannas). Termites are especially abundant in tropical savannas of the world, and their tall termitaria are conspicuous elements of the savanna landscape. Termites are important in soil formation and their termitaria provide shelter for other animals. Termites are also important food sources for anteaters (endemics of the neotropical zoogeographic province) and aardvarks and pangolins in African savannas.

Tropical savannas are inhabited by the world's proportionately large and rapidly growing human population, and a majority of their rangelands are dominated by livestock. Large population size associated with the prevalence of poverty increases the demands for immediate environmental resources. Overharvesting of timber and nontimber products, introduction of exotic species of plants and animals, overstocking, deliberate massive killing of wildlife, and fire suppression in some areas have significantly degraded tropical savannas, changing their scenery. There is likelihood that with climate change impacts, tropical savanna might be completely replaced by a different landscape.

African Tropical Savannas

African savannas are grassland-dominated vegetation mixed with scattered and isolated trees found between latitudes 15 degrees north and 30 degrees south and longitudes of 15 degrees west and 40 degrees east. Savannas in Africa are found in several countries including Guinea Bissau, Sierra Leone, Liberia, Ghana, Somalia, Nigeria, Ethiopia, Ivory Coast, Benin, Togo, Central African Republic, Cameroon, Chad, Tanzania, Sudan, Democratic Republic of Congo, Ethiopia, Kenya, Angola, Malawi, Burundi, Uganda, Rwanda, Zimbabwe, Zambia, Mozambique, Botswana, Madagascar, and South Africa. The savannas occupy areas such as the Maasai steppe ecosystems in northern Tanzania and southern Kenya within the Somali-Maasai regional center of endemism. Savannas in Africa also include the widely spread miombo (mopane) woodlands spanning from southwestern Tanzania to Zambia, Malawi, Mozambique, Zimbabwe, parts of Botswana and Namibia, and Angola. Miombo is an extensive tropical savanna falling mostly under the Zambezian regional center of endemism, and mostly characterized by such plant species in the genera *Isoberlina*, *Jubernadia*, and *Brachystegia*.

Flora and Fauna

In terms of plant species, savannas in Africa are characterized by such species as *Acacia*, miombo/mopane, candelabra trees, baobabs (*Adansonia digitata*), sausage trees (*Kigelia africana*), amarula plants (*Sclerocarya birrea*), *Combretum* species, star grass, elephant grass, Bermuda grass, and many more.

African savannas are known for their diversity of both invertebrates and vertebrates. Millions of animals are known to coexist and are spatially or temporally specialized to environmental resources in savannas. There are carnivores, including cats (lions, leopards, cheetahs, serval cats, caracals, and civets); dogs such as wild dogs and jackals, hyenids (aardwolves, spotted and striped hyenas); and mustelids (honey badgers). Herbivores include elephants, hippopotamuses, kudus (greater and lesser), giraffes, elands, topis, sable antelopes and roan antelopes, hartebeests, black and white rhinoceroses, wildebeests, gazelles, impalas, gerenuks, aardvarks, waterbucks, bushbucks, reedbucks, buffalo, zebras, oryx, dik diks, lemurs, and sunis. These herbivores utilize different parts and types of plants, thus reducing competition, and hence increasing coexistence between them.

There are also a large number of birds such as ostriches, vultures, chaffinch, doves, pigeons, eagles, shrikes, starlings, shoebills, storks, ducks, hammerkops, hornbills, and many others. Reptiles include snakes (venomous and nonvenomous), crocodiles, lizards, and tortoises. There are also amphibians such as frogs and toads, and many invertebrates, with insects accounting for the majority. Biota in African savannas interacts in a complex way and produces a complex food web. Some animal species such as lions, elephants, beetles, and termites are keystone species maintaining diversity of the biomes. For example, elephants play the roles of ecosystem engineers, maintaining and creating vegetation structure, which increases usable habitat for other species.

Economy and Culture

Wildlife abundance and the multitude of culture from the people residing in these areas make them attractive destinations for cultural tourism. Bushmen (San in Namibia and South Africa, Baswara in Botswana, and Hadzabe in Tanzania) and pastoralists (Maasai, Tutsis, and Ankole) persistently maintain their traditional ways of life. Other communities are composed of agropastoralists (peasants and small-scale livestock keepers). Preferred types of crops are maize, beans, peas, wheat, millet, cassava, coconuts, groundnuts, sugarcane, and cashew nuts, and the common livestock are an indigenous breed of cattle, sheep, and goats.

Threats to African Savannas

The majority of community livelihood in African savannas relies on the use of the immediate environmental resources for survival. People in these areas lack alternative livelihoods, which places direct pressure on natural biological resources, posing a greater threat to their survival. Indigenous trees are unsustainably harvested for construction, fuelwood, and charcoal production, while animals are illegally overhunted or captured for food and trade to satisfy the desires of the growing population in nearby towns and abroad for food and trophies. The tragedy of the ivory trade in recent years has increasingly jeopardized the survival of elephants.

"Tropical savannas denote communities with an expanse of grass layer and scattered woody vegetation (shrubs and trees)."

Human-wildlife conflicts are prevalent in the African savannas. Conflicts between pastoralists on one side and African wild dogs and cheetahs on the other side have driven these wildlife species to the danger of extinction. On the other hand, contact between wildlife and domestic animals creates an outbreak of zoonotic diseases such as rabies, canine distemper, and anthrax, claiming the life of many wildlife, domestic animals, and human beings. Trampling by moving domestic animals creates loose and bare soils, which erode easily during rainy seasons but also suppress fire recurrence. The growing human population requires large areas for expansion of settlement and agriculture; this creates encroachment into protected areas and blocks wildlife migratory corridors and dispersal areas. Ultimately areas with wild animals (for example, protected areas) remain as ecological islands in the sea of human-modified landscapes.

Climate change and associated impacts invariably affect tropical African savannas resulting in changed climatic patterns in some areas. Frequent drought occurrences in the Horn of Africa (Kenya, Sudan, Ethiopia, and Somalia) dramatically reduce food production for humans and kill livestock (a major asset for most people) as a result of the loss of pasture and water.

Conservation Status

African savannas comprise major protected areas of the world, differing in their conservation status, ranging from fenced national parks such as Kruger National Park in South Africa to the free-ranging wildlife in the major natural ecosystems such as the Serengeti in Tanzania. Other notable protected areas falling under the African savannas include Selous, Rungwa, and Moyowosi Game Reserves; Ruaha and Mikumi National Parks in Tanzania; Maasai Mara, Amboseli, and Tsavo National Parks in Kenya; Luangwa Valley in Zambia; Gonarezue in Zimbabwe; and the Great Limpopo Transfrontier Park straddling the borders between Mozambique, Zimbabwe, and South Africa. These areas preserve pristine habitats for wildlife and play a crucial role as representative areas for biodiversity conservation in the region annually visited by millions of tourists. Many countries in this region are members of international conservation treaties regulating management of biological resources. These treaties include the International Union for Conservation of Nature (IUCN), Convention on International Trade in Endangered Species of Wild Fauna and Flora (CITES), Convention on

Biological Diversity (CBD), and the Ramsar treaty for wetland management.

South American Tropical Savannas

Tropical savanna also extends to South America in llanos and cerrado, but with relatively few species existing. Savannas occupy almost 965,255 square miles (2.5 million square kilometers) of Brazil, Colombia, and Venezuela, an area equivalent to one-quarter the size of Canada. Animals dispersing from the neighboring biomes play a key role in maintaining diversity and persistence of this region.

Llanos in South America

Flooded annually by the Orinoco River, the llanos of the Orinoco Basin of Venezuela and Columbia contain plants adapted to growing for long periods in stagnant waters. They span in a northeast direction from the foothills of the Andes Mountains in Colombia along the Orinoco River to its delta in the Atlantic Ocean. The vegetation in these areas is dominated by grass species in the genera *Melinis, Panicum, Homolepis, Scleria, Paspalum, Trachipogon, Gymnopogon,* and *Axonopus* with shrub species in the genus *Casia*. Fauna in the llanos includes capybara and marsh deer (mammal species adapted to a semi-aquatic life). The area also harbors the Orinoco crocodile, an endangered endemic crocodile species. Other animals include the world's extant largest boa anaconda, Orinoco turtle, giant armadillo, giant otter, black and chestnut eagle, swallow-tailed kite, broad-winged hawk, neotropical migratory shorebirds, and many species of cat fishes.

Brazilian Cerrado

The Brazilian Cerrado is an extensive open woodland savanna of short twisted trees composing 21 percent of the country. The Cerrado savanna extends from the Amazonian forest margin to outlying areas in the southern states of Sao Paulo and Parana, to Bolivia and Paraguay. The Cerrado is a moister tropical savanna with rainfall ranging from about 31 to 79 inches (800 to 2,000 millimeters), with vegetation less adapted to waterlogging compared to its northern counterparts, the llano of Venezuela and Columbia. The Cerrado has several endemic plants and animals, and is a biodiversity hot spot consisting largely of savanna ecosystem. Plants in the Cerrado are estimated at around 5,000 species, with dominant grassy level vegetation five to seven times the tree and shrub layer. Similar to other savanna, it is characterized by remarkably long dry seasons supporting drought and fire-adapted plant species. Common genera are *Anisacanthus, Dicliptera, Geissomeria, Hygrophila, Jacobinia, Justicia, Lophoslachys, Mendoncia, Poikilancathus, Ruellia, Stauurogyne, Stenandrium,* and *Thunbergia*.

The Cerrado is characterized by indigenous animal species such as edentates (tamanduas, anteaters, and armadillos), marsupials (opposums), latyrrhine monkeys (marmosets, howlers, and capuchins), rodents (agoutis, pacas, capybaras, and many mouse-sized species) and maned wolf. The Cerrado has great diversity of animal life, though not in comparison to its counterpart in Africa.

Economy, Culture, and Threats in the Cerrado

In the past, the Cerrado domain was sparsely populated by backwoodsmen and Indians, typically Brazilian countrymen, while llanos were shielded from human settlement and other threats. In most cases the countrymen kept livestock, cultivated crops, and utilized natural resources on a subsistence scale. In the past three decades, neoliberal changes in government policies have commercialized agriculture in the Cerrado leading to encroachment of many areas for farming. Crops like soya, maize, rice, and cassava are commonly preferred. Invasive plant species have increased in number as a result of tree plantation and livestock fodder improvement. Construction of Brazil's new capital city (Brasilia) and its associated development also created demand for the growing urban population, which increased pressure on natural resources. With livestock numbering around 48 million, it is estimated that 35 percent of the Cerrado has been destroyed.

Another threat to Brazilian savannas comes from the Brazilian steel industry, which solely depends on charcoal as source of energy. This places a remarkably high pressure on trees and is the second-largest threat after agriculture.

One of the views of the extensive, tree-dotted savanna in Kenya's Samburu National Reserve. African savannas, which are grassland-dominated vegetation mixed with scattered and isolated trees, are rich in animal diversity and home to many species of mammals, reptiles, birds, and insects. This abundance supports traditional pastoralists and bushmen while also attracting worldwide tourism. (National Science Foundation)

The industry gets 71 percent of its charcoal from native trees. On the other hand, expansion of agriculture has led to dramatic changes in water use in llanos, diverting water to irrigation areas and drying up wetlands. Crops commonly cultivated are palm oil and rice, leading to fragmentation of the once continuous ecosystem. Oil exploration and extraction together with construction of dikes and water channels exacerbate habitat fragmentation. Furthermore, agrotoxins poison the ecosystem, leading to fatal consequences.

Conservation Status in the Cerrado

The value of the biodiversity-rich Cerrado and llano has always been undermined. When compared to other biomes, Cerrado is considered to be poor, and not even considered by the Brazilian constitution as a national heritage. Although only 1 percent of the area is currently conserved, local conservation organizations collaborating with the World Wildlife Fund (WWF) are advocating for conserving and rescuing the biome. On the other hand, llanos have been poorly represented in protected area systems. Again, WWF, working with a local conservation agency, has proposed more than 63,000 acres in Colombia to be the first llano protected area. Increasing private areas for conservation and collaboration with the U.S. Fish and Wildlife Service are other efforts that are in place to rescue these ecosystems and their wildlife populations.

Australian Tropical Savannas

Australia's tropical savannas are landscapes of dense grass and scattered trees covering approximately one-quarter of the continent, about 733,594 square miles (1.9 million square kilometers). They cross the continent from Rockhampton on the east coast, across the Gulf of Carpentaria, to the Top End of the Northern Territory, and over to The Kimberley in western Australia. Australian savannas fall within the tropical latitudinal ranges 10 degrees to 20 degrees south with medium to low mean annual rainfall (16 to 47 inches or 400 to 1,200 millimeters). Climates are characterized by two distinct seasons: the wet and the dry, with high daytime temperatures and high evapotranspiration rates. The dry season lasts five to six months, usually from May to October. The wet season lasts five to six months from December to March. Australia's tropical savannas are also referred to as the *monsoonal tropics* and the *wet-dry tropics.* The soils of Australia's tropical savannas are generally poor, with the exception of the southeast sector of the tropical savannas, which is more fertile.

Australian Fauna and Flora

Australian tropical savanna is one of the major biodiversity reservoirs which is less intensively developed compared to the temperate areas of the world, and this accounts for its relatively intact habitat compared to that of southern Australia. The savanna is home to hundreds of species of native plants, mammals, birds, reptiles, and amphibians and tens of thousands of different species of invertebrates. Many species in these groups are found nowhere else in the world. This part of Australia also has World Heritage areas like Kakadu National Park and Purnululu National Park.

Australian savannas are dominated by marsupials, mammals whose young are born undeveloped. These animals include the echidna, eastern gray kangaroo, the koala agile and whiptail wallabies, wallaroos, possums, gliders, the northern quoll, and the golden bandicoot. These animals live in or near the few trees in this area, utilizing them for shade, food, and water. Reptiles dominate in the other parts of the tropical savannas. The saltwater crocodile, which is found here, is the world's largest reptile, growing up to 26 feet (8 meters long). Associated with these are many species of birds, amphibians, and invertebrates, many of which are endemic to the region.

Open woodlands with grassy understory are the major vegetation type in Australia's tropical savannas. *Eucalyptus* trees of restricted height (49 to 82 feet or 15 to 25 meters) and open canopy dominate Australia's tropical savanna woodlands in areas receiving over 23 inches (600 millimeters) mean annual rainfall, with acacia or *Casuarina* species inhabiting the drier, and less fertile, areas. *Eucalyptus* species grow in warm climates and are the most important tree for timber, oil, and gum in Australia.

Australian Socioeconomic Activities

Australian savannas have a rich culture from the Aborigine people, who have a long association with the land and much traditional knowledge of land management. The didgeridu wind instrument and the band Yothu Yindi come from the tropical savannas. Tourists and locals also value the Australian tropical savannas for their wild and spectacular natural features. The main areas of employment include the livestock, conservation, tourism, mining, and horticultural industries. Until a few decades ago, livestock was the main economic base of the tropical savannas, accompanied by lesser but more intensive uses such as mining, agriculture, or urban development.

More recently, however, mining and tourism have become the dominant economic industries. Like many other savanna ecosystems, Australian savannas are home to a pastoral industry that includes some of the most extensive operations in Australia and manages the largest area of land of any group in the northern part of the country. The livestock industry is the major user and manager of land in the tropical savannas. It contributes hundreds of millions of dollars to regional economies across northern Australia and provides employment for thousands, both directly and indirectly via service industries

Australian tropical savanna is experiencing significant population pressure and land-use changes and is vulnerable to rapid land degradation. Other threats include habitat fragmentation, deforestation, climate change and variability, as well as frequent fires.

Asian Tropical Savannas (Terai-Duar Savanna)

Asian savannas are estimated to cover an area of 13,400 square miles (34,706 square kilometers) spreading through Nepal, India, and Bhutan. These savannas are dominated by subtropical grasslands and shrublands in the Terai-Duar Savanna. The Terai or "moist land" is actually the belt of marshy grassland, savanna, and forest located at the base of the Himalayas, about 15.5 miles (25 kilometers) wide, but it is very long. The area is mainly found in the semi-arid and semi-humid climate regions of subtropical and tropical latitudes. Rainfall ranges between 12 and 60 inches (300 and 1500 millimeters) annually, with climate significantly influenced by the topography, especially the great complex of mountains that flanks the Himalayas.

The Asian savannas have originated over generations from woodland ecosystems through deforestation, abandoned cultivation, and burning and are maintained at a subclimax by repeated grazing and burning. These mixed forces have led to the formation of mosaic types of savanna communities, depending upon the age, mode of origin, and the intensity of biotic disturbance. Savanna communities at different seral stages tend to differ in their biotic composition (that is, species composition, productive potential, and nutrient cycling).

Asian Fauna and Flora

The Asian savannas are home to diverse and abundant invertebrates and vertebrates. These include diverse amphibians, birds, and reptiles such as the gharial crocodile (*Gavialis gangeticus*). There are also mammals, including tigers (*Panthera tigris*), pygmy hogs (*Sus salvinius*), swamp deer or barasinghas (*Cervus duvauceli*), and the greater one-horned rhinoceros (*Rhinoceros unicornis*). Asian savannas are home to three endemic birds: the spiny babbler (*Turdoides nipalensis*), gray-crowned prinia (*Prinia cinereocapilla*), and Manipur bush-quail (*Perdicula manipurensis*).

There are a number of plant species in the region including wild sugarcane (*Saccharum spontaneum*), 10-foot tall tropical reeds (*Phragmitis kharka*), and silky kangaroo grass (*Themeda villosa*). Smaller species include cranberry grass (*Imperata cylindrical*) and six-weeks three-awn (*Aristida adscensionis*). Many of the grass species found in the Terai-Duar Savanna serve as fodder for livestock and wildlife species such as elephants and rhinoceroses, as well as cover for the predatory species such as tigers.

Agriculture is the principal economic activity in the region. Other major employment sectors include livestock keeping, nature-based tourism, cultural tourism, and natural resources. The main threats to the region include poaching for wildlife species, overgrazing, clearing for cultivation, population growth, irrigation projects, and water diversion.

Alex W. Kisingo
Henry K. Njovu
Gideon Alfred

Tundra Ecosystems

A circumpolar band of roughly 5 million acres of treeless country called the arctic tundra lies between the forests to the south and the Arctic Ocean and the polar ice caps to the north. Smaller yet ecologically similar regions found above the timberline on high mountains are called alpine tundra. The Russian word *tundra* came from the term *tunturi*, meaning "treeless plain." This name gives a brief but effective description of the plant species that grow on the northern flanks of the latitudinal tree line in the Northern Hemisphere, and above the altitude of the vertical tree line. Based on the temperature and environment, tundra can usually be categorized as short-term (hours, days, or a half month), seasonal (a half month to several months), and permafrost (years to thousands of years). Geographically, short-term, seasonal frozen soil and permafrost areas occupy about 50 percent of land area on Earth, while the permafrost area can take up to 25 percent of the land area. Nearly half of the territory of the former Soviet Union and Canada is permafrost, and 85 percent of the land in Alaska is permafrost. In addition, even near the equator, permafrost is found on the peak of Mount Kilimanjaro.

Among many factors that affect the distribution of ecosystem types, the temperature is a very important one. The higher the temperature is, the more energy is available for plants to use in photosynthesis and to be stored in the biosphere. When we go forward north from the equator, the temperature generally decreases. Also, as altitude increases, the temperature declines. So there is a general similarity between latitude and altitude in determining the distribution of ecosystems. Bearing this in mind, arctic tundra and alpine tundra are described and compared below.

In tundra, cold is the number-one limiting factor. Temperature often ranges from 60 degrees F (15.5 degrees C) in the summer to minus 130 degrees F (minus 90 degrees C) in the winter. Precipitation is also low, but water is not a limiting factor because of the very low evaporation rate in the tundra region. The growth of woody plants is essentially restricted due to the lack of warm summer at high latitude and altitude. Therefore, vascular plants that can survive in these cold environments must complete their annual growth rather quickly during the short but cool summer, and need to have the ability to endure the long yet extremely cold winter. Consequently, the species diversity in the tundra is lower compared to most other biomes. In fact, tundra can be considered as a cold marsh or a wet grassland that is frozen for a portion or even most of the year.

Tundra Temperature and Soil

The tundra regions at high latitude areas are dominated by cold, dry, polar wind for most of the year. Air with limited moisture only enters the tundra region in summer when the sea ice begins to melt down. The typical summer temperature is usually below 50 degrees F (below 10 degrees C), and the temperature stays above the freezing point for only two to six months in a year. In the short summer, the sun remains above the horizon for most of the day, and does not even set for several weeks at extreme latitudes. But in the winter, the sun rises only briefly or not at all. Alpine tundra is similar to arctic tundra in terms of having a rather short growing season but a long and cold winter. Otherwise, the climates are quite different. Alpine tundra has less extreme photoperiods, more precipitation (mainly as snow), stronger winds, and greater day-night temperature fluctuations. Therefore, in contrast to arctic tundra, the weather in alpine is very variable. In addition, alpine tundra differs from arctic tundra in having higher ultraviolet radiation and lower oxygen concentration.

In addition to temperature, soil is an important environmental factor for the formation and function of ecosystems. Many types of soil are present in tundra areas, ranging from moist deep peats to coarse rocky materials where only mosses and lichens are able to grow. Usually only the very top layer of soils can thaw every year. The length and depth of thawing depends primarily on the amount of solar radiation received. However, factors that affect surface thermal properties, including soil color and moisture, vegetation cover, snow depth, and so on, will also play a role. The thawing layer can range from about three feet (one meter) to only a few centimeters in depth. In most areas, lower layers can remain frozen throughout the year. Consequently, drainage is usually poor and many soils remain waterlogged in summer. In contrast, alpine tundra soils generally lack permafrost but are thin-

A caribou antler rests among wildflowers in the arctic tundra in the Bering Land Bridge, Alaska. In Alaska, 85 percent of the land is permafrost. Some of the estimated 48 animal species include arctic fox, ermine, and grizzly bear. (National Park Service)

ner and immature, mostly classified as cryic (cold) entisols or inceptisols. Thus, they are better drained than arctic tundra soils. Because of the good drainage, late-summer drought can occur, limiting plant growth. Therefore, rocky sites with little plant cover are fairly common in alpine tundra. Solifluction (also known as soil fluction), a type of mass wasting in which waterlogged sediment moves slowly downslope over impermeable material (in tundra's case, a deeper frozen layer or bedrock), occurs in both arctic and alpine tundra, especially in alpine tundra because of the slope. Solifluction tends to produce an undulating landscape.

Tundra Fauna and Flora

To survive in extremely cold environments, plants must have special morphological and physiological characteristics. In arctic tundra, the dominant plants are dwarf shrubs and perennial herbs. These plants grow very close to the ground surface from which they get more energy during the rather short growing season. But in extreme environments, other than lichens and mosses, very few plants can survive. These plants usually have the ability to go dormant during severe drought or cold and resume growth when conditions become better. Some plants can even survive under continued snow cover for years. In addition, plants must be able to tolerate short and cool summers when temperatures are only slightly above the physiological limit for plant growth. They have to grow fast because the summer is very short in tundra areas. Also, many tundra plants have perennating organs underground to help them survive in extreme habitats.

"Tundra can be considered as a cold marsh or a wet grassland that is frozen for a portion or even most of the year."

Trees are almost entirely lacking except in very limited areas near rivers, or on sheltered slopes, where the unfrozen soil can be relatively deep. The herb layer is in fact common in tundra areas and there are quite a few different species of plants, especially grasses and sedges. Surprisingly, the lichen layer covering exposed rocks is better developed than in any other biome. Alpine tundra is often similar to arctic tundra in terms of plant species. However, there are exceptions. For example, the tundra of the Sierra Nevada shares only 15 to 20 percent of its plants species with the arctic tundra.

Therefore, the ecosystems in arctic and alpine are similar, and both can be placed into the category of "tundra." However, there are many obvious differences between them. Therefore, we must formally call them arctic tundra and alpine tundra separately to show the differences.

Although extremely cold, there are animals in the tundra area. It has been suggested that approximately 48 species of land mammals are found on arctic tundra, but surprisingly the abundance of each species is not low. Most noticeable mammals and birds in arctic tundra are arctic fox, caribou, ermine, grizzly bear, harlequin duck, musk ox, polar bear, and snowy owl. There are not many species of insects in the tundra, and the commonly found ones are black flies, deer flies, and mosquitoes. The surviving strategy of mosquitoes is to keep themselves from freezing by replacing the water in their bodies with glycerol, a chemical that works like an antifreeze and allows them to stay alive during cold winter. Also, it should be mentioned that arctic and alpine tundra share few animals.

In addition to the commonly recognized arctic tundra and alpine tundra, Antarctic tundra occurs on Antarctica and on several Antarctic and sub-Antarctic islands. Most areas in Antarctica are very cold and dry, and most of Antarctica is covered by ice fields, so plants are not supported. However, limited areas of Antarctica, particularly the Antarctic Peninsula, have coverage of rocky soil that can support vegetation. Compared to the arctic and alpine tundra, the Antarctic tundra has much less mammal population, largely because of its physical separation from other continents. Seabirds and sea mammals such as penguins and seals have their habitats nearby the shore. Some small mammals such as cats and rabbits occasionally present in Antarctic tundra, and are usually considered to be introduced by humans to some sub-Antarctic islands.

Tundra and Global Warming

Soils in tundra are considered to have the highest carbon density (21.8 per square meter) among all biomes, and the total soil carbon in tundra ecosystem is also the highest (191.8 times 10^{15} grams) among all specific ecosystems. In fact, tundra is recognized as one of Earth's three major carbon dioxide sinks. A carbon dioxide sink is a living entity (basically biomass) that takes in more carbon dioxide than it releases. Therefore, when discussing tundra, one thing cannot be neglected in this century: global warming. Global warming began to draw researchers' attention in the 1990s. It is a function of the greenhouse effect and caused by many kinds of greenhouse gases, such as carbon dioxide, methane, ozone, and nitrous oxides, of which carbon dioxide is thought to be the most important one. Carbon dioxide (CO_2) levels in the atmosphere are increasing, which may be mainly due to continued use of fossil fuels. It is estimated that during the 21st century, the carbon dioxide concentration could double and, consequently, simulations from general circulation models suggest that global average temperature would rise between 1.5 and 4.5 degrees C (34.7 and 40.1 degrees F). The gradual warming is accompanied by the melting of inland glaciers. One fundamental concern is that higher temperature in tundra and taiga, in which around 30 percent of the total soil-bound carbon found on our planet's surfaces are stored, will accelerate the degradation of peatlands, layers of permafrost, and forest litter, and thus lead to a large release of CO_2 and methane, exceeding the speed of biomass production. The second fear may be that a hotter, drier climate will bring more natural fires to areas in tundra where dead vegetation and peat have accumulated, which will also increase the amount of CO_2 in the atmosphere. In addition, not only the density of vegetation but also the composition of plants will be affected by climate change. Animal communities will be affected indirectly through their food plants.

From another viewpoint, the melt water from permafrost provides a direct source of water for plants in a severe drought summer, and keeps surplus water in the soil until the next summer. Global warming may disturb this permafrost system and, consequently, plants in tundra ecosystems could be negatively impacted. Also, freezing and subsequent thawing of soil in tundra may cause an increase in microbial activity, which is an important factor to keep the soil ecosystem active. Unfortunately, climate change might break the balance. However, climate warming, which brings increased availability of nitrogen and longer growing seasons, will likely increase biomass production in tundra ecosystems; this is what we should not ignore.

Zhiqiang Chen

Urban and Suburban Systems

An ecosystem is a community of living organisms interacting in complex ways with the physical environment, such as soils, geography, chemistry, and weather, to form a relatively cohesive functional unit. This definition suggests a certain degree of similarity across a given area such that what happens in an ecosystem is fairly predictable. Cities meet that definition and, in fact, there are several urban ecosystems being studied under the international Long Term Ecological Research Network. This article describes urbanization and explores characteristics, issues, and some solutions common to urban ecosystems.

The simple definition of a city is a large and densely populated urban area. But how large or dense, and what do we mean by "urban?" These questions have been surprisingly hard to answer because there are many different land uses in urban areas and there isn't a clear or widely accepted division between urban, suburban, and rural.

Smaller cities, for example, those with 5,000 to 50,000 people, typically include residential, business, and industrial areas and have formal governance such as a city council and mayor. In a large city there is a gradient of urbanization, from high-rise downtown areas transitioning slowly—or sometimes more abruptly when an urban growth boundary is in place—to low-density, sparsely populated rural areas, with other towns and cities within the area of influence. Therefore, it is useful to think about urban ecosystems in terms of a metropolitan area, where concentrations of people live in large cities, suburbs, and the satellite cities and towns close enough to provide the jobs, goods, services, and cultural experiences important to people.

Over many centuries, humans shifted from lower populations to the current and rapidly growing global population of about 7 billion people, with a corresponding shift from hunting/agriculture to urban areas. Prior to the 18th century, 3 percent of people lived in cities; as of 2008, for the first time half of the world's population lived in cities. By 2050 about 70 percent of all people are expected to live in cities. That's a great number of people, acres, and impacts on the environment, so it warrants some attention.

Key Characteristics of Urban Ecosystems

Ecologically, urban ecosystems have both positive and negative aspects. On the good side, concentrating people in one area reduces time and expense in commuting and transportation while improving opportunities for jobs, services, education, housing, and transportation. Concentrating human population can also reduce the impact on the rest of the environment. On the other hand, urbanization takes a heavy toll on air and water quality, fish, wildlife, and habitat. The key is to reduce these impacts without substantially increasing the urban footprint. Success doesn't mean that a metropolitan area resembles the original natural environment. Rather, it accommodates the needs of people, provides contact with nature, and conserves the biological resources and diversity so important to our own survival. These goals can be achieved using a strong foundation of science, the art of social and political compromise, and a variety of tools including urban planning, conservation, and regulation.

Compared to natural ecosystems, characteristics shared by most cities include: changes in land cover (less vegetation and more hard, or impervious, surfaces), changes in natural disturbance regimes, air pollution, warmer air and water temperatures, water quality and quantity issues, changes in the amount and type of habitat, invasive species issues, and wildlife communities where generalist species prevail. This article will explore the causes and effects of these changes, offer some solutions, and present a case study from the Portland-Vancouver metropolitan area in the northwestern United States.

Land Cover Changes

Land cover is material at the surface of the Earth, such as trees, grass, pavement, or water. Converting natural habitats to urban land cover is the overarching, and ecologically overwhelming, reason cities are similar to one another and different from other ecosystems. Much of the original landscape, whether forest, desert, prairie, or some other type of ecosystem(s), is now characterized by significant impervious land cover such as roads, parking lots, driveways, sidewalks, and rooftops.

In cities the air, habitat, and water quality are products of the cumulative effects of past and present human constructs and activities. Not all urban areas are the same, however. Table 1 describes some general urban land uses, the typical characteristics, and pervious and impervious surfaces.

Altered Disturbance Regimes

Disturbances are events such as fires, floods, landslides, or wind storms that disrupt and can change an ecosystem or community. Ecosystems are adapted to certain types of disturbances that occur relatively predictably over time and space—a disturbance regime. When land cover changes, it changes the timing and spatial characteristics of disturbances and introduces new types of disturbance. Eventually, the ecosystem stabilizes under a new, relatively predictable disturbance regime associated with urbanization. Now it is an entirely different kind of ecosystem.

For example, larger river systems include extensive floodplains to accommodate increased water during the wet season. There are different floodplain levels, or "benches," adapted to floods occurring annually and less frequently, perhaps inundating an area every 10, 50, 100, or 500 years on average.

Frequently flooded areas are characterized by fast-growing plant species such as grasses, sedges, and shrubs, as well as species such as cottonwood and willow that can physically withstand the force of the floodwaters and survive underwater for periods of time. A 50-year floodplain can sustain longer-lived species less adapted to flood disturbance, and so forth. The floodwaters deposit sediments, nutrients, rocks, and woody debris on the floodplains as the water slows and recedes. The substantial sediment deposits can form some of the richest farm soils in the world.

Urbanization has often occurred in floodplains because they are flat and close to shipping channels, a key means of transporting goods for import or export to support concentrations of people, industries, and jobs. However, the paving, vegetation removal, dikes, dams, levees, and floodwalls that come with urbanization alter the natural disturbance regime of the floodplain. Most times, the water is intentionally confined in the river channel; when it is not, substantial economic and structural damage, and sometimes loss of life, occurs.

Other types of natural disturbance are intentionally disrupted in metropolitan areas. For example, whether the original ecosystem was forest, desert, shrub, or grasslands, fire suppression is nearly universal, particularly around the fringes of a metropolitan area where significant natural habitat remains, along with homes and other rural uses. Fire suppression in these areas reduces danger to humans and economic damage and blocks the fire from spreading closer or into the urban area.

Nonnatural types of disturbance also characterize urban ecosystems. The process of changing land cover is a disturbance, but the new land cover is not; humans disturb the ecosystem in predictable ways. For example, every so often a building might be demolished in a downtown area and rebuilt. While this is not likely to happen again at that particular site for decades, it will happen in various other places in the city. Demolishing a building creates substantial noise for days or months, creates extra construction traffic, reroutes car and transit routes, and creates many tons of waste. This disturbance is regular within the system. A more short-term type of disturbance is freeway or light rail traffic, in which rush hour sets a regular pattern of higher disturbance. Ball games in a lit stadium, people walking in a park, blasting at a rock quarry, plane traffic around an airport—these are all part of the urban ecosystem's disturbance regime. They influence the ecosystem and organisms living there.

Often an ecosystem has been altered from its original state prior to urbanization. The most typical example is conversion from the original mix of habitats to agriculture, and then to an urban area. In such an area, essentially three different ecosystem types have characterized exactly the same area, often over a few decades to a few hundred years: the original ecosystem type, an agricultural ecosystem and, finally, an urban ecosystem.

Urban areas often have high levels of phosphorus, nitrogen, carbon dioxide, and other nutrients. For example, nitrogen and carbon dioxide concentrations are high near busy roadways. These nutrients can allow certain plants to thrive at the expense of others. In a study near San Jose, California, an endangered butterfly species—the Bay checkerspot—declined in numbers near busy roads because invasive plants thrived on higher nitrogen levels, pushing out the butterfly's host plant. In this case, limited cattle grazing may be a solution to return nitrogen to more natural levels conducive to maintaining native prairie habitat.

Air Quality and the Urban Heat Island Effect

Anyone who has walked or ridden a bike on a hot sidewalk or roadway and stopped to cool off under a shady tree already knows something about urban ecology: cities are warm places. Replacing cool, moist, natural vegetation with dry buildings, roads, and other urban constructs translates to higher air temperatures, called the urban heat island effect. The effect is most intense on hot summer days. For example, cities with a million or more people typically average 1.8 to 5.4 degrees F (minus 16.8 to minus 14.8 degrees C) warmer than nearby rural areas during the day, with peak intensities often reaching 18 to 27 degrees F (minus 7.8 to minus 2.8 degrees C) higher. Cities are warmer at night, too, as the heat stored during the day is

Table 1. Common metropolitan land uses and characteristics

Metropolitan area land uses	Typical characteristics	Pervious/impervious characteristics
Rural residential ("exurban": outside but near the urban area)	Large lots or properties One dwelling unit per one or more acre(s); may include outbuildings Associated with farms, forests, hard to build areas	5–10 percent impervious Pervious land cover typically natural, agriculture, lawns, landscaping
Low density residential (suburban)	Two to four dwelling units/acre Large lot, single-family homes Often associated with suburbs and areas outside the city center	20–35 percent impervious Pervious land cover typically landscaping and lawns with some natural areas
Medium density residential (suburban/urban)	Six to 12 dwelling units/acre; single-family homes, some single-family attached (e.g., duplexes) 20 dwelling units/acre—attached single and some multifamily	40–60 percent impervious Pervious land cover typically landscaping and lawns
High density residential (urban)	22-plus dwelling units/acre Mostly attached including condos or apartments; may have local shops, businesses mixed in	75–90 percent impervious Pervious land cover typically small landscaped areas, street trees
Mixed use	Mixture of commercial, high-density residential, small medical and business offices, other May include a "town center" within walking distance, transit station, etc., for a more self-contained community; less reliance on cars Fairly common in Europe, less so in United States	75–90 percent impervious Pervious land cover includes small landscaped areas, street trees, occasional open space for community gatherings
Commercial and industrial	Typically large properties; includes concentrations of retail businesses, shopping malls, factories, mills and industry, medical buildings, large retail stores, parking areas May have significant on-site natural areas under regulatory protection (e.g., wetlands, streams.)	75–90 percent impervious Pervious cover typically small landscaped areas in parking lots or adjacent to buildings, street trees; usually close to key transportation corridors.
City center	Most intense urban land cover with multistory buildings and high-rises; residential buildings may have hundreds of dwelling units/acre	85–100 percent impervious, high street density; may include rooftop gardens.
Parks, greenways, open spaces	Virtually all metropolitan areas have some designated parks and open spaces; may be large swaths of green (greenways), small pocket parks, playgrounds, ball fields, or golf courses	0-30 percent impervious

released to the night air. Even a desert city that has more cooling plants than the surrounding landscape is likely to be warmer because the impervious surfaces still store a great deal of heat.

Interestingly, the urban heat island effect may offer extra opportunities to offset warmer temperatures due to climate change. Because lack of vegetation makes cities warmer, "regreening" can reduce temperatures. In the most urban areas this can be accomplished by increasing street and parking lot trees, commercial and industrial landscaping, and green roofs (ecoroofs). Dark, hard surfaces absorb more heat than light ones, so lightening the color of roadways and rooftops will help. These activities can especially target the neighborhoods and areas that are hottest in the summer, when they would have the most effect in reducing air temperature. In terms of climate change there's a double added bonus: fewer air conditioners running reduces energy use, plus trees and vegetation store carbon.

Urbanization also changes air quality through emissions from industry, power plants, motor vehicles, wood-burning stoves, and a myriad of other causes. These activities increase pollutants (which can also be nutrients) such as nitrous oxides, which are produced during combustion and harm human health; ozone, which is protective high in the atmosphere but hard on the respiratory tract near the ground; heavy metals, such as highly toxic mercury from industrial emissions; and particulate matter that may harm the heart and lungs. Combustion also produces carbon dioxide, a key greenhouse gas. A haze of pollution over metropolitan areas is often visible from far away, and it is carried by wind currents to other areas.

Water Quality and Quantity

In cities, the amount and timing of water delivery are critical. The hydrologic cycle relates to the occurrence, pattern, timing, and distribution of water and its relationship with the environment. Impervious surfaces, combined with loss of natural soils and vegetation to slow and capture water, interrupt the hydrologic cycle, alter stream structure, and degrade the chemical profile of the water that flows through streams. These changes to water storage and delivery harm the environment in a variety of ways, and are cumulative within watersheds. The cumulative effects are products of an altered hydrologic cycle, or altered hydrology.

Water quality responds predictably to changes in land cover, typically declining as vegetation is replaced with impervious surfaces. In metropolitan areas, hydrology is most altered from the central city, with the impact declining as land cover becomes more permeable. However, numerous studies demonstrate that even low levels of imperviousness, in the range of 5 to 10 percent, are enough to damage stream channels and water quality. The following are some of the common effects of altered hydrology due to urbanization:

- Streams become "flashier"—higher flows during storms, but less water in the dry season; some streams that had year-round water prior to development dry up.
- Stream channels widen and deepen to accommodate the higher flows, damaging stream banks, stripping vegetation, and increasing sediments in the water and on the stream bed.
- Deepening the stream channel can cut into the groundwater table, partially draining it. Groundwater is typically what keeps a stream flowing during dry periods.
- Impervious surfaces stop water from percolating down to the groundwater, so the water table drops.
- Disconnection with large floodplains.
- Locally, flooding becomes more common and severe, particularly in heavily urbanized areas, because too much water enters the stream too quickly and it overruns the banks.
- Water temperature rises because impervious surfaces are warm, whereas trees and vegetation shade slow and cool stormwater. Unnaturally warm water is one of the most ubiquitous water quality problems in cities, and has contributed substantially to the decline of salmon and other cold-water organisms.

In older, developed areas, hydrology may reach a different but stable condition. Streams and other water bodies won't necessarily support the same plants and animals, but the new system can support longer-lived species or those that require predictable environments. The new system may have more generalist species and more individual organisms, at the expense of specialists such as cold, clean water specialists or those that need a rocky stream bottom.

For example, freshwater mussels can be good indicators of system stability because they are relatively slow-growing and adults are sedentary, while juveniles travel by attaching to specific host fish species. Some species can live more than 150 years. If only young mussels are present, the system is probably still changing, whereas older mussels have probably been there a while, indicating some level of stability. In North America, nearly three-quarters of all freshwater mussel species are imperiled and about 35 went extinct in the last century. Altered hydrology is undoubtedly a contributor to these species' declines.

Habitat Changes

At the local or site level, human habitat management leads to loss of structural complexity. Structural complexity and the total amount of vegetation are well-known contributors to wildlife species richness in forested areas. Humans tend to like a parklike setting with trees and grass, all green and alive. Dead trees and fallen branches are removed and leaves are raked. Unfortunately, this "clean" habitat fails to meet the needs of many wildlife species. Most birds feed on insects, seeds, and berries from shrubs, which also provide a great deal of cover and nesting habitat. Salamanders, centipedes, and salmon rely on dead wood, on the ground or in the water. Standing dead trees, or snags, are required nesting or roosting habitat for scores of birds, mammals, reptiles, and amphibians. Woodpeckers, swallows, bats, and bluebirds nest in snags and control pest insects. Leaves on the ground attract insects and their predators, such as towhees and thrushes. Sparrows, voles, bugs, snakes, lizards, and amphibians make use of rocks, brush, and wood piles. Leaving or adding some of these features to a yard will make it more natural and attract more native wildlife species. These elements also help build healthy soil.

At a larger scale, habitat fragmentation is the process of breaking apart large areas of natural habitat into multiple smaller disconnected patches. The term is generally used in the context of forested areas, but also applies to other habitat types such as wetland, shrub, or grassland. Fragmentation is widely recognized as an overarching threat to wildlife and ecosystem health and is closely linked to habitat loss, loss of habitat connectivity, and invasive species. Habitat fragmentation is characteristic of urban ecosystems.

Three basic characteristics—habitat patch size, isolation or connectedness, and size—heavily influence wildlife diversity in fragmented urban landscapes. Native habitat loss and conversion are first and foremost, with metropolitan areas retaining only a small fraction of the original habitat. The remaining natural areas are often converted to other habitat types, sometimes as a result of altered disturbance regimes. For example, in many cities native prairie or grasslands are converted to shrub or forest habitat due to fire suppression and invasive species such as Himalayan blackberry. Around the urban fringe, agriculture often replaces more natural habitats. The combination of habitat loss and changes threatens native species, particularly those that specialize on specific habitats.

When habitat is fragmented the patches become increasingly isolated. Two patches that are close together typically contain more species than two that are further apart, and if there are connecting corridors to other patches, even more species are present. In completely isolated habitats animals are essentially trapped or in danger if they leave the habitat patch. Isolated patches lose species over time, and without connection to other habitats, the species cannot come back.

Identifying important wildlife movement corridors and providing viable connectivity between remaining habitat patches can help reduce many of the ecological impacts of habitat fragmentation. Urban areas often protect streams through zoning and regulation and streams tend to connect habitat patches, therefore stream corridors sometimes offer the best options for wildlife movement. In addition, the amount and placement of a few key landscape features, such as trees and shrubs, significantly influence the types of wildlife that can survive in urban areas. Studies show that landscaping and street trees increase wildlife connectivity in measurable ways.

"Converting natural habitats to urban land cover is the overarching, and ecologically overwhelming, reason cities are similar to one another and different from other ecosystems."

Patch size is another critical factor, because larger habitats support more species per acre than smaller patches. Species requiring a large habitat patch (area-sensitive) may become locally extinct in fragmented habitats. In larger patches, area- and disturbance-sensitive species can find refuge in the middle of such patches away from disturbance, and the habitat quality may be better away from the patch's edge. Some studies suggest certain thresholds, depending on species or geographic area, at which area-sensitive species begin to appear. For example, a lower threshold, but one that may be particularly important in urban areas with smaller habitat patches, may be around 30 acres in naturally forested areas.

Large habitat patches benefit many of the region's sensitive species, but small habitat patches increase the mobility of wildlife in a landscape. Urban areas with trees and shrubs scattered throughout, combined with larger natural areas connected by corridors, are likely to hold more species and more animals than large patches and corridors embedded within an entirely urban matrix. Backyards, street trees, right-of-ways, and green roofs can all provide valuable opportunities to increase permeability. For these more urban solutions, an emphasis on native plants will help maintain native animal diversity. Numerous studies link native wildlife, bees, and butterflies to native plants.

The edge of a natural habitat patch is an ecotone, or a place where two types of habitat (for example, forest and urban) meet. Larger patches have less edge habitat than smaller patches, and patch shape also influences the amount of edge habitat. A long, narrow habitat patch has relatively more edge than a round patch of the same size. Edge effects are the changes that occur at the edge of a habitat area, and fragmented urban habitats have a lot of edge. Because ecotones host species from each habitat type, the number of species is often higher than inside a habitat area, and some species such as turtles and amphibians require more than one habitat type. Elk require forest for cover plus fields and shrubby areas for feeding. Fifty years ago, when world population was quite a bit lower and more natural habitat remained, biologists counted edge effects as a plus.

However, too much of a good thing can be bad, particularly in urban ecosystems where roads, buildings, cars, foot traffic, cats, and dogs can be quite hostile for native wildlife. If a deer steps out of a forest onto a busy roadway, wildlife-vehicle collisions, economic damage and injuries, and human and wildlife mortality may occur. Other well-studied negative edge effects include jays, crows, and small predators stealing bird eggs and nestlings along edges, where such high-protein food is poorly hidden; loss of disturbance-sensitive species near roads and heavy-use trails; and increased problems with invasive species because seeds are carried along edges by birds, car tires, wind, and other means.

Invasive species are typically nonnative plants or animals that negatively affect habitat and biodiversity, often with serious economic consequences. In 2004, the estimated costs of damage by and control of invasive species in the United States was estimated at $120 billion per year. Many nonnative species are not invasive, and many landscaped areas include such species. Invasive species typically share certain traits, including the following:

- Fast growth and rapid reproduction
- High dispersal ability, for example, wind-carried seeds or rhizomes carried downstream
- Generalist species that are able to tolerate or adapt to a wide range of conditions

- Early successional species—those that colonize an area after disturbance such as fire, such as grasses and thistle
- Association with humans
- Lack of natural controls such as predators, competitors, and disease

Infamous examples of invasive species examples in the United States include Japanese kudzu and English ivy, which can take down an entire forest. On the other hand, rhododendrons native to the northwestern United States are an invasive problem in Europe. There are literally thousands of other examples throughout the world.

Fish and Wildlife

In general, species best adapted to urban environments are those not limited to a single habitat type, those with populations easily maintained by outside recruitment, and those that can exploit the more urban habitats. For example, in the western United States habitat generalists or edge-loving species such as scrub jays, American robins, and European starlings are abundant, and chimney-nesting species are increasing. European starlings are particularly harmful to native cavity-nesting birds along streams because they nest early and often, and they aggressively remove other species to occupy nest cavities. However, they can be controlled by increasing tree cover and the width of riparian forests.

Predators are at the top of the food web in essentially all ecosystems. Urban areas see an increase in small and medium-sized predators such as cats, raccoons, and coyotes, and a loss of top predators such as cougar, bear, and wolves to control the smaller predators. Smaller predators prey on smaller animals to the detriment of many birds, small mammals, and reptiles. Birds that nest on or near the ground tend to decline rapidly in newly urbanized areas. Backyard bird feeders and other supplemental feeding may increase birds but also favor smaller predators.

Long-distance migratory bird species typically decline in urban areas across the Northern Hemisphere. The reasons are unclear; studies in the Netherlands linked disturbance from road noise to bird declines, and studies elsewhere show that some bird and frog species change the pitch of their song to be heard over road noise. There are probably other urban-related reasons as well. A disproportionately high number of neotropical migratory birds (those that nest in the United States and Canada, but travel south of the United States-Mexico border to winter) are habitat specialists, area-sensitive, or both. Many rely on shrubs and complex vegetation structure, although grassland species, which tend to require large habitat areas, are also rapidly declining.

Many migratory birds are sensitive to human disturbance. In fact, some so-called area-sensitive species probably need to avoid disturbance more than they need large habitat patches. There is little doubt, however, that human disturbance from roads, trails, industry, housing, and other development harms wildlife through noise, sound, light, and human and pet impacts. The species most sensitive to these impacts die, fail to reproduce, or leave.

Land-Use Planning: Portland, Oregon

The ecological problems brought about by urbanization are cumulative, but occur at a variety of

The Green Street Rain Garden in Portland, Oregon, is one of the results of the Portland City Council's 2007 Green Street resolution to promote and incorporate the use of green street facilities in public and private development. (U.S. Environmental Protection Agency)

spatial scales. So do the solutions. This section describes some effective tools used in the Portland, Oregon, metropolitan area to reduce the impacts of urbanization on wildlife, habitat, and water quality.

Comprehensive land-use planning with nature in mind can reduce negative impacts. Comprehensive planning helps a community identify goals and aspirations, always including development, housing, jobs, services, and transportation but often pertaining to nature as well.

In 1977 the Portland, Oregon, metropolitan area implemented the first Urban Growth Boundary (UGB) in the United States, designed to protect high-value farm and forest land from urban encroachment. A UGB motivates efficient use and redevelopment of urban lands and thoughtfully planned infrastructure such as roads and sewers. The Portland metropolitan area's current population is about 1.6 million people. There are 25 cities and the urbanized portions of three counties within the UGB, each with a different governing body. An elected regional government, Metro, serves to bring these governments and citizens together to plan major transportation projects, UGB expansions, and other projects to meet the needs of the population.

The region faces federal Clean Water Act and Endangered Species Act issues due to water quality and salmon declines. In 2005, Metro Council passed an ordinance requiring increased habitat protection along the region's streams. The regulation is implemented by the local jurisdictions. Regulation was limited to the most important water-related habitat, but the ordinance also called for a review of cities' development codes to identify and remove barriers to nature-friendly development practices and proposed voluntary measures including natural area acquisition, restoration, and environmental education. Local jurisdictions have stepped up to the challenge and, despite a growing population, the region is looking more and more green.

Metro also has a substantial green spaces program through two citizen-funded bond measures. Metro has acquired and is restoring more than 12,000 acres of natural areas to date, with a focus on large, healthy, natural areas and connections between habitats. Some of the bond measure funds go to the cities and counties to meet more local habitat and park needs, and some funding goes to acquiring regional trail easements to promote nonvehicular travel.

The adjacent Portland and Vancouver, Washington, metropolitan areas provide homes for more than 300 native wildlife species. These animals must be able to navigate the intricate network of roads, parking lots, backyards, and barriers to survive and thrive. The region is expecting significant population growth in coming decades—about a million more people by 2025. Further, anticipated changes in temperature and weather patterns will impact habitat and wildlife in ways that are not yet known.

The Portland–Vancouver metropolitan region teamed up to create The Intertwine, a collaborative organization designed to help improve and connect the region's system of parks, trails, and natural areas. The Intertwine created a voluntary biodiversity plan that outlines existing conditions, habitat types, and major ideas, concepts, and challenges to conserving the region's nature. It provides a way for people to look at larger-scale issues such as habitat connectivity, but it is specific enough to identify specific habitat areas where preservation and restoration would most benefit wildlife and water quality. It represents shared knowledge so that the entire region can work toward some of the same goals.

There are many other environmental efforts in the Portland–Vancouver metropolitan region, including highly successful recycling programs, incentives for green roofs on businesses, financial and technical assistance for landowners to deal with stormwater onsite, and a major backyard habitat certification program led by the local Audubon Society and a land trust. There is always more to be learned and more to do, but the region has made a good start toward a more sustainable urban ecosystem.

Lori A. Hennings